Tyron Hoppe

Willford Press
www.willfordpress.com

Published by Willford Press,
118-35 Queens Blvd., Suite 400,
Forest Hills, NY 11375, USA

ISBN: 978-1-64728-479-4

Cataloging-in-Publication Data

Composite materials for aircraft structures / Tyron Hoppe.
p. cm.
Includes bibliographical references and index.
ISBN 978-1-64728-479-4
1. Airplanes--Materials. 2. Composite materials. 3. Airframes--Materials.
4. Aerospace engineering. I. Hoppe, Tyron.
TL699.C57 C66 2023
629.134--dc23

For information on all Willford Press publications visit our website at www.willfordpress.com

Willford Press

Contents

Preface

Composite materials are a combination of two or more constituent materials that have a unique set of properties. They are versatile in nature and are widely used in aircrafts and spacecrafts, hot air balloon gondolas, gliders, passenger airliners, fighter planes and space shuttles. The application of high-performance composite materials in aircraft structures offers many advantages such as weight reduction, low specific heat, high thermal stability, high impact resistance, and high damage tolerance that improves accident survivability. There are three main types of composite materials having different mechanical properties. These include carbon fiber, glass, and aramid-reinforced epoxy. Carbon fiber is the most widely used composite fiber in aerospace applications. The structural applications of composite materials range from complete airplanes such as the Beech Starship to wing assemblies, helicopter rotor blades, propellers, seats, and instrument enclosures. This book is a vital tool for all researching or studying the applications of composite materials for the construction of aircraft structures. It will provide incredible insights to readers with respect to emerging trends and concepts in this area of study.

Various studies have approached the subject by analyzing it with a single perspective, but the present book provides diverse methodologies and techniques to address this field. This book contains theories and applications needed for understanding the subject from different perspectives. The aim is to keep the readers informed about the progresses in the field; therefore, the contributions were carefully examined to compile novel researches by specialists from across the globe.

Indeed, the job of the editor is the most crucial and challenging in compiling all chapters into a single book. In the end, I would extend my sincere thanks to the chapter authors for their profound work. I am also thankful for the support provided by my family and colleagues during the compilation of this book.

Tyron Hoppe

1

Adhesive Bonding Technology and Quality Assessment

Michael Noeske, Welchy Leite Cavalcanti, Hauke Brüning, Bernd Mayer, Antonios Stamopoulos, Apostolos Chamos, Thomas Krousarlis, Paweł H. Malinowski, Wiesław M. Ostachowicz, Konstantinos Tserpes, Kai Brune, and Romain Ecault

Abstract The first chapter highlights the relevance of both adhesive bonding technology and in-process quality assessment for mastering twenty-first-century challenges in joining functional and lightweight materials like carbon fibre reinforced polymers. The ongoing developments of the relevant technological and regulatory procedures and frameworks are hereby outlined, following trends for data-driven innovation and standardisation. Advances from monitoring process variables towards the in-depth and objective Extended Non-destructive Testing (ENDT) of material-related features are presented, based on methodological and technological innovation and insights from recent European joint research projects like Horizon 2020s ComBoNDT—"Quality assurance concepts for adhesive bonding of aircraft composite structures by advanced NDT". Introducing ten heuristic principles for quality assessment in bonding processes, a concept is demonstrated for establishing empirically consolidated sets of quantitative material and process-specific correlations between design-relevant joint features and quality data measured during the manufacture or repair of adhesive joints using ENDT. Each correlation is obtained by systematically introducing disturbances of relevant process features identified by experts and is levelled once by linking findings from standardised mechanical tests

---

M. Noeske (✉) · W. Leite Cavalcanti · H. Brüning · B. Mayer · K. Brune
Fraunhofer Institute for Manufacturing Technology and Advanced Materials IFAM, Wiener Str. 12, 28359 Bremen, Germany
e-mail: michael.noeske@ifam.fraunhofer.de

A. Stamopoulos · A. Chamos · T. Krousarlis
EASN Technology Innovation Services, Terweidenstraat 28, 3440 Budingen, Belgium

P. H. Malinowski · W. M. Ostachowicz
Institute of Fluid–Flow Machinery, Polish Academy of Sciences, Fiszera 14, Gdańsk 80231, Poland

K. Tserpes
Laboratory of Technology & Strength of Materials, Department of Mechanical Engineering & Aeronautics, University of Patras, 26500 Patras, Greece

R. Ecault
Airbus Operations S.A.S., 316, Route de Bayonne, B.P. D4101, 31060 Cedex 9 Toulouse, France

with ENDT results obtained for joints that have intentionally been manufactured or repaired in an off-specification way. Subsequent chapters will demonstrate the suitability of the broadly applicable process.

Keywords Adhesive bonding technology · Quality assessment · In-process data · Extended non-destructive testing (ENDT) · Heuristic principles · Concepts from ComBoNDT project

1.1 Introduction

Adhesive bonding was already being applied more than 35,000 years ago in the Upper Palaeolithic in the Near East using naturally occurring bitumen [1] as well as in Europe during the contemporaneous last Ice Age [2] and starting from the Middle Palaeolithic to the Iron Age using wood or birch-bark tar [3, 4]. Adhesive joints based on biomaterials are also known from the Swabian Aurignacian [5] in the Upper Palaeolithic. For example, hollowed-out adherends consisting of flint and split animal bones or mammoth teeth were re-joined and re-sealed using an airtight resin glue (an organic mastic that was possibly birch tar obtained using fire [6]) to manufacture a single piece of work that contributed to the social cohesion inside fire-heated caves inhibited by modern humans or even Neanderthals. These joints created musical instruments, highlights of humanity's cultural repertoire, and their product quality was tested by the musicians and their audience. As early as the Neolithic, inorganic sandwich composite adhesives consisting of fillers and binders were used to manufacture poly-layered joints as ornamental artefacts [7]. In the Taklamakan Desert in north-western China, early evidence of adhesives and their development around 3500 years ago was recently collected in the form of a bone sculpture-inlaid wooden artefact [8]. The development of adhesives and bonding applications was further advanced in ancient Greece and the Roman Empire [9]. In fact, throughout human history, adhesives have been among the most widely applied materials [8] and their use was one of the earliest known transformative technologies [10], yet many of these historical advances have only recently been revealed through archaeological excavations and investigations.

In Europe, the modern era was initiated by the re-discovery of Greek and Roman cultural and knowledge heritage, triggering the three Industrial Revolutions, which facilitated machine-driven mass customisation and computer-driven digital data processing. The fourth Industrial Revolution is currently ongoing, encompassing exchangeable decentralised datasets and individualised information available anytime and anywhere and joining the biological, physical, and digital fields to form cyber-physical systems. The advent of Industry 4.0 is updated music to the audience's ears—be it students or professionals, engineers or natural scientists, and it will connect more than seven billion humans worldwide. Such extensive communication will be based on documenting and preserving, sharing and high-speed exchanging trackable digital data or even knowledge, a development that is entering site-comprehensive production chains step by step. This advancement also challenges modern adhesive bonding for the series production of multifunctional materials for the simple reason that quality assessment and quality management in production technologies are not based on archaeological methods or tools. Nonetheless, we suggest that making relevant material-based aspects accessible is a crucial endeavour, both in archaeology and in forward-looking production technologies, and that this perspective will spawn the analysis of augmentable domain-specific data sources that are virtually unbounded in both cases.

We consider modern tools for quality assessment in adhesive bonding technology worth the readers' attention, and we present recent progress in the research, development and application of process-integrated monitoring technologies for joining composite structures. We systematically introduce our readers to extended non-destructive testing (ENDT) and provide a compendium for quality or process engineers throughout their professional lives as well as for students as an application-oriented introduction to the interdisciplinary topics at hand, comprising physicochemical, material and process engineering aspects.

In this section, we give a short introduction to ENDT and quality assessment in adhesive bonding processes relevant to the manufacture or repair of composite structures.

Why are these topics so highly relevant at present? Fundamentally, we share expectations that in the twenty-first century, progress in joining functional or lightweight materials and components will essentially be driven by innovations based on adhesive bonding technology and by developments that will enable manufacturers to safeguard the quality and optimise production [11–13], integrating these three drivers into a factory-of-the-future approach for composite manufacture, e.g. in the aerospace or automotive industries [14–16]. With the aim of achieving cost efficiency and increased flexibility with many degrees of freedom [17, 18], computer-integrated manufacturing will be established for all process steps, including those related to adhesive bonding. In this way, the increasing market demand for new and highly customised products with high quality [19] will be met and a small time-to-market delay will be achieved. In view of challenges like shorter response opportunities and flexibility in a turbulent and unpredictable market environment while preserving product quality [19], knowledge capturing and processing profiting from formal methods, e.g. taxonomies and ontologies [20], are gaining importance and may greatly contribute to rule-driven manufacturing control [21]. In addition to the software and information technology infrastructure presently being developed, hardware that facilitates the gathering of precise in-process data from all process steps will be needed as a third pillar [18, 22, 23] within less than a decade. These intelligent hardware elements of machines and devices such as sensors and actuators contribute to the "field-level" base of the industrial automation pyramid [19, 24], which highlights the numerous components and pathways of industrial communications, as sketched in Fig. 1.1. Within an automation pyramid, the upper levels are characterised by a deeper understanding and a higher responsibility for the performance of the complete overall system. The deeper levels within the hierarchy are distinguished by increasingly detailed knowledge about individual processes and process steps or

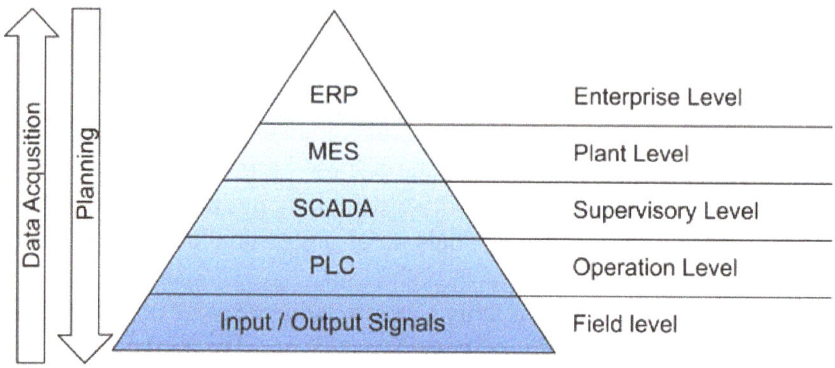

Fig. 1.1 Hierarchical model of an industrial automation pyramid based on field-level sensing. The first level or "field level" comprises the physical devices such as actuators and sensors. The second level or "operation level" includes logical devices such as Programmable Logic Controllers (PLC) or PCs usually found in complex machinery. The "supervisory level" (third level) corresponds to the supervisory control and data acquisition systems (SCADA) used to monitor and operate individual processes. The manufacturing execution systems (MES) are responsible at a higher "plant level". The top of the pyramid ("enterprise level") comprises the company's integrated management system (ERP) controlling the company's global operations. Providing, implementing and evaluating hardware that enables precise in-process data to be gathered within a comprehensive concept for quality assurance is a key target of the present book

technologies. In matters of smart manufacturing knowledge management, the basic data form the foundation for the higher levels, which are targeted towards manufacturing information, scientific understanding (e.g. providing predictive models), and knowledge [25]. In the future, for some systems, e.g. for monitoring bonded repair of composite aircraft structures, highest level standalone and autonomous opportunities appear accomplishable from a scientific point of view [26].

Bearing these expectations and challenges, but also the self-limiting frame of this book, in mind, we will introduce one comprehensive concept, ten pathbreaking heuristic principles, and more than a dozen tools which are accessible for interoperability and facilitate implementing and operating a quality assessment (QA) system. These form a profound base for quality assurance within a superordinate and nevertheless interlinked quality management system. We expect that sensing systems comprising ENDT tools and concepts [27, 28] as well as sensors and sensor concepts [18] will thus provide a solid foundation both for this approach following Industry 4.0 and for quality assessment as a technical pre-requisite for quality assurance.

1.2 Technological and Regulatory Framework

In this section, we will introduce a succinct description of adhesive bonding processes and render some aspects of quality assurance and monitoring before describing the ongoing advancement of quality assessment with a focus on adhesive bonding processes. After detailing ten heuristic principles for quality assessment, we will integrate methods and tools for ENDT. Finally, we will present a concept for ENDT and quality assessment in adhesive bonding, which will then be further elaborated in the subsequent chapters of this book.

1.2.1 Adhesive Bonding Processes

The track record and success story of adhesive bonding technology are based on establishing and safeguarding reliable joining processes. The early one and a half decades of the twenty-first century highlighted that quality assurance "for correct adhesive application and documented via certification" contributes to minimising faults, saving money, generating trust, promoting the wider use of adhesives, and sustainably improving the image of adhesive bonding, which may be clouded by observations that "bonding errors are often still encountered" [11]. So, let us first and foremost grasp a clearly arranged and predominantly technical image of this promising joining technology.

According to DIN EN 923, adhesives are non-metallic substances capable of joining materials through surface bonding (adhesion), with the bond possessing adequate internal strength (cohesion) [9]. The result of applying bonding as a material joining technique, i.e. the product of the bonding process, is an adhesive joint. Concerning the feasible lifecycle of an adhesive joint, such a bonding process may occur during manufacture [12] or during repair [29].

Characteristically, an adhesive bonding process may be divided into consecutive phases of the preparation of the constituent materials (such as adherends, [optionally] prepregs and the adhesive system comprising [optionally] primers and the adhesive), the application of the adhesive system, the assembly/lay-up, the curing and the final finishing. Each of these phases may be subdivided into further steps and finally strung together to form the process chain [30].

In the framework of these technical processes (or process chains), the adherend and adhesive materials may be described as operands, i.e. objects with initial relevant properties that are changed by the effect of one or more factors [31]. Following a procedure presented by Mattmann [31] for describing a product lifecycle, these factors are provided by technical systems denoted as operators. Correspondingly, at the end of a bonding process, the final state of the operands is different from the initial state, with the difference being described by the new, process-induced properties. In most cases, the bonding process is a successive multi-step process which may be described as a process chain with a chronologically defined sequence of process steps grouped in process phases [30]. Other operands in addition to the adherends and the adhesive will often need to be considered, e.g. in the case of a multi-layer adhesive system.

Therefore, in a short formal description consistent with Mattmann's approach [31], we suggest the diction for the adhesive bonding process as comprising the time-dependent (state and property) transformation $T(t)$ of several operands starting from an (overall) initial state $S_i(t_0)$ prevailing at the onset of the process at the point in time t_0. When the process is completed at time t_f, a final state $S_f(t_f)$ is achieved. As expressed by Eq. (1.1), the states $S_i(t_0)$ and $S_f(t_f)$ are described by property vectors at the defined points in time t_0 and t_f, and the achieved change and difference resulting from the process. Within a process chain, the operand is unequivocally described by the entity of its properties (in three spatial dimensions) at any time (as the fourth dimension) for all consequentially feasible intermediate process states.

$$S_f(t_f) = T(t_f) \bullet S_i(t_0) \tag{1.1}$$

In a technical engineering approximation highlighted by Eq. (1.2), among the operand properties, process-relevant time-dependent features, such as $s_A(t_1)$ and $s_B(t_1)$, may be identified based on the requirements to be met. Thus, the related feature vector $S(t_1)$ is a quantitative descriptor for the prevailing state of the operands as governed by the effect of the, respectively, performed set of operations, represented by the transformation $T(t_1)$, having been executed for (i.e. until) the point in time t_1.

$$S(t_1) = (s_A(t_1), s_B(t_1), \ldots) \tag{1.2}$$

Subsequently, the initial and final states of the operands will be technically described by the feature vectors $S(t_0)$ and $S(t_f)$, respectively. Clearly, the complete description of these states using the respective property vectors may comprise additional properties which are not significantly modified during the course of the adhesive bonding process. In this way, the pursued concept for systemically approaching the features will determine the accuracy of this approximation, which in practice will crucially depend on the iteratively achieved process and material know-how.

In a nutshell, following this diction the adhesive bonding process is described by the time-dependent procedure and changes of both the operator and operand states, as highlighted in Fig. 1.2. In more detail, when implementing such process representation as described by Eq. (1.1), the estimated operator-operand interactions may either be neglected, i.e. considered small as compared to the main effects of the operations affecting the operators or the operands, or they may be included in the concept of the operator or operand.

Visualising the first approach, a process in a controlled environment—with small and controlled deviations from a known and understood procedure described by $(T + \Delta T)(t) \bullet (S_i + \Delta S_i)$—may be assessed based on the knowledge of the main effects in the "reference" process, e.g. a qualified process. As a first example, in a wet cleaning step, the accumulation of known auxiliary materials of preceding process steps within the cleaning bath is maintained within evaluated parameter intervals $[(T - \Delta T)(t), (T + \Delta T)(t)]$ through the tailored measures of the process control. As a second example, identifying a deviation of the adherend surface cleanliness exceeding the evaluated parameter interval $[(S_i - \Delta S_i), (S_i + \Delta S_i)]$ may result in deciding to perform a further, e.g. repeated, cleaning step rather than deciding to, e.g. change the adhesive system of the following bonding steps.

Visualising the second approach, the material-dependent effects of processes may be attributed to respective material features in a material-specific process described

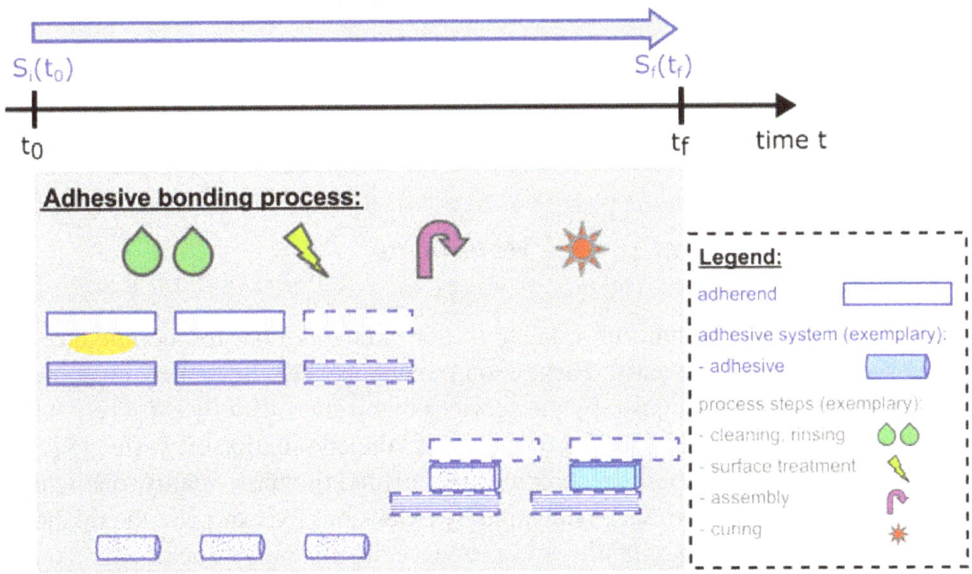

Fig. 1.2 An example of a process chain in adhesive bonding, with operands (with a blue outline) being successively changed through a time-dependent transformation $T(t)$ from an initial state $S_i(t_0)$ to a final state $S_f(t_f)$ due to the effects of the operators (with a grey outline). In this example, the top adherend is partially covered by contamination (shown in yellow) at the beginning of the bonding process

by $T(t, S_i)$, i.e. promptly adapting the operator action following the detection of deviations from an "expected" initial operand state. For instance, in the curing oven of an adhesive bonding process, the temperature reached by the pre-connected adhesive system and adherends within a given time may be consciously adjusted following the heat capacity and heat conductivity of the involved materials and devices. Therefore, using different operands will mean applying a different, "new" process tailored to the materials and the geometry of the joint to be manufactured.

From our point of view and following the perception highlighted in Fig. 1.1, technically facilitating the first approach is currently feasible and in the near future will increasingly be substantiated based on access to manufacturing information. We expect that the second approach will require a more profound understanding, which will prospectively be provided by longer term information-related data evaluation.

When geometrically extended specimens or even devices are to be adhesively bonded, the comprehensive prevailing feature vector $S(t_1)$ may be complex and comprise regions of the devices which are insignificantly affected by the considered bonding process. In such cases, a reduction of the topological complexity of the respective process description may be necessary and can be achieved by predominantly embracing the relevant conformational elements, e.g. those characterising the design regions comprising the bonding areas of the adherends. As adhesive bonding is based on adhesion, the feature vector describing the operands "adherend1" and "adherend2" shall necessarily comprise the surface properties of these solids in the bonding region. Meanwhile, the time-dependent feature vector describing the adhesive system shall comprise the interaction with the adherend surfaces.

In the following sections, we will demonstrate the application of these aspects for a described bonding process.

1.2.2 Quality Assurance and Monitoring

According to a strict definition, quality is first and foremost the compliance of services with the requirements. These requirements can be created by customers, manufacturers or users, but also by the services themselves. Briefly, with respect to competitiveness, an essential aspect of quality is "the elimination of waste" [32].

However, there is also a far-reaching aspect behind the term quality, one which is often associated with product safety and by extension, for example, the financial success of a company. The reliability of a product can often only be achieved through good quality. Moreover, high-quality products provide the basis for manufacturing in the high-wage countries of the EU, as these reach higher market prices. Nevertheless, quality control is a challenge for many manufacturing companies. House-made standards and testing standards usually help to achieve internal quality goals, but in order to demonstrate quality to external customers, it is often necessary to refer to norms and known standards or even be certified according to a recognised standard. Therefore, proof of compliance with a standard is provided by a certification process, followed by the issuance of a temporary certificate by independent certification bodies. One of the best-known quality standards is ISO 9001, which specifies a quality management system that an organisation (e.g. in the frame of a manufacturing process) must meet in order to comply with the quality requirements. Among other aspects, it introduces the concept of the so-called "special process", which is a process "in which the result cannot be fully verified (checked) by subsequent monitoring and measurement or non-destructive testing of the product" [33]. This includes processes such as welding or bonding. With these "special processes", a strategy for avoiding errors must be developed through a complete planning of the process, whereby all error-influencing factors must be identified and defined. In the production phase, all parameters must be checked and finally documented for the feasibility of the considered process. Even if this standard only defines the minimum requirements for a quality management system, the basic idea is ingenious. Secured processes prevent errors at specific points in the process, and over the entirety of the individual secured process steps, the quality of a product is controlled [34]. An example detailed by Espie et al. [32] highlights that "adhesive bonding can be a more complex procedure to control than other joining methods", demonstrated by the fact that "on a car assembly line spot welding is the responsibility of one station, but up to five points of the line can contribute to success or failure of a bonded joint". This indicates of course that the basic idea of ISO 9001 must be put into practice in a technology-specific way, since it is very general and the quality management system only conducts specific checks. It does not contain any further information on specific application techniques such as joining technology, e.g. bonding. Based on the core idea of ISO 9001, standards such as DIN 2304 [35] apply and specify the requirements in a technology-specific manner. DIN 2304-1 specifies requirements for the quality-assured execution of structural, i.e. load-carrying, adhesive bonds along the process chain of bonding—from development to manufacture and repair—and thus provides a basic framework for achieving high-quality bonded products.

Quality management thus comprises the product which the customer is buying, the process to manufacture or deliver this product, and the organisational system behind it [32]. Aiming at effectively implementing the continuous improvement of the product, the process and the system, the monitoring of events is desirable in addition to establishing a controlled environment [32]. Perceived as a tool, "the essence of monitoring is to look at trends and changes (or the lack of them) *over time* to reveal actions necessary to be taken with processes and the system, or to confirm that all is well" [32].

One immediate effect of monitoring in the age of digitalisation is that data are not only measured but also stored in great quantities, making them accessible for extensive evaluation and analysis [23]. Formally, the acquired data contribute to a complex materials characterisation data space. The precise format and architecture of such data are subject to ongoing research and industrial initiatives, e.g. considering concerted taxonomies and ontologies for contextual data. For example, Allotrope Foundation, founded in 2012, "is developing advanced data architecture to transform the acquisition, exchange, and management of laboratory data throughout its complete lifecycle" [36]. Within the Allotrope Taxonomies Domain Model, an entity of data is composed of the five domains of material (e.g. sample), equipment (e.g. spectrometer), process (e.g. method), result (e.g. spectrum) and property (of the data type or object). A similar approach for laboratory data appears feasible for structuring in-process monitoring data. Indeed, for the evaluation of monitoring data as a tool of quality control, mathematical, e.g. statistical, approaches are currently being discussed in the literature, which deals in-depth with structuring, e.g. clustering, the data and identifying their relevance to the quality of a manufactured product. Some key aspects are the following:

- Monitored data, i.e. the data vector or matrix obtained, may include on the one hand measurements of process variables related to the manufacturing process (and formally $\mathbf{T}(t_1)$), and on the other hand measurements of quality variables related to the manufactured product (and formally $S(t_1)$) [37, 38]. For example, statistical process monitoring (SPM) is based on both process variables and quality variables, while the focus of multivariate statistical quality control (MSQC) literature is on the monitoring of quality variables [37]. As highlighted above, the measurement process itself also consistently requires process monitoring and documentation.
- Process variables are often measured frequently and come in large quantities, while quality variables are measured at much lower rates and often come with a significant time delay [37].
- Root causes of potential quality problems may sometimes be related to a set of certain process variables [37], which is why two [39] or more [40] sub-blocks

for process data have been suggested, depending on their correlation with quality variables which are characterised by, e.g. mutual information (MI) values [39].

- Smart manufacturing which links physical and cyber capabilities will profit from managing large amounts of information and will facilitate improved diagnostics and prognostics, e.g. for fault detection or predictive maintenance [41]. Manufacturing operations and product quality may be improved by implementing process analytics delivering high-quality data and by "incorporating subject matter expertise in solution design" [41]. Combining such domain knowledge like subject matter expertise in analytics with the process and material expertise appears especially relevant to approaching the relationship and possibly the correlation between the measured data vector and the material feature vector $S(t_1)$.
- Additional information for quality control purposes may be generated by a combination of multiple information sources (that provide data) using information (data) fusion, e.g. by combining non-destructive testing and simulation [42]. Berger et al. laid out that "concurrent", "complementary" and "cooperative" integration types may be considered, depending on the amount and types of sensors that are being applied in combination. Following these authors, the method of combining data from a sensor network depends on the spatial and temporal relationship between the sensors.

The authors of this chapter forecast that in addition to regulatory requirements the availability of appropriate non-destructive testing devices for monitoring will in the medium term at the latest boost the frequency of their application, for instance, in adhesive bonding technology and especially in view of quality variables. We recommend identifying, monitoring and collecting high-quality data relevant to product quality. In the subsequent chapters, and especially in Chaps. 3 and 4, we will draw on subject matter expertise to characterise and categorise monitoring techniques and devices.

As depicted in Fig. 1.3, our quality monitoring approach will follow a description of a material analytical process considering an impacting probe P_i, a detected probe P_d and their interactions with the investigated sample material. The P_i-sample

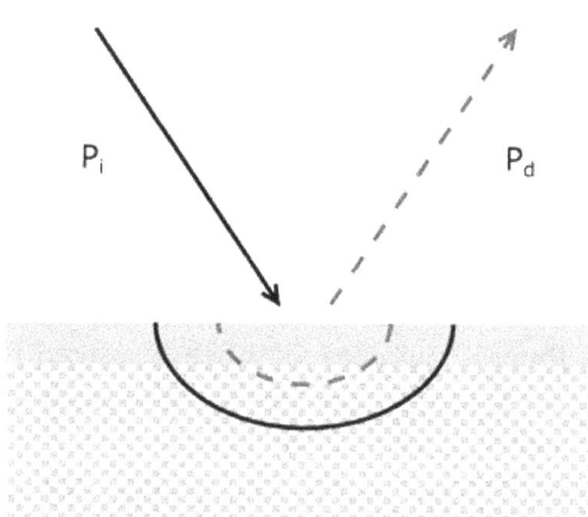

Fig. 1.3 The material analytical process considering an impacting probe P_i (solid line) and a detected probe P_d (dashed line) as well as their respective interaction volumes with the investigated sample material comprising the sample surface

interaction volume may be larger than the information volume produced by the P_d-sample interaction, but due to causality not vice versa. Finally, the focus will be on knowledge-based monitoring, which needs to answer the key question of this book: "How significantly does the state of the detected (set of) probe(s) P_d depend on the state of the monitored operands?" Evidently, monitoring the state of the operators is equally essential.

We observe that advanced information in combination with cyber-physical systems is currently establishing the fourth generation of manufacturing [41]. Indeed, in the past two decades the assessment that "monitoring adds cost but no value and may save cost at a later stage" [32] may have even hampered the speed of innovation in quality monitoring techniques, since "it is very hard to get companies to invest in something of which the added value is vague" [43]. Based on interviews with representatives from the NDT sector, C. Wassink spotted that companies looking for NDT solutions appeared to do so at a rather low aggregation level (plant-by-plant basis), on a rather small time scale of weeks or a few months, and by predominantly addressing technical issues. Subsequently, he suggested that a new and alternative innovation mindset should be applied at the industry level and on a time scale of several years, advancing innovation by multiple iterations and improvements and by widening the focus from mere defect detection to safety and risk reduction considering the economic value and social acceptance. Such an approach was to be followed by "mixed teams of practitioners and scientists" that were formed "to launch and improve new innovative solutions" and to establish a shared vision and innovation model comprising the active role of NDT service providers.

Following this perception, we intend the present book and the presented work to contribute our subject matter expertise in analytics and in adhesive bonding technology to a vision implementing extended non-destructive testing, thereby embracing the far-reaching aspects underlying the concept of quality.

1.2.3 Quality Assessment for Adhesive Bonding

We dedicate this section to the advancement of selected aspects in quality assessment for adhesive bonding technology over the past quarter of a century. We decided to tackle this agenda by first inviting the reader to engage in some time travel to the past decade of the past millennium, to about 1990. The idea is that we intersperse numerous citations from the comprehensive report of the EUREKA research project EU716 "Quality Assurance in Adhesive Technology" authored by Espie et al. [32], which was already touched upon in Sect. 1.2.2 Quality Assurance and Monitoring, thereby highlighting the awareness and vision of 20 years ago. The reader will thus be given the opportunity to compare their experience and perception of the challenges and perspectives with ours, which will be detailed subsequently and can be summarised as follows: Basic requests for in-process QA in adhesive bonding technology have been persistent for the past three decades, and these have been expedited with increasing intensity from several aspects. It is expected that, with the achieved progress described in the following chapters of this book, they will advance quickly.

As outlined in the previous section, Espie et al. highlighted that quality management in adhesive bonding relies on two major concepts [32]:

(1) "The control of joint design and specification of materials and processes".
(2) "The process monitoring and/or inspection".

In line with this, over one decade later a typical QA program was said to be composed of three parts, including the aim of applying the QA concepts and reference to criteria for the acceptance of operands [44]:

(1) "Establishing limits on bonding process factors that will ensure acceptable joints and product".
(2) "Monitoring the production processes and quality of bond in joints and product".
(3) "Detecting unacceptable joints and product, determining the cause, and correcting the problem".

Nowadays, QA is assigned an even wider mission, embracing the service life of the product resulting from the bonding process. For example, sustainability is a central environmental, economic and social concern on the "adhesive bonding roadmap", which was recently published by Dechema (Society for Chemical Engineering and Biotechnology) and the Joint Committee on Adhesive Bonding (GAK), supported by the German Welding Society (DVS), the Research Association on Welding and Allied Processes, FOSTA (the Research Association for Steel Applications) and iVTH (International Association for Technical Issues related to Wood) [13]. This wider sense, based on the future-oriented public and technological perceptions, makes further "increasing the trust in adhesives" the essential caption on the frame of this roadmap. The three pillars for the roadmap are based on "managing production processes", "understanding ageing" and "computer-aided bonding (CAB)". Within the first pillar, aspired targets are "quality assurance using non-destructive testing methods" using standardisation, guidelines and training and with—on the horizon in the year 2025—the "introduction of health and monitoring systems".

Thus, concerning quality assurance, we nowadays perceive that widening the focus to include economic value and social acceptance (as highlighted by Wassink [43]) is indeed a common sense in adhesive bonding technology. Moreover, it appears that after mastering static aspects, the upcoming decade will focus on assessing time-dependent changes within the operand materials during bonding and the application of adhesive joints. Following the up-to-date "adhesive bonding roadmap", managing a production process will embrace non-destructive testing, 100% monitoring in production, networked systems and sensors, quality assurance and practical NDT [13].

What foundation has been laid in this regard over the past three decades? What contributions have promoted the progress beyond that which this book and its authors intend to highlight? In 1989, Light and Kwun described in a state-of-the-art report "the bonding process, the destructive methods used to measure bond strength, and the various NDE methods that have been evaluated for determining the quality of a bond. These NDE methods include sonics, ultrasonics, acoustic emission, nuclear magnetic resonance, X-ray and neutron radiography, optical holography, and thermography". They concluded that with respect to non-destructive evaluation (NDE), "each of these methods has shown some limited success in detecting debond conditions", and that "at the present time" partially a "potential capability to differentiate *qualitatively* the gradations between a good bond and a debond" is ascertainable, which may "provide a correlation to bond strength" [30]. Approximately ten years later, in the EUREKA EU716 project [32] it was claimed that in adhesive bonding processes "continuous monitoring and compliance with documented procedures are required to provide assurance of quality" because adhesive bonding is a special process. The "application of general quality management systems already in place in manufacturing industry" was one of the aims of this 3-year collaborative project between the Centre for Adhesives Technology at The Welding Institute (TWI), Cranfield University and the Department for Trade and Industry (DTI). The project "identified that highlighting design and production issue during very early stages of design" and "well before a component reaches the production stage … enabled potential problem areas to be recognised and avoided". We will return to this latter (design) aspect and begin by reporting the details observed by researchers two decades ago.

In contact with eleven enterprises and institutions, exemplary manufacturing process checklists were completed in the EUREKA EU716 project by following the subsequent aspects/factors for describing and documenting the (two) adherends of adhesive joints:

- Description of the part,
- Manufacturer,
- Grade,
- Incoming specification,
- Supplier QA status, e.g. ISO 9001,
- Acceptance test(s),
- Sampling basis,
- Key attributes (critical factors), e.g. physical form, chemical composition, mechanical properties,
- Shape (critical factors), e.g. dimensions, tolerance,
- Surface condition (critical factors), e.g. as received, known contamination (like oil, grease, moisture, mould release agents, dust, dirt), existing coating,
- Required surface condition prior to bonding,
- Pre-treatment(s), listing process, materials, monitoring methods.

Among the required surface conditions prior to bonding, several of the following aspects were typically indicated by the contributing manufacturers:

- Not specified,
- No damage, e.g. intact peel-ply,
- Clean (we comment that from our viewpoint this could be "cleaned", i.e. with a cleaning process having been performed), grease-free, dirt-free,

- Free of loose cement, gravel or dirt,
- Dry,
- No contamination from preceding processes (steps), e.g. free from abrasion debris,
- Untreated, or with pre-treatment (e.g. passivation, abrasion, as-processed, primer application) and optionally with calibration within 1 h of bonding,
- Surface tension >56 mN/m,
- Less than a maximum number of pin-holes per length (or area),
- Sterile.

Among the monitoring methods for pre-treatments, typically one aspect among the following optional methods was specified by the manufacturers:

- None,
- Operator control (for a solvent cleaning process),
- Visual examination, e.g. colour, side to be bonded, no evidence of contamination, clean and dull appearance of (abraded) surfaces, reflectivity, thickness control, optionally with specified illumination (e.g. strobe light),
- Surface temperature,
- Surface tension, e.g. ink,
- Non-oiled, handling kept to a minimum.

Further aspects recommended in the EUREKA EU 716 report to be considered for adherend materials checklists were appearance, surface energy, exposure, handling, storage, and despatch. We would like to highlight in this context that with a technological background the (informative) Annex A within "Adhesive bonding of railway vehicles and parts—Part 2: Qualification of manufacturer of adhesive bonded materials, English translation of DIN 6701-2:2015-12" [45] comprises an overview of relevant aspects ranging from assessing the main function of the bond, surface preparation, type of adhesive used, testing and degree of mechanisation.

The numerous and often rather qualitative selected aspects concerning the state of the adherend surface(s) before the application of the adhesive system reflect concerns that are common in adhesive bonding technology and which address the area that will contribute to the bond line of the resulting joint. The integrity of this region often is considered "a significant 'Achilles heel' in the outright acceptance of adhesive bonding in structural engineering" [46]. Following M. Michaloudaki, who refers to the situation prevailing in 2005, "the predominant strategy to quality assurance is based on destructive testing of the bonded joint with subsequent statistical evaluation". She points out that "this procedure is combined with high costs and does not allow 100% controlling of the components or a repair of defects occurring during manufacturing" and, moreover, that such "testing itself or process mistakes during manufacturing (e.g. false applicator nozzle positioning) inevitably lead to product waste". Essentially at the same time, M. Davis, a Principal Research Scientist at the Directorate General Technical Airworthiness of the Royal Australian Air Force attending a workshop of the Federal Aviation Administration in 2004 [47], reported some observations considered characteristic for applying adhesive bonding technology for the construction and repair of (military) aircraft structures. The author considered surface preparation to be the "most significant factor in long term bond durability", and claimed that failures are often "caused by ineffective processes not just contamination". He concluded that "a clean surface alone is not sufficient" and that "process specifications are useless unless properly validated". Among the "causes of service bond failures", the author listed "inappropriate quality assurance tests". With this respect, he reported that "NDT only tells of bond-line gaps", which may be a reason why "you never hear reports about good bonds". The situation that "some OEMs claim good bonds, blame failures on operators" might be overcome by a quality management that including as best practice "to manage quality through the process, not just to measure it after bonding". The author highlighted that with respect to repair bonding, "requirements are the same as construction" but "the processes are different", e.g. with regards to surface preparation or the heating and pressurisation because heater blankets and vacuum bags are often used instead of autoclaves during production. We would like to highlight the essential aspect of quantifying the process quality implied in these observations—an approach that has typically been based on mechanical characteristics in the past decade. Glancing at adhesive bonding as a substantial and complex technology, Niermann et al. [48–50], when reviewing and discussing quality assurance aspects, outlined the distinct phases and respective process steps, finally flowing into the manufacture of a well-designed adhesive joint, i.e. the planning, concept, design and final development as well as the production and the usage phases, which require cross-process quality assurance measures. A guiding mechanical principle was highlighted for proving usage safety: the load capacity throughout the service life must be greater than the expected loads. The authors stressed that in production, processing parameters are to be defined by manufacturers for cleaning and pre-treatment products, primers and adhesives, and—above all—that these must be observed. Any change in the parameters should occur (after being authorised) only after testing. Certified training courses in quality assurance measures for bonding technology were identified to help recognise and prevent errors from the beginning, and these were highlighted as an essential tool for quality assurance in adhesive bonding [50].

In this context of a complex technology based on numerous process steps, from a current viewpoint we would like to highlight again the relevance of the process chain characteristics when manufacturing adhesively bonded joints. Interfaces for handing over the operands from one process step to the next need to be as carefully addressed as the interphases between each adherend and the adhesive system. Global sourcing from multiple sources may result in process steps being performed at different locations and with a certain time delay, accompanied by storage, conditioning or transport operations [51]. For example, a cleaning step preceding the bonding steps is contained in most adhesive bonding processes. The process management in parts cleaning aims at ensuring "sufficient parts cleanliness as required for the respective follow-up process" with a minimal consumption of resources [52]. Consequently, within the bonding process chain an interface-comprehensive quality assurance concept is required, e.g. involving expertise from cleaning specialists and bonding specialists. Moreover, the exchange of quality-related information on the state of the operands is expected to be especially intensive at such interfaces, and in practice an all-over monitoring is aspired to, reliably linking the process steps to literally form a chain.

One accepted approach for assessing quality-related information about the state of the operands is to introduce a process-control specimen which (i) accompanies the production (or repair) parts throughout the phases of cleaning, assembly and cure [45, 53], and (ii) remains accessible in the cured state for destructive testing, thus documenting the effects of the performed bonding process. A second approach is highlighted in this book and is based on extended NDT, comprising (i) in-process monitoring of the actual operands by performing time-dependent control within the very regions of technological relevance and (ii) post-process characterisation of the resulting adhesive joints.

Generally speaking, we consider the objective in applying techniques to monitor materials in the frame of manufacturing processes the same as two decades ago: contributing to collecting and documenting facts [32] which support safeguarding the compliance of the state of the material with the requirements, which are typically set during the process qualification. This objective indispensably holds true for the material state, which corresponds to the product of the manufacturing process. Concerning quality assurance in adhesive bonding technology and visualised in a pyramid model in Fig. 1.4 (which is based on Fig. 1.1), considering additional elements of upstream quality assurance is highly recommended. Technical provisions for pre-process

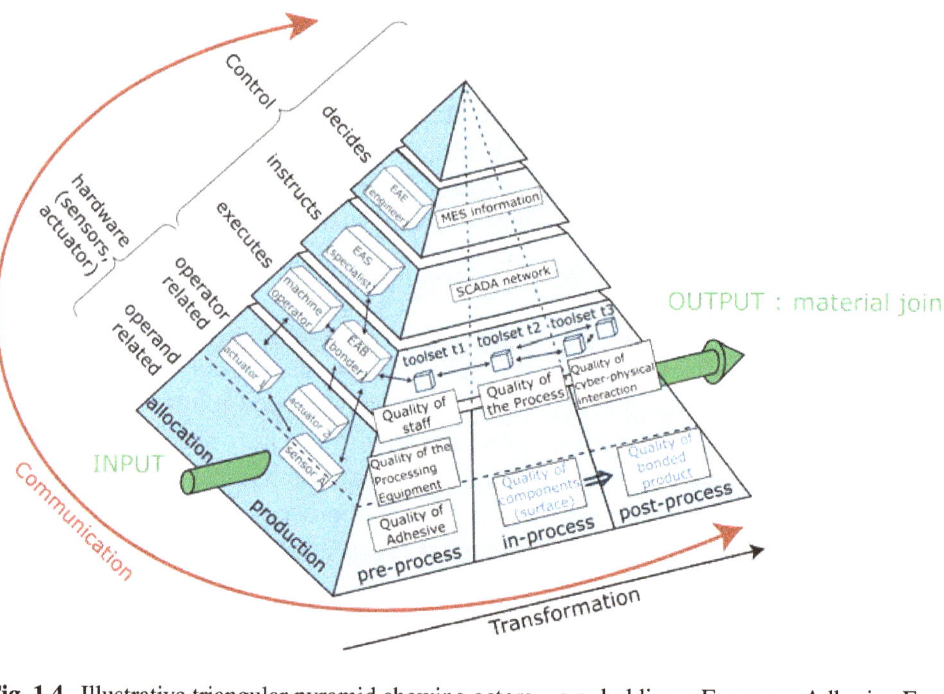

Fig. 1.4 Illustrative triangular pyramid showing actors—e.g. holding a European Adhesive Engineer (EAE) qualification—as well as material and process-related aspects contributing to quality assurance in adhesive bonding technology during manufacture or repair. The hidden and yet meaningful rearward face of the tetrahedron is related to ecological aspects. Quality topics in blue characters are the focus of the following chapters

quality assurance, e.g. incoming inspection [54], and in-process quality assurance, e.g. after each production step [9], are requested. This is because the closer to the error source within the production chain the technologically relevant amendments for clearing material state deviations from the requirements are performed, the easier, more precise, faster, more sustainable and altogether cheaper they can be. In addition for the "continuous monitoring and control of process parameters", qualified operational staff and compliance with documented procedures, as claimed by Espie et al. [32], strongly contribute to the quality of adhesively bonded products. As highlighted in Fig. 1.4, great emphasis must be placed on employee qualifications, processes, reliable testing equipment and first-class communication [51]. We recommend considering interface-comprehensive communication among humans, between humans and machines (HMI or also human-computer interaction—HCI), and exchange within cyber-physical systems (CPS) in this context.

From the analytical point of view, the implications for the monitoring techniques and processes are manifold. A suitable monitoring should facilitate a comparison between the actual states and the target states of the process and the material, i.e. the states of the operators and the operands, and relevant deviations from the qualified target state should be significantly and reliably be indicated. With the target state typically being defined by a data interval, the monitoring process shall facilitate a differentiation between, on the one hand, states corresponding to the boundaries of that interval and, on the other hand, states corresponding to the centre of that interval.

In regard to monitoring in adhesive bonding technology, the analytical requirement for an in-process monitoring process is, therefore, much less complex than predicting the extent of property deviations for the ultimately manufactured adhesive joint or predicting properties which are not accessible without destructive testing, like the initial (or even the final) joint strength (respectively fracture toughness), which is often a fundamental design specification. In other words and in view of the formalism represented by Eq. (1.1), the monitoring is clearly not intended to contribute to assessing $\mathrm{S}'_\mathrm{f} = \mathbf{T}'(\mathrm{t_f})\mathrm{S}'_\mathrm{i}$ in the case that the initial feature vector $\mathrm{S}'(\mathrm{t_0}) = \mathrm{S_i}(\mathrm{t_0}) + \Delta\mathrm{S_i}$ deviates from the qualified feature vector S_i or in the case that the process characteristics $\mathbf{T}'(\mathrm{t_1}) = \mathbf{T}(\mathrm{t_1}) + \Delta\mathbf{T}$ deviate from the qualified ones $(\mathbf{T}(\mathrm{t_1}))$ Rather, monitoring is intended to contribute to revealing whether $\Delta\mathrm{S}(\mathrm{t_0})$, $\Delta\mathrm{S}(\mathrm{t_1})$ or $\Delta\mathbf{T}(\mathrm{t_1})$ are acceptable based on the qualified corridors, i.e. the parameter ranges which conform with the qualification. Therefore, the analytical requirement for a monitoring process is "to look at trends and changes (or the lack of them) *over time* to reveal actions necessary to be taken with processes and the system, or to confirm that all is well" [32]. Concerning the surface properties of adherends before the application of the adhesive system in a bonding process, the lack of changes or differences as compared to the qualified process may reveal that the adherends are "ready to bond" based on the requirements of this particular process, the material combination and the application by customers. Clearly, among the required surface conditions listed above an attribute like "bondable" was not indicated, possibly due to the fact that it can hardly be considered a metric, numerical or steadily continuous material feature. In contrast, "ready to bond" is considered to refer to the feedback given to a technical operator or worker within the bonding process and is accounted for by the bonding supervisor in charge [11]. Such feedback is based on comparative monitoring and evidence of compliance with the quality/requirement-relevant data obtained for a qualified benchmark system. Finally, from the user's point of view the monitoring should be plausibly applicable in-line, in a non-destructive way, and executed at all positions relevant to the technical properties of the product.

Further aspects intended to spur the approach between monitoring system providers and users in adhesive bonding technology will be developed in the subsequent sections.

1.2.4 Ten Heuristic Quality Assessment Principles for Adhesive Bonding Processes

In this section, we present some qualitative aspects for discussion which we consider relevant for assessing quality assurance in adhesive bonding processes [55]. We call them heuristic principles and understand them here as a kind of set of pragmatic rules of thumb which will need to be refined by ongoing research; nevertheless, we have formulated and recorded them to support the reader in directing quality assessment in the direction of understanding and interpreting quality-related data rather than merely gathering them. Inherently, these principles may become a starting point for developing a guide for recommended practice. These heuristics are also intended to be principles in the sense of a starting point for iteratively improving the steps of the QA system. This improvement will contribute to further developing a superordinate quality management system that contains targeted actions in cases where the QA indicates quality-related issues. Based on experience (e.g. from the ENCOMB project [27], detailed in the subsequent section), discussions (e.g. within the ComBoNDT team [28] or with colleagues at Fraunhofer IFAM), ongoing literature studies (e.g. aspects from Nielsen's contributions concerning the usability design of user interfaces [56], or the methodology for discovery described by Kleining and Witt [57]), we propose a consideration of the following ten heuristic QA principles when targeting user-friendly quality assurance for the discovery of possible errors in adhesive bonding processes.

Heuristic Principle 1: "QA during an adhesive bonding process shall comprise the initial state of the operands, i.e. the adherends and the adhesive system, as well as their final state, i.e. the adhesive joint".

- An incoming component inspection is to be performed.
- At the start of the bonding process, i.e. at the initial time t_0, the initial state of the operands shall be characterised and documented. At best, this may include properties beyond the quality-relevant features.
- During the bonding process, the complexity of the operands' property vector may be reduced by following a feature vector, which considers surface and bulk

features. The data resulting from monitoring these features are documented within the data vector.
- At the end of the bonding process, i.e. at the final time t_f, the final state of the operands shall be characterised and documented. At best, this may include properties beyond the quality-relevant features.

Heuristic Principle 2: "QA shall comprise the time-dependent features describing the state of the operands and the acting operators".

- The process-relevant features shall be identified and then monitored during the bonding process.
- Interactions between operands and operators are only accessible if the features of all operands and operators are measured, especially close to the bonding region.
- Time intervals between measurements shall be chosen reasonably; they ultimately govern the dimensions of the data matrix which comprises the data vectors obtained at distinct points in time.
- Referring to adhesive bonding technology, the period during which QA is performed may definitely go beyond the manufacture of the joints. This means that it may encompass the manufacture of the operands themselves in addition to the application of the joint (or non-destructive testing of the obtained joint when applying application-specific operational demands).

Heuristic Principle 3: "QA shall be part of each step of an adhesive bonding process".

- The bonding process is considered a process chain of subsequent process steps, with the chronological sequence being relevant ("non-commutative steps" that follow causality).
- The time resolution of an assessment embracing the initial and final states of the operands shall be improved by pursuing QA for each step within each phase of the bonding process.
- The monitoring shall comprise process parameters (characterising the operators) and quality-relevant features (characterising the operands), and the acquired data shall be evaluated and rated following the QA approach.

Heuristic Principle 4: "The status of the QA monitoring system shall be made perceivable".

- The monitoring system shall be regularly calibrated using reference calibration standards and reference materials, and the determined status of the instrumentation shall be displayed.
- The monitoring system and the subsequent data evaluation procedure shall indicate to the adhesive bonding supervisor and to the respective worker the level of quality of the performed measurement. Measurements of low quality shall be rerun.

Heuristic Principle 5: "As a result of a QA inspection, the QA system shall give digital documentation/reports to the adhesive bonding supervisor and indications shall be passed to workers".

- The inspection system indicates information about the state of the bonding process; decisions are taken and imparted by (interacting) persons who are part of the adhesive bonding staff.
- An in-process and real-time availability of such information is often desirable during the manufacture and repair of adhesive joints.

Heuristic Principle 6: "The position inspected by the monitoring systems shall be promptly linked with the progressive position within the operand".

- User-friendly representations of spatially resolved monitoring data should be referenced to visual documentation (e.g. by scaled photographs) considering the magnifications.
- For user-friendly representations of several discrete spatially resolved quality-related data, the monitored regions of operands should be located considering their relative orientation.
- The time of the monitoring should be indicated and it should be referenced to the moment of the process start.
- Besides the spatial relationship between several monitoring devices and monitored regions, the temporal relationship between the acquired data also needs to be considered.

Heuristic Principle 7: "Consistency with the DIN 2304 standard and the qualified bonding process shall be assured".

- These are the benchmarks to be achieved during a manufacturing or repair process. Specifically, not only must non-bonding situations, such as a (local) lack of adherend pre-treatment or adhesive application as well as kissing bonds, be avoided, but also good bonds must be safeguarded and documented.
- Consider all process-relevant and application-relevant factors (in a controlled system), involving the adhesive bonding supervisor in the QA process.

Heuristic Principle 8: "QA inspection shall help to optimize the bonding and the QA processes".

- The monitoring of quality data and process data for the operands and operators as well as their evaluation and rating shall be documented.
- Workers shall be continuously educated and trained to contribute process-relevant perceptions. Subjective observations and information given by workers shall be stimulated and taken seriously and shall be documented and evaluated, aiming to make them objective.

Heuristic Principle 9: "QA data shall be collected under the paradigm of the maximum structural variation of perspectives"

- During the qualification of the manufacturing or QA processes, the expected variability of well-defined initial sample states shall be considered by scheduling intervals of acceptable feature values.
- When performing monitoring, different (modes of) ENDT techniques shall be considered. In detail, several mono-modal ENDT techniques may be applied or multi-modal techniques shall be used.
- Time-dependent or stimulus-responsive interactions between ENDT probes and the materials to be investigated shall be considered, both when documenting the parameters of the ENDT data acquisition and when elaborating ENDT procedures.

Heuristic Principle 10: "Analysis of QA data shall embrace the discovery and quantification of similarities".

- A first question to be quantitatively answered by the data analysis shall be "How big is the similarity to the qualified operand state?"
- When an error is detected, a second question to be quantitatively answered shall be: "Is the error pattern similar to known operand states which were deemed necessary to reject during qualification?"
- Check the observed changes in process parameters and quality parameters for common time-dependent trends, patterns or correlations.
- Basically, this aspect assesses part of the Industry 4.0 cyber-physical connection and smart manufacturing because it aims to gain information by understanding patterns and rooting causes to their situation and use cases [24, 37].

We anticipate that the thus assessed quality-related data sets will be amply accessed for documentation, reporting and evaluation purposes. In particular, capturing the features describing the actions of relevant operators contributes to preserving often proprietary manufacturing domain knowledge, which presently may only reside in the heads of engineering staff [21]. Evaluating the monitored operand-related features in the framework of manufacturing control will form the basis of rule-driven decisions [21], for example, determining if a certain time-dependent operand state is to be classified "in order" (or "not in order"), if actions are to be taken, or which actions are expedient. On the one hand, we expect that such rules and the recipes to be followed will be based on human reasoning in the decades to come, but the availability of binary data will help to enhance and refine the criteria upon which these rules work. On the other hand, documenting such rules themselves, the human reasoning and strategies behind them as well as the decisions taken and the formalised recipe to be followed is a task for the steady optimisation of the interacting QA and quality management systems. We assert that this task will be supported by ontologies. Therefore, we recommend applying a clear-cut taxonomy for the elements of a production or repair process, and the things or concepts which are subset to "operator" or "operand" will be specified with greater clarity, more concretely, and customised for each manufacturing site.

Assessing sets of binary data which represent quality-relevant operand features will be detailed in the following chapter. We will highlight, on the one hand, CFRP adherends and adhesive joints thereof as exemplary and descriptive materials and, on the other hand, NDT tools to monitor the quality-relevant features of CFRP adherends and/or adhesive joints thereof in the frame of quality assessment. Afterwards, we will introduce a concept for quality assessment in adhesive bonding which is based on these environments, our experience and human reasoning.

1.2.5 Extended Non-destructive Testing for Bonding CFRP

In this section, we do not comprehensively survey all ENDT approaches, but instead focus on the characterisation of composites based on fibre-reinforced polymers (FRP), especially carbon FRP (CFRP), distinguished by layers made from electrically conductive long fibres. We may highlight that the performance of adhesively bonded joints manufactured from such composite materials depends on the intensity of the operational loads to which the adhesive bond is exposed during in-service operation (e.g. of an aircraft), on the density and size of defects such as debonds, pores and delamination, and on the physico-chemical properties of the adhesive bond. While the operational, environmental and mechanical loads are considered in the structural design, the question remains as to how issues regarding the quality assessment of the joints contemplated here were considered and tackled at the end of the first decade of the third millennium.

In a nutshell, we may state that the defects in the joint area could (and still can) be detected by means of conventional NDT. However, there were no methods available to test the physico-chemical properties of adhesive bonds. In more detail, we present the complex line-up starting with the requirements for the said joints and the main parameters affecting the product quality, which comprise the surface treatment, joint configuration, geometric and material parameters, and failure mode [34, 58]. In their recent review, Budhe et al. stress that the durability of adhesive joints is governed by environmental factors such as temperature and moisture (including pre-bond moisture). Especially in the manufacture of automotive parts made from CFRP, variations in material quality (e.g. the thickness of the fibre-covering polymer matrix) or a variety of different contaminations, to some extent with considerably varying surface concentrations [16, 59, 60], call for the implementation of in-process surface quality monitoring, e.g. using scattered light technology. Concerning the repair of FRP structures, as for any substrate "skilled repair technicians, good surface preparation, well-designed repair procedures and the use of first-rate materials" are required [61]. The National Composites Network Best Practice Guide further highlights "stringent quality control encompassing reliable damage detection, surface cleanliness and texturing examination, drying to known limits, undertaking work within permitted temperature and humidity envelopes, and controlling resin cure to manufacturers' recommendations", "followed by NDT inspection of the finished repair or destructive testing of sample coupons or bars". During a typical repair procedure, thorough cleaning and degreasing may be succeeded by a water-break test and thorough drying [61]. After finishing the repair, inspection and certification of the resulting outcome are recommended [61].

Considering the quality assessment in more detail, the available NDT tools which can be applied during the production of FRP composites comprise facilities for monitoring the flow front, curing degree, void content, and possible delaminations (between matrix and fibres), for which established and various techniques are available, e.g. based on the electromagnetic, optical, mechanical or thermo-dynamical properties [17]. We may summarise that quality assurance processes for adhesively bonded CFRP primary structures that are not load-critical existed and were applied. Adhesively bonded structures were (and are) inspected by means of such conventional NDT in order to detect defects like pores, debonds or delaminations in the joint area. The materials (e.g. adhesives, prepreg materials) and process parameters (e.g. surface treatment, curing) were also controlled and monitored. In addition, specimens that had run through the complete manufacturing cycle were tested by both non-destructive and destructive methods to identify systematic process failures. However, in order to ensure the performance of adhesively joined load-critical CFRP structures, technologies suitable for the detection of the adhesion properties of bonded components were required [27]. Driven by central challenges within the aeronautics industry and with the above-mentioned requirements set, the ensuing development and adaptation of ENDT methods for the pre- and post-bond inspection of CFRP aircraft structural components is ongoing [62], and it is being expedited as a basis for establishing a reliable quality assurance concept for adhesive bonding. Briefly, as introduced in the ENCOMB project [27], the principle of such ENDT methods is based on the detection of selected physico-chemical properties which are important for the performance of an adhesive bond. Within the ENCOMB consortium, leading experts in aeronautics research and development from ten European countries cooperated to provide advanced non-destructive testing methods for reliable quality assurance of adhesive bonds in CFRP structural components, respecting the fundamental aspects most relevant from a manufacturer's and ENDT user's point of view. The constellation is schematically depicted in Fig. 1.5.

Two essential questions to be answered during the ENCOMB project were the following:

- From the point of view of the manufacturing process: "Which changes in which physico-chemical properties of the adherend surfaces and adhesives (i.e. the discrete operands) result in which changes to the properties of the adhesive bonds (between the joined operands), such as mechanical strength?"
- From the point of view of a feasible monitoring process: "Does a deviation in the state of the monitored operands influence the state and distribution of the detected (set of) probe(s) P_d?"

Concerning the physico-chemical properties of adherend surfaces, the degree of contamination or the type and level of surface pre-treatment, for example, needs to be considered. The physico-chemical properties of adhesives depend on a range of conditions, from the curing parameters and age of the adhesive, to the application parameters and ambient conditions. The adhesion, the morphology of the interphase and the cohesion of the cured adhesive are a direct product of these properties and

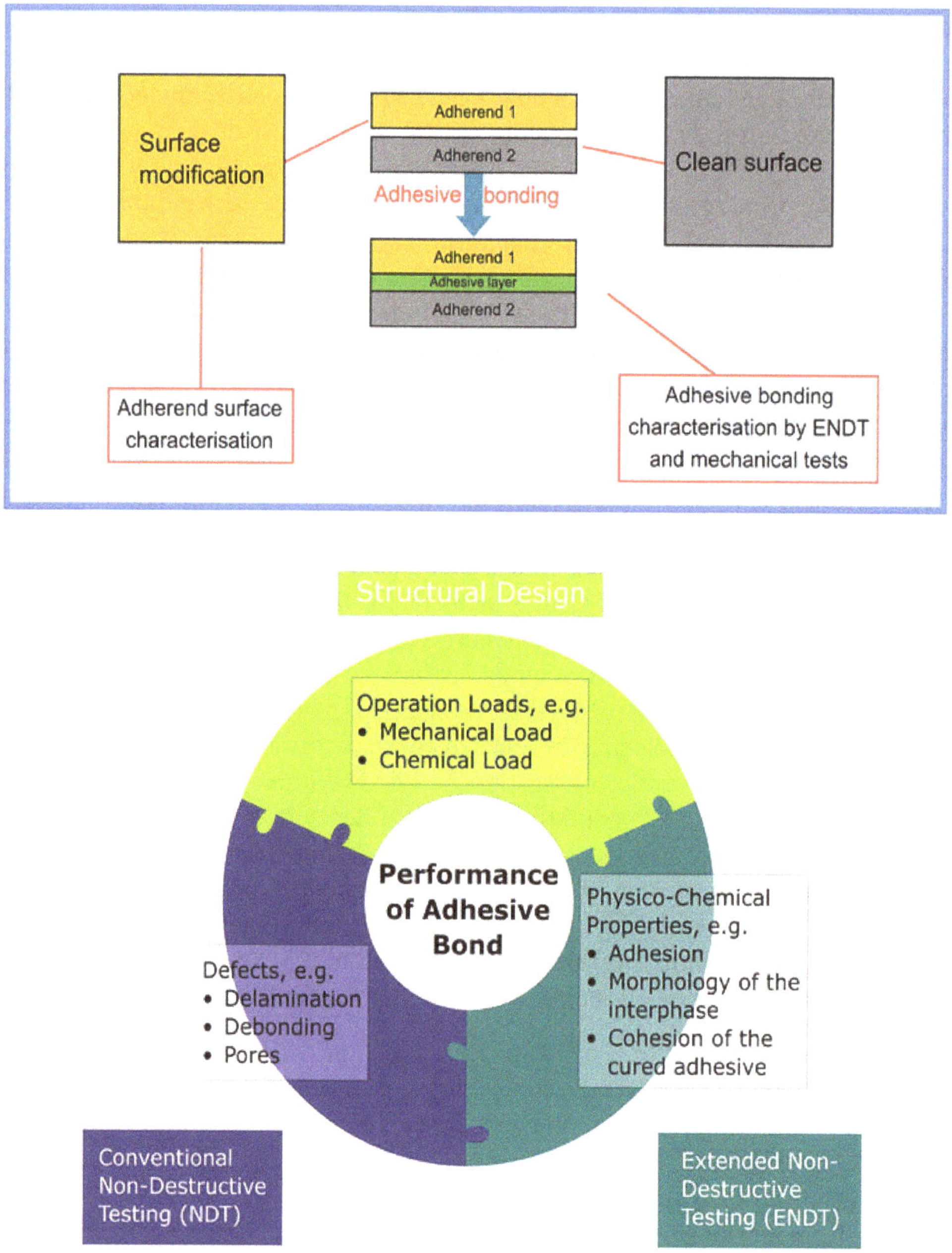

Fig. 1.5 The target and procedure of the ENCOMB project [27]. The focus of this project was to identify, develop and then adapt methods that are suitable for the assessment of adhesive bond quality, comprising investigations of CFRP adherend surfaces and adhesives. Not only the detection capabilities but also the sensitivity of measuring techniques were tested, evaluated and improved in order to achieve analytical results that could be quantified

are fundamental to the strength and durability of the adhesive joint. If the physico-chemical properties of adhesive bonds are not sufficient, then adhesion failure, weak bonds or bonds that weaken in-service can occur. Based on this rationale, ENCOMB identified and provided promising and developable non-destructive testing (NDT) methods for the pre- and post-bond inspection of CFRP aircraft structural components. State-of-the-art NDT techniques were screened, with the most suitable being adapted to important application scenarios with regard to aircraft manufacturing and in-service repair; finally, the performance was validated.

From the point of view of joint manufacture, the physico-chemical properties of CFRP adherend surfaces and the quality of the adhesive bonds were affected by intentionally applying different contamination levels down to the threshold levels of an insignificant impact on bond strength. From the point of view of identifying capable ENDT technologies, a screening was performed from among 31 technologies. Technologies facilitating a differentiation between treated specimens and a clean reference specimen were then adapted and validated in five different application scenarios that had been identified as being of primary importance for aircraft manufacturers. For each of the application scenarios, several techniques were developed that were able to detect different contamination levels and that had passed the validation step. Furthermore, several techniques with good potential were also developed further to comply with the requirements.

With these advancements in mind, we may state that research and development in extended non-destructive testing have been ongoing for over a decade, and the advances are increasingly providing tools and procedures for approaching the technical aspects of quality assessment in adhesive bonding technology. The trends we perceive in terms of progress in monitoring and the growing impact facilitated by ENDT are highlighted in Fig. 1.6.

Details of several promising ENDT techniques as well as their present performance and future potential in adhesive bonding technology are presented in the subsequent chapters. The contributions in this book highlight the development status which, as compared to ten years ago, clearly exceeds the prototyping stage, as will be substantiated by the assessment of the respectively accomplished technology readiness level (TRL). That being said, we are aware that the development of ENDT techniques is at present very dynamic, and we are confident that further progress will be achieved over the decade to come, motivated by the increasing interactions and exchange of views between specialists in the fields of monitoring, quality assurance and the manufacture and application of adhesive joints, accompanied by increasing standardisation.

To exemplify the technological perspectives of ENDT for the decade to come, we will focus on two of the surface-sensitive ENDT techniques, namely the aerosol wetting test (AWT) and optically stimulated electron emission (OSEE), and deduce advancements that may be achieved based on evaluating the dynamics of the respective measuring processes. Presently, key information is gained from the signals measured after a certain fixed period of time, starting with the deposition of primary

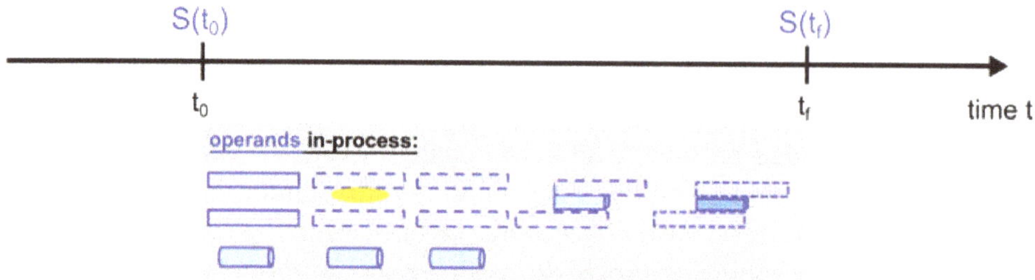

Inspection					
pre-process		**in-process**		post-process	
frequency	non-destructive	**frequency**	**non-destructive**	frequency	non-destructive
in 2010 (before ENCOMB project):					
random	**0** (laboratory: +)	random	- (visual: +)	up to 100 %	**0** (strength: -)
in 2018 (after ComBoNDT project):					
100 %	+	**100 %**	**+ (extended NDT)**	100 %	+

Fig. 1.6 The fields of application for the inspection of operand-related and process-relevant material features, considering the schedule of an adhesive bonding process and thus, comprising incoming inspection, manufacturing (or repair) inspection, and final inspection. The domain for ENDT inspection is shaded blue, and the basic contributions achieved in the ENCOMB [27] and ComBoNDT [28] projects are highlighted. We evaluated the availability and distribution of the respective inspection tools and procedures qualitatively and indicate our rating by "+" (widespread), "0" (temporary), or "−" (rather rare)

liquid droplets or with the ultraviolet light illumination of the solid adherend surface, respectively. This observation implies on the one hand that these time periods need to be carefully documented among the ENDT measuring conditions in the respective metadata set. We expect that this will call attention towards standardising formats for documenting these metadata, e.g. for achieving interoperability, for instance, following the approach developed by Allotrope Foundation [36]. On the other hand, any observed time-dependent effects on the data may provide additional information about the adherend surface state and may even be purposefully triggered or stimulated. For the examples of AWT or OSEE based surface inspection, additionally accessible adherend surface properties might be the rate of liquid (droplet) evaporation or the electrical charging upon electron emission. Prospectively, the optically stimulated electron emission of an adherend surface may be purposefully modified by the stimulus, which is temporarily effective during an AWT measurement, i.e. by the deposition of aerosol droplets. A stimulus-responsive ENDT (SR-ENDT) approach may be realised in this way [63].

As anticipated in Fig. 1.6 and as a first-hand example of advancing ENDT towards a new and hopefully brilliant horizon, the following scope of this book is attributed to the progress and outcomes of the three-year Horizon 2020 European research project ComBoNDT [28].

1.2.6 Concepts for ENDT and Quality Assessment in Adhesive Bonding

In this section, as a consequence of the above-mentioned considerations for ENDT and quality assessment in adhesive bonding, we exemplarily describe for the reader a feasible concept which complies with the ten heuristics and the systematics upon which the approaches described above are based, while also providing a comprehensive and customisable toolbox and schedule for implementing monitoring-based decisions in a quality assurance system.

Following this concept, sets of several empirically obtained correlations will have to be compiled between, on the one hand, destructively tested quantitative and design-relevant joint features and, on the other hand, operand-based quality data measured during the manufacture or repair of adhesive joints using ENDT. Each correlation is obtained by systematically introducing disturbances of one operator-related process feature (as performed within the ENCOMB project [27]), or even by introducing disturbances of several process features together in a set (as performed within the ComBoNDT project [28]). For example, a disturbance of one operator-related process feature might be due to contacting the adherend surface, such as when the finger of a worker touches freshly cleaned CFRP material. As a consequence, a deviation of the material state from the requirements may result because during the contact time some material may have been transferred from the finger to the CFRP surface. Strictly speaking, the process characterised by accidentally touching the surface may be referred to as a contamination (or a contamination operation), while the material transferred as an effect of this process is a contaminant (or contamination agent) affecting the CFRP surface quality. While bearing in mind the differences in meaning behind the words, the contemplated phenomenon might simply be referred to as a contamination. However, the exemplified approach based on semantics highlights that consciously recognising the cause-and-effect chain during a process chain is a basic element of the quality assurance process, which may be built on monitoring causes or effects (or even both) in the course of the quality assessment.

The working hypothesis behind the concept that we introduce to the reader is the following. If all feasible (or pragmatically all imaginable) disturbances of process features are identified, implemented and their effects tested, then such a set of all feasible (or imaginable) empirically obtained correlations (based on the respectively measured ENDT data sets and the data sets resulting from testing quantitative and design-relevant joint features) will reveal

- Whether applying selected ENDT investigations will to a full extent provide quality data that allow an identification (during the manufacture or repair process) of all feasible (or imaginable) joints that will not fulfil the design-relevant joint requirements as given by the qualification procedure;
- Whether one or several ENDT methods (or the measurement modes of multi-modal ENDT tools) will be necessary to obtain a set of quality-related data which covers and provides significant quality-relevant information for all feasible (or imaginable) disturbances of process features;

- Which measures of quality assurance are required in order to clear the detected distinctive quality-related operand features of the particular joint;
- What the consequences are when potentially clearing quality-assurance actions are taken in response to detecting a "not in order" state of the operands (already during manufacture or repair);
- Which operator-based measures of quality assurance are required in order to avoid a recurrence of the observed variation in quality.

The concept thus relies on evaluating experimentally acquired data sets, and we intend to devise a way towards an elaborate data acquisition process on the basis of the following steps:

- The identification and quantification of feasible (or pragmatically all imaginable) disturbances of process features;
- The separate or combined implementation of the identified and quantified disturbances in the fabrication of specimens, e.g. adhesive joints; testing and quantifying the effects of an implemented quantified disturbance applying;
  - An in-process quality assessment of operand-related features (e.g. using ENDT approaches and methods), specifying both the limit and the probability of detection (POD);
  - A post-process assessment of quantitative and design-relevant operand features (e.g. performing destructive testing to determine the strength or fracture toughness, including an inspection of the fracture pattern);
- Plotting the data set obtained by post-process assessment against the data set resulting from in-process assessment, highlighting the significance of the considered operand feature for the quality of the product, with the obtained relationship being specific for each implemented disturbance.

We are aware that applying (at least) the integral parts of such a concept is presently becoming increasingly widespread and is entering the phase of research, development and technology [15, 16, 63, 64]. In this book, we highlight the usefulness of our application-oriented approach in a descriptive and concrete way.

The subsequently discussed set of sketches shown in Fig. 1.7 displays how an application-relevant "not in order" statement which is based on destructively testing quantitative and design-relevant operand features (denoted as Y), may be translated to a threshold criterion related to an ENDT data set that comprises information on quality-relevant operand features (denoted as y). In Fig. 1.7 we imply that instead of the directly measured quality-relevant operand feature itself, its deviation from the value obtained during the process qualification is considered. In an example of use based on non-destructive quality assessment in adhesive bonding manufacturing, we note on the one hand that following a common qualification procedure the design-relevant operands feature Y may be joint strength, as obtained in a defined test bench for as-manufactured or aged adhesive joints. On the other hand, following an expedient ENDT monitoring procedure, a quality-relevant operand feature y may be the surface wetting behaviour of one adherend, as obtained with a process-integrated

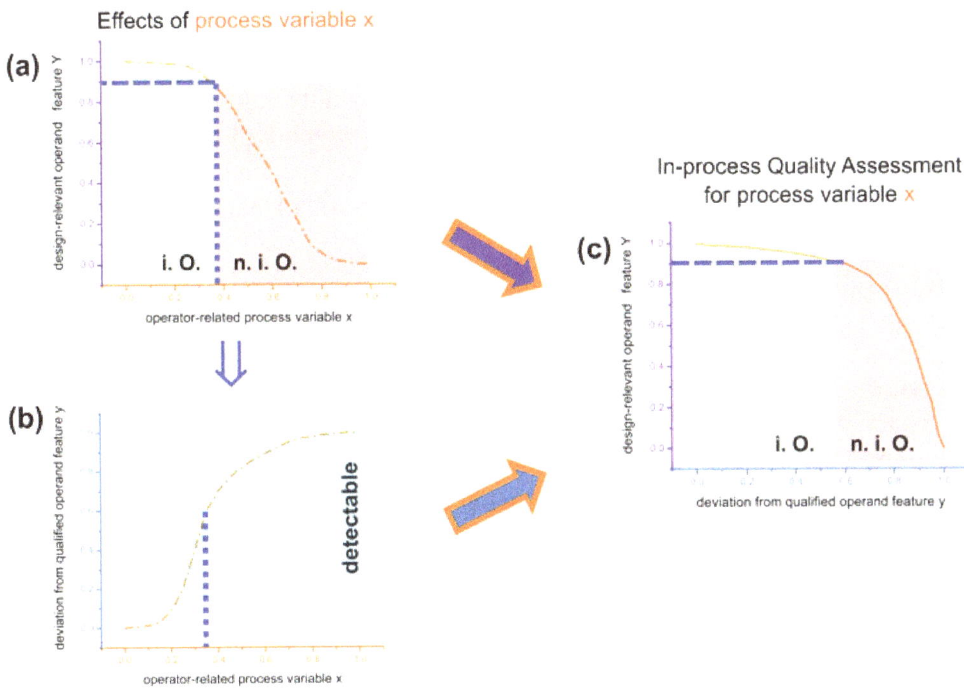

Fig. 1.7 Schematic introduction to the concept of ENDT and quality assessment in manufacturing processes, e.g. in adhesive bonding. For exemplification, the two functions y(x), and Y(x) were selected as one-to-one functions in a way that for each value of operator-related process variables, i.e. x, one value results in each one of the operand features y or Y. Y represents a design-relevant operand feature, and y represents a quality-relevant operand feature. On the left column of the plots, the functions Y(x) and y(x) are displayed, and on the right column the correlation $Y(y)|_x$ is shown, tied with the boundary condition that the regarded process variable is x. More details can be found in the text

setup prior to the bonding step. Basically, the correlation $Y(y)|_x$ links two observable responses of testing procedures, with a Y and y pair being obtained for the same disturbance x of the qualified bonding process.

We would like to accentuate that this set of sketches in Fig. 1.7 illustrates the following:

- Figure 1.7a shows that the accuracy of the experimental approach may iteratively be optimised aiming at finally introducing any disturbance in a way that its grading (e.g. represented by levels of intentionally applied contamination) is particularly fine around the crucial value of the regarded process variable (denoted here as x), which corresponds to the "in order"/"not in order" transition point of the design-relevant operand feature Y; we would like to highlight here that such an optimisation was one of the advancements achieved in the ComBoNDT project as compared to the ENCOMB project;
- Figure 1.7b shows that variations around the crucial value of the regarded process feature x should manifest in significant signal alterations upon an in-process assessment of quality-relevant data y (e.g. concerning the signal obtained when

applying ENDT; a minimum requirement is thus that the respective feature is detectable by the ENDT investigation);

- Figure 1.7c shows that the result of an in-process quality assessment might anticipate the "in order"/"not in order" categorisation which will be revealed by the post-process inspection $Y(y)|_x$ of the manufactured specimen only if precisely that investigated process feature x is the only uncontrolled feature of the inspected process;

We would like to accentuate that the set of sketches in Fig. 1.8 aims at a more elaborate contemplation and illustrates the following:

- Figure 1.8a corresponds to Fig. 1.7c and is related to effects resulting from the impact of process variable x.
- Figure 1.8c, d, which are similar to Fig. 1.7a, b but address a different process variable (namely $X \neq x$), show that the progression of the characteristic line y(X) is sketched differently from y(x), which is done to indicate that it is usually specific, e.g. for the respective intentionally applied contaminations;

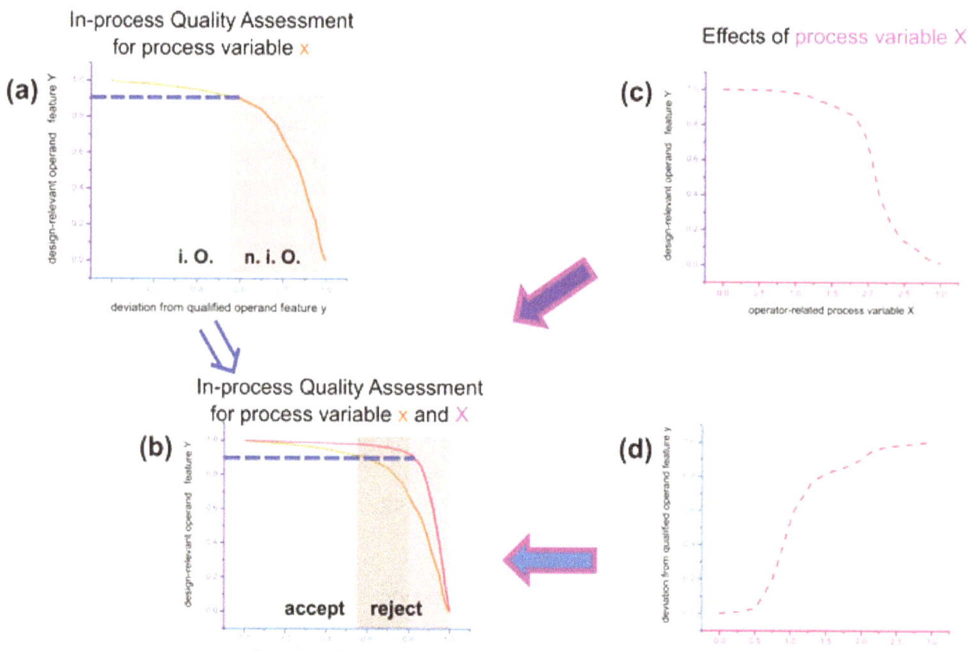

Fig. 1.8 The elaborated concept of ENDT and quality assessment in manufacturing processes, e.g. adhesive bonding, which is central to this book. For exemplification, the four functions y(x), Y(x) as well as y(X) and Y(X) were finally selected as one-to-one functions in a way that for each value of operator-related process variables, i.e. either x or X, one value results in each one of the operand features y or Y. Y represents a design-relevant operand feature, and y represents a quality-relevant operand feature. In the left column of the plots, the functions $Y(y)|_x$ and finally both the functions $Y(y)|_x$ and $Y(y)|_X$ are displayed (in plots a and b, respectively). More details can be found in the text

- Figure 1.8b, which is the aggregation of the sketches in Fig. 1.8a, c, d, shows that to avoid false positives, the decision to set a threshold value in a binary in-process categorisation is governed by the most sensitively detected process disturbance whenever more than one process variable may not be controlled in-process.

We prominently applied the presented concept in the ComBoNDT project [28], which is introduced in the next section.

1.3 Recent Joint Research for Advancing QA in Adhesive Bonding

In the following subsections of the present chapter, we will introduce the reader to the framework of the joint research project "Quality assurance concepts for adhesive bonding of aircraft composite structures by advanced NDT" (ComBoNDT) [28] and address its characteristics as well as the beating heart of this project, namely our consortium. This project received funding from the European Union's Horizon 2020 research and innovation programme under grant agreement No. 636494 and our consortium was active from 01-05-2015 to 30-04-2018 to achieve substantial progress beyond the starting position described in the previous chapter. The present book is based on the advances for quality assessment in adhesive bonding technology which were achieved over the course of the EU Framework Programme for Research and Innovation aspiring for smart, green and integrated transport. Briefly, the input and the output provided by the ComBoNDT project is essential but not confining for this book, and all authors of this book contributed to this project.

Subsequently, we will detail the overall concept, the goals pursued in this book, the step-by-step planning towards the achievement of the goals, the strategies followed as well as the intended impact according to the project workflow.

1.3.1 Objectives and Rationale

The context in which we were working had previously been defined by the key performance indicators (KPI) of the European aeronautics industry as defined by Flight Path 2050 "Maintaining and extending industrial leadership" [65]. Among these KPI, "process time reduction 80%, cost reduction 70%, and competitiveness of products produced in Europe compared with those produced in low labor cost countries" were the key factors inspiring the activities performed early in the second decade of the third millennium. To this end, we aimed at the development of ENDT tools for CFRP, adherend surfaces and bonded CFRP joints that could potentially cope with the technological problem in two ways:

- Safety improvement and cost reduction for building this kind of structure due to more reliable and longer lasting adhesively bonded joints
- Augmentation of the competitiveness of the European aeronautics industry by adopting such lightweight structures and joints, which may lead to a further cost reduction and greener air transport

More precisely, even though there is a strong need to exploit the potentials of lightweight CFRP structures in the aeronautic industry alongside the application of adhesively bonded joints, their adequacy for primary aircraft structures remains low. Despite the advances made in this sector as well as previous attempts, the shortcomings observed were caused mainly by the absence of adequate quality assurance processes. The corresponding requirements concerning the quality assurance of the manufacturing process of load-critical CFRP structures are particularly high, as potential failures could directly affect the overall safety of an aircraft.

Considering the above-mentioned aspects, the implementation of reliable adhesive bonding processes through advanced quality assurance would lead to the increased development of highly integrated structures with an optimum combination of advanced composite materials, which would, in turn, minimise the use of the rivet-based assembly. Consequently, metallic assembly concepts would potentially be surpassed through a redesigning of the structures. The benefits of such a procedure can be crucial, especially in the field of weight saving, which may be expected to amount up to 15% in the case of the fuselage airframe. This could have further positive effects on the size and weight of other aircraft parts, such as the engines or the landing gear, causing in parallel a reduction in both the fuel consumption per passenger-kilometre and the operating aircraft costs. Furthermore, a significant reduction in greenhouse gas emissions (CO_2) would contribute to the mitigation of climate change and further improve the environmental performance of the transportation sector.

The already established quality assurance processes for adhesively bonded CFRP non-critical load structures are based on measurements using methods suitable for the detection of potential defects (pores, debonds, delaminations) in the joint area. However, using such conventional NDT methods does not facilitate a detection of any further defects of interest, such as kissing bonds, nor does it assess the weakening of a geometrically intact bondline. Thus, the development of quality assurance processes which, on the one hand, provide a correlation to the physico-chemical properties of the probed adherends and adhesives and, on the other hand, could make the adhesion properties of bonded components accessible had to be spurred. This goal also met the EASA (European Aviation Safety Agency) certification requirements for structural bonding [66]. Our research and development (R&D) in ENDT techniques aimed at establishing reproducible and reliable non-destructive inspection tools in order to ensure the manufacture of joint structures that reliably feature the required strength.

In a nutshell, our overall objective was to develop a quality assurance concept for the adhesive bonding of load-critical CFRP primary aircraft structures, which could be applicable within the whole life cycle of the aircraft to overcome the current limitations regarding the certification of composites. Therefore, we established a detailed approach regarding in-line ENDT for CFRP materials at an increased technological readiness level (TRL), both in-process and post-process.

Applying ENDT technologies, we aimed to overcome the limitations of conventional NDT methods and allow the reliable assessment of the surface state of CFRP adherends as well as the quality of the final adhesive joint. We focused on the implementation of previously developed ENDT techniques, their assessment and, potentially, the increment of their TRL. The integration of these techniques into future adhesive bonding process chains, quality assurance concepts and material testing for maintenance processes could pave the way for the safe and reliable integration of lightweight structures into aircraft design. Applying the bonding of complex lightweight structures and the replacement of the traditional riveting method may overcome the use of CFRP as a "black metal".

Among the most important topics to which the book aims to contribute is the establishment and strengthening of the production and material testing processes at all stages of an aircraft's lifecycle using the previously mentioned quality assurance concepts. A fast and precise detection of surface contaminations and defects like kissing bonds in bondlines could help save time (up to 70% time savings using ENDT) during production, maintenance, overhaul, repair and retrofit. This way, aircraft parts could be checked for contaminants without disassembly or time-consuming steps. Also, parts of the aircraft could be replaced or fixed when necessary, resulting in up to 50% higher cost efficiency for ground operations. All of the above will significantly contribute to the competitiveness of the European aircraft sector.

The advancement of highly promising ENDT technologies was necessarily tested and demonstrated exemplarily in the frame of two fields of application, namely aircraft manufacture and in-service bonded repair. These fields of application determine the requirements in terms of the detection capabilities (e.g. of unknown contamination), applicability and robustness (i.e. TRL) that need to be met by the ENDT technologies. The maturity of the techniques will also involve approaches concerning automation and industrialisation, which means that aspects like the mobility of the measurement setup, objective and unambiguous data evaluation and interpretation were also addressed. In more detail, among the main objectives for our research and development was the incrementation of the current maturity level (TRL 3–4) to a TRL of 5–6. The aim of an increased TRL was addressed both as a measurable project result and a ground-breaking step towards the implementation of the developed ENDT techniques in real application scenarios.

The improvement of material testing during manufacturing as well as ground operations (overhaul, maintenance) will allow the automation of processes that are currently performed manually. The resulting time savings (in combination with more reliable results) should also be utilised to obtain measurable results by comparing the state-of-the-art process with the newly developed techniques. An important step was therefore the determination and improvement of the performance of ENDT in terms of the speed of inspection and data evaluation (aim: 10 min/m^2 of the inspection area at three to five times faster than the current state of the art).

In summary, a successful R&D process would enable

- The reliable and reproducible detection of unknown and potentially multiple contaminations on adherend surfaces;
- The reliable and reproducible detection of poor bond quality in bonded adhesive joints;
- A robustness of methods and a suitability for field measurements in aircraft manufacturing and repair environments in terms of detection limits and measuring speed;
- ENDT technique(s) which are validated in the relevant environments (TRL 5–6).

1.3.2 Concept and Approaches

Regarding the overall concept, the aircraft manufacturers within our project consortium provided the other partners with their demands and targets, aiming at the successful exploitation of the research activities. The sketch displayed in Fig. 1.9 summarises our overall conception of this book and its chapters:

We greatly benefited from the outcome of the EU FP7 project "Extended nondestructive testing of composite bonds—ENCOMB" [27], the results of which were detailed in the previous chapter. Launched back in 2010, ENCOMB involved a screening of potentially suitable techniques for the characterisation of CFRP adherend surfaces and adhesive bonds, whereby over 20 different non-destructive technologies were tested. Several contaminations or defects (e.g. a silicone-based release agent, hydraulic fluid, moisture, thermal degradation or poorly cured adhesive) were introduced to adhesively bonded CFRP joints in order to adapt the measuring techniques in such a way that the different contaminants and the resulting effects on the bond quality could be detected down to relevant threshold values

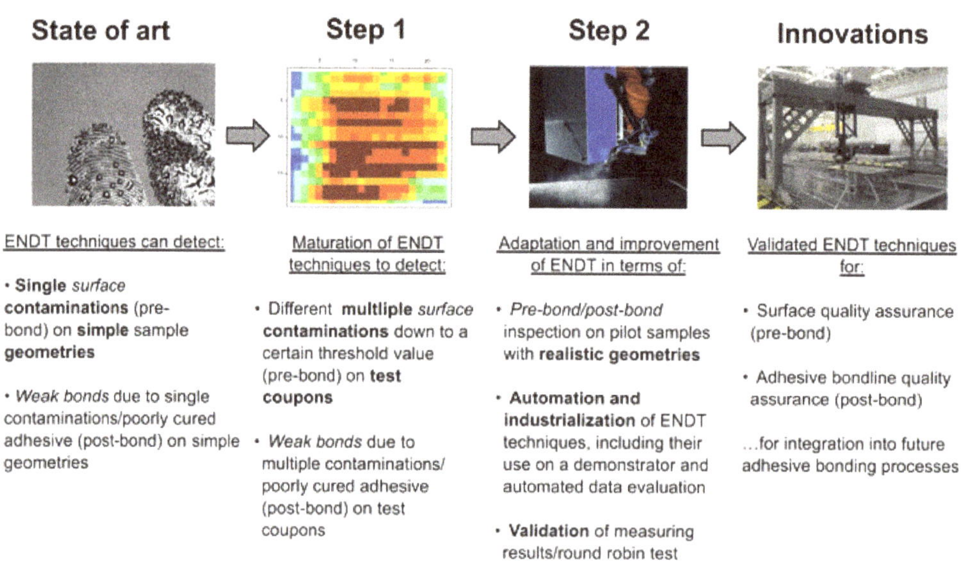

Fig. 1.9 The conception of this book and its chapters

(e.g. related to surface concentrations). Additionally, relationships were established between the degree of the pre-bond contamination and the observed degradation and the mechanical performance of the resulting adhesive joint. We briefly recall here that substantial progress was made with regard to providing the basis for an integrated approach to the quality assurance of adhesive bonding processes. Against this background, when planning the activities reported in this book, it was of strategic importance that the R&D work reported here technically and methodically enhances the most promising quality assurance concepts identified within the ENCOMB project and advances them to a TRL of 5–6.

Moreover, we aimed at additionally exploiting the knowledge gained from a series of further aeronautics and NMP projects while interacting with ongoing projects in the field that were in a parallel state of progress, e.g. CleanSky [67] and BOPACS [68]. To accelerate the further development of ENDT technologies for integration into future adhesive bonding process chains as well as allow the assessment of surface quality before bonding and the quality of the finished adhesive joint, the work within the ComBoNDT project was performed alongside test scenarios extracted from the fields of application of manufacturing and repair. Besides their relevance for manufacturing and repair, the outlined test scenarios represented as yet unanswered scientific questions, such as the effect of adherend surface contaminations on the adhesion properties, the overall bonded joint performance and the joint durability. To achieve these goals, we addressed an increase in the degree of automation in connection with a high reproducibility and an adequate measurement speed, an increase in the detection capabilities and the sensitivity of the techniques, a decrease in the costs and an adequate simplicity of handling, in particular with regard to quantifiable results.

Considering all of the above, the ENDT methods identified for further maturation in the frame of the R&D work reported here were carefully selected with respect to their suitability for the corresponding research approach. The techniques used in the project already existed in the form of laboratory setups, and their applicability was previously tested in the frameworks of other research projects. As summarised for our readers in Table 1.1, feasibility studies were demonstrated in the early phase at the beginning of this decade. To serve the book's scope, as addressed in Sect. 1.3.1 Objectives and Rationale, the selected techniques to be maturated were selected because they had previously demonstrated great applicability and effectiveness. Up to the starting point of our recent R&D work, the ENDT techniques tested met the basic requirements of a TRL up to 3 or 4. In some previously conducted projects, feasibility and accuracy were tested for some techniques, which also demonstrated capabilities of detecting and quantifying contamination and indicating failure scenarios of CFRP relevant to aeronautics (ENCOMB, ABiTAS). Among these methods, only the most promising and reliable were chosen for further investigation and improvement. Considering the objectives of this book, their accomplishment was in selecting the most promising techniques developed in previous research, as described above. Hence, the ambitious technical goal of the ComBoNDT project was to transfer the selected technologies from their current TRL state to a higher level (5 or 6). We addressed this goal through activities performed in different fields:

Table 1.1 Survey of national and international research and innovation activities connected to the ComBoNDT project

Research and innovation acronym and status	Main topics	Relation to ComBoNDT
ABiTAS [69]	Characterisation of adherend surfaces with conventional NDT with the aim of developing a robust, flexible and economic process chain for structural assembly by adhesive bonding. Surfaces after pretreatment were screened, tested and optimised	The results of in-process surface quality control of composite surfaces with the aim of achieving an increased TRL level were applied for ComBoNDT
ENCOMB [27]	Screening of more than 20 different technologies suitable for the characterisation of adherend surfaces and adhesive bonds, also those with a low TRL (1–2). Distinct contamination and degradation scenarios were introduced to the samples to test the eligibility and versatility of the techniques	Much of the knowledge from ENCOMB regarding suitable, highly promising NDT for the development of ENDT was transferred to ComBoNDT. The ENCOMB project finished in April 2014, so the full benefit of its results provided a foundation for ComBoNDT
CleanSky [67]	The influence of composite surface properties on bond durability in repair applications was studied. The pre-bond inspection of surfaces was a task in this context	The studies within ComBoNDT went beyond the work carried out in CleanSky, especially concerning the characterisation of adhesives and the testing of adhesive bonds by ENDT
BOPACS [68]	The aim of BOPACS was the weight reduction of primary aircraft structures by developing boltless adhesively bonded joints. The focus lay on the understanding of crack growth and debonding expansion mechanisms in adhesive joints with an aim of developing specific design features capable of preventing crack growth	Some partners from BOPACS were also members of the ComBoNDT consortium (Fraunhofer IFAM, Airbus Group, LTSM University of Patras). Thus, it was ensured that the findings within BOPACS were implemented in ComBoNDT

- The first step included the **test scenarios** for the project's scope as well as the sample preparation and overview of the sample measurements as previously defined by both the industrial end users and the ground operation service providers.
- The second step included the **sample preparation**, which complied with the requirements defined by the industrial partners.
- The next step was the **characterisation of pre-bond and post-bond quality** by reference methods using lab-based analytical methods. (The second and present steps were of paramount importance as a high quality of samples was imperative

to ensure reproducibility and comparability because the results of the ENDT technologies would be compared to the results of the lab methods).

- The next step included the conduction of **mechanical tests** to address the influence of contaminants on the bond strength using both established and innovative mechanical tests in samples with different levels of intentionally applied contaminants and, thus, different influences on the bond strength.
- After the completion of the mechanical tests, the next step concerned **ENDT for the quality assessment of adherend surfaces** (pre-bond inspection). The performance of the ENDT technologies was adapted and optimised regarding their sensitivity to detect the physico-chemical properties of adherend surfaces with multiple pre-bond contaminations as well as their suitability in quantifying the measuring results. This work included the further development of the technologies with regard to sensitivity, reliability, automation and mobility, the performance evaluation within real manufacturing or repair processes, and approaches for industrialisation. Another activity was performed in parallel which dealt with **ENDT for adhesive bondline quality assurance** (post-bond inspection). Within this activity, the evaluation and development of appropriate technologies with high sensitivity formed the primary interest in order to determine the bondline quality influenced by multiple pre-bond contaminations as well as to quantify the results. The sensibility, reliability, automation and mobility of the techniques were assessed by their implementation in real manufacturing or repair processes.
- A further step was the **demonstration** of the adapted and improved ENDT techniques in real application environments for both manufacturing and repair in order to reveal their suitability for future use in aeronautical industry applications.
- Another step was the **validation and technological assessment of ENDT** methods and investigations. In this part, the work was aimed specifically at the comparative evaluation of the produced innovations regarding their suitability for measurements in production or in-service environments through an assessment of their TRL (including simplicity of handling, time for data processing, detection capability, costs, lifetime, etc.), in-line capability, mobility and robustness. This activity also included probability of detection (POD) studies to evaluate the detection performance of each technique. In the frame of validation of the techniques of the project, a round-robin test was used.
- Finally, the broad **dissemination and exploitation** of the results, in parallel to the protection and safeguarding of the intellectual property rights of the partners involved in the project, was ensured, as the reader may infer from the present book.
- On the way to all these achievements, the overall **project management** was dedicated to all activities, ensuring efficient project coordination towards achieving the project's objectives.

Following the conceptualisation and throughout the course of our R&D work, distinct levels of pre-bond contamination, the influence of which was addressed via mechanical testing, needed to be previously detected and preferably discriminated by the ENDT techniques and quantitative evaluation of the quality-related data.

Table 1.2 Approaches pursued within this book, characterised by different types of test scenarios addressing the fields of application for in-plant manufacturing and in-service repair based on adhesive bonding

Sample geometry and surface state	*Fields of applications investigated*	
	ENDT for pre-bond (a) and post-bond (b) quality assurance for *manufacturing*	ENDT for pre-bond (a) and post-bond (b) quality assurance for *repair*
• Test coupons (flat surface) • Multiple contaminants	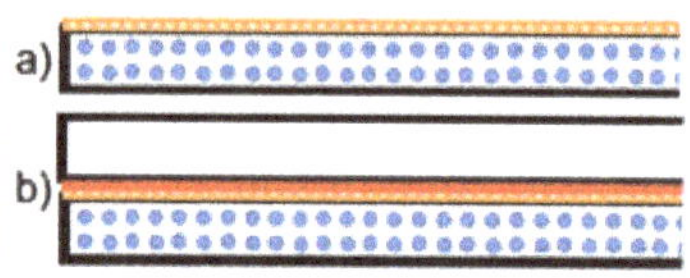	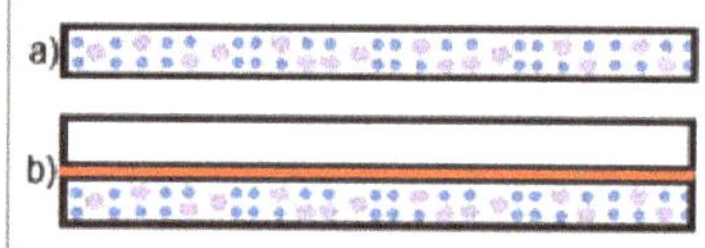
• Pilot samples (complex geometry) • Multiple contaminants	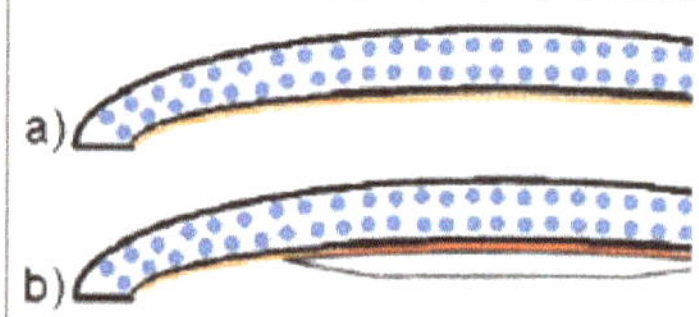	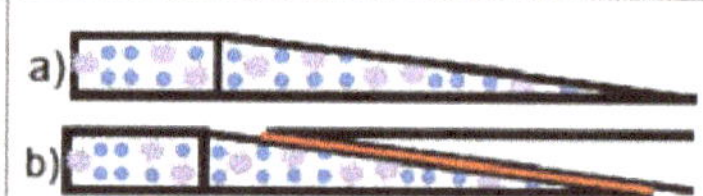
	moisture contamination release agent contamination	Skydrol© contamination adhesive bondline

From a manufacturing perspective, in the production scenarios, a contamination of adherend surfaces with, e.g. release agent and moisture was considered to be highly relevant. A variation due to the impact of moisture, hydraulic fluids such as Skydrol©, kerosene or de-icing fluid may also occur during aircraft operation. Therefore, corresponding exposures were taken into consideration for the repair test scenarios. These scenarios with contaminations, relevant for the respective field of application from manufacturing and repair, were applied to samples with increasing geometric complexity to mimic technologically relevant situations, as displayed in Table 1.2.

1.3.3 *Aims and Key Aspects*

The exploitation of lightweight CFRP materials for further use in aeronautic applications presupposes adhesive bonding as a very appropriate joining technology for load-critical primary structures [62] and the compliance with the appropriate certification (e.g. by the European Aviation Safety Agency, EASA). Due to the absence of an adequate quality assurance concept which can guarantee the safety of adhesively bonded joints and enable a corresponding certification, the previously mentioned exploitation was not possible. Such a quality assurance concept for adherend surfaces and the contacting bondline is crucial for load-critical bonds and must be effective. However, quality assessment exclusively involving the already existing NDT testing methods is not considered sufficiently reliable [70]. As the major part (80–90%) of any inspection is performed visually, it is very important for advances in quality inspection that the technologically relevant effects of deviations from qualified procedures, e.g. damage, which are not accessible to human sensory perception can be sensitively detected and revealed in-process. Taking this proposition into account, a number of research projects targeted addressing and developing innovative NDT technologies in laboratory environments. These trendsetting technologies should be able to monitor the adherend surfaces and bonded joints and assess their quality-relevant features which may be affected by surface contaminations or manifest in kissing bonds. To this end, the effect of single contaminants was assessed for simple sample geometries. Nevertheless, these NDT techniques had not been tested in scenarios of multiple contaminations or on realistic three-dimensional and complex geometries.

In the frame of this book, we will highlight these tests comprising tools and procedures for in-process quality assessment using pilot samples charged with multiple contaminants; moreover, the techniques' operation and their application are demonstrated on real manufacturing and repair components. The aim is to prove their suitability and reliability for surface and bondline quality detection and compilation. This work is performed with the ultimate goal of the certification of CFRPs as the primary material in critical structural applications. Another issue is the maturation of innovative NDT methods in terms of the TRL, considering the fact that these basically had only been tested in a laboratory environment. To this end, our R&D work aimed not only at the increment of their TRL but also at building up a catalogue of criteria for an assessment of ENDT techniques applicable to adherend surfaces and bondlines of load-critical CFRP structures. In addition, within the developed concept of assessing technologically relevant quality-related features, mechanical tests of the resulting bonded joints were highlighted as necessary in order to determine the influence of contaminants up to a contamination level that shows a risk. Considering that the existing standardised methods entail a high cost and work effort, focused destructive tests providing selected specific bond strength parameters and statistics were performed. Despite this issue, the mechanical tests chosen and performed were those that appeared to be the most time and work-efficient, in addition to the most innovative ones (i.e. the centrifuge test).

Mechanical testing

More precisely, when evaluating the influence of the intentionally varied adherend surface state (clean, single or multiple contaminations) on the mechanical properties, established standardised mechanical tests like mode-I, mode-II and mixed-mode fracture toughness tests are widely used. These tests are time- and cost-consuming due to complex sample preparation, destructive single-sample testing and manual evaluation of the mechanical load and of the two obtained fracture surfaces. These limitations for standardised testing have been recognised in many previous research projects such as ENCOMB and ABiTAS. Although these significant standardised mechanical tests were knowingly chosen to be performed in the frame of our R&D work, we additionally used a new multiple sample test (centrifuge-testing) because it was ascribed great potential to overcome the limitations of standardised tests within the BOPACS project. With this novel centrifuge test, up to eight samples can be measured within five minutes. The measured mechanical properties have previously shown accuracy with a very good precision and reproducibility. The novel testing is cost-efficient, fast and reliable, and it indeed increased the information value of mechanical testing compared to the results from standardised mechanical tests.

TRL assessment

TRLs are commonly used to evaluate the maturity of technologies (e.g. NDT techniques, pre-treatments) regarding their degree of development and applicability in industrial processes. There are several definitions of TRLs, e.g. from the EC for H2020, from the U.S. Department of Defense, NASA and ESA; however, these definitions are more general without including specific criteria for special applications. With regard to the fields of application addressed within this book (i.e. manufacturing and repair), different criteria are relevant for TRL assessment.

Establishing a satisfactory catalogue of criteria for the assessment of TRL for pre- and post-bond ENDT technologies to be used in the fields of manufacturing and repair for CFRP primary structures is a major challenge that will be elaborately assessed in this book, especially by our co-authors from industrial consortium partners experienced in manufacturing and ground services. They worked on the creation of a catalogue of requirements tailored specifically for the TRL assessment for ENDT in the chosen fields of application. With this catalogue, a distinct determination of TRL will be possible for the specific fields of application investigated within this book. With a distinct and comprehensive TRL assessment, the comparability of the test methods and the TRL improvements achieved within the project were measurable and became very clear. TRL assessment was performed according to this catalogue at the beginning (initial TRL) of our R&D work and at its end (final TRL achieved).

Monitoring of adherend surface conditions and bondline quality by in-line techniques

During the manufacturing and repair processes of CFRP materials, the quality assurance of adherend surfaces has, up to now, been performed using the water-break test for the large-area inspection of wettability properties [71]. Hydrophobic surface areas (originating from residues of, e.g. release agents or lubricants) are detected by changes in the wetting behaviour. The test is performed manually and its evaluation is done visually with an individual pair of human eyes and is, therefore, is subjective and error-prone. The water-break test is followed by time- and work-consuming drying and cleaning steps for the investigated specimens. Furthermore, water-soluble contaminants cannot be detected by this procedure (though they might have been present in the non-investigated surface regions as well), which is another disadvantage. For the inspection of small areas, contact angle or also surface energy measurements and test-inks are commonly employed. These tests only allow a monitoring of very small areas of the sample and require an additional cleaning step before bonding, and it is even recommended that they be performed adjacent to the intended bonding area, not inside it. Figure 1.10 shows a demonstrative example of applying the water-break test (left) and an ultrasonic picture of a delamination in CFRP (right).

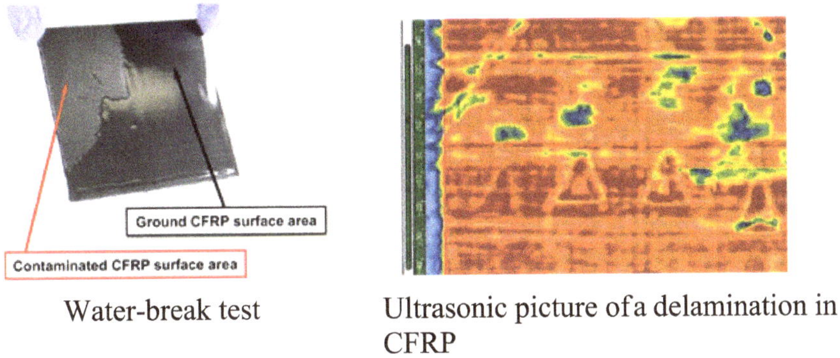

Fig. 1.10 The water-break test (left) and a state-of-the-art ultrasonic picture of the delamination (right) of CFRP

Adhesively bonded structures are also inspected by means of conventional NDT, e.g. visual inspection and audible sonic testing (tap test), to detect damages by comparing the local perception to one obtained in the vicinity or with a reference specimen. These tests are performed manually and are therefore subjective as the results depend to a great extent on the concentration, skills and experience of the operator. Furthermore, damage that is non-perceivable by unaided human senses will not be detected. More complex measuring techniques like ultrasound, thermography, shearography and radiography often need well-trained personnel and require time-consuming spectrum evaluation steps. Current conventional NDT techniques allow the detection of defects like pores, resin starvation/richness, wrinkles, discolouration (e.g. due to overheating, lightning strike), disbonding and delaminations in the joint area [72]. In addition, for ultrasound testing a couplant material such as water must be applied between the sensor and the investigated surface. Expensive re-drying and cleaning processes are therefore necessary and only single-point measurements are possible [70]. The state-of-the-art devices do not provide the necessary information for the quality assurance of adhesive joints and still have many disadvantages and shortcomings. The aforementioned state-of-the-art methods and devices are furthermore limited by the fact that defects like kissing bonds and a weakening of bonded joints cannot be detected.

Based on the results obtained within former research activities and extracted from a literature review, we carefully selected NDT technologies based on their state-of-the-art performance for further investigation in order to address these limitations. The identified technologies proved their capability to successfully detect contaminants relevant in aircraft manufacturing and in-service repair as well as to assess adhesive bond quality. All of the selected NDT techniques had the potential to reach a TRL high enough for their use in manufacturing and repair environments in order to provide quality control systems for surface and bondline inspection. The ENDT technologies that are relevant for the R&D work within this book are listed in Table 1.3 together with their main limitations in the early 2010s and the progress that was aspired to be realised within the project.

Table 1.3 State of the art, limitations of technologies and progress beyond limitations concerning ENDT for surface and bondline quality assessment contributing to quality assurance

Technology	State of the art (early 2010s)	Limitations (early 2010s)	Progress (shown in this book)
ENDT for surface quality assessment			
OSEE Optically stimulated electron emission	OSEE is a fast, contactless NDT technique to detect contaminants on surfaces. With lab setups, a detection of different contaminations is achieved	• Lab-scale for small samples; • Investigations limited to single contaminations; • Distance sensitivity and influence of surface topography unknown	Mobile detection unit for large-scale CFRP surface evaluation, influence of factors (topography, distance, mixture of contaminants) on signal rated
LIBS Laser-induced breakdown spectroscopy	Established for bulk metal and surface quality analysis Contaminations detected by spectral peak evaluation of tracer elements	• Restricted to elemental peak evaluation; • Limited to small sample sizes due to laser protection requirements; high costs	Evaluation of molecule bands including automated spectra evaluation, downsizing of the technique to reduce costs
Electronic nose	E-nose is under rapid development for gas mixture analysis, classification and quantification of components. Its suitability for ENDT purposes has been demonstrated	• Unsuitable to cope with moisture contamination; • Mixtures of contaminants not yet investigated; • High costs per device; • No onboard data analysis	Detection of contaminant mixes and water; Minimal set of sensors to lower the costs; Data analysis tools
AWT Aerosol wetting test (bonNDTinspect)	Determination of surface wetting behaviour with AWT was successfully applied for the detection of surface contaminants. Existing lab setup	• Detection of multiple contaminants not tested; • Large-area scanning and automation not yet performed	Development of automated AWT System (bonNDTinspect) for robotically operated AWT for large area scanning; Measurement of surfaces with multiple contaminants

(continued)

Table 1.3 (continued)

Technology	State of the art (early 2010s)	Limitations (early 2010s)	Progress (shown in this book)
Full-field vibrometry	Detection of cracks and delamination in CFRP demonstrated; applicable to larger samples	Applicability for surface contamination and weak bonds not yet investigated	Improvement of signal processing techniques for full wave field measurements; Signal analysis in wavenumber domain; separation of propagating wave modes; Adaptive wavenumber filtering
FTIR Fourier transform infrared spectroscopy	Used for qualitative detection of different materials; First approaches for quantitative evaluations of IR spectra by multivariate methods	Penetration depth in CFRP is restricted to the first fibre layer No experience with a combination of different contamination scenarios	Collecting absorption spectra of contaminated surfaces and evaluating the data with a multivariate method
ENDT for adhesive bondline quality assessment			
SWAT Shock waves adhesion test	Shock wave propagation used to induce localised and adjustable tensile stress within assemblies Laser shocks have proved to be efficient in a variety of samples	• Availability of shock wave generation technologies; • Difficulty to generate traction at a selected interface; • Single shock duration tests not suitable for strong bondlines	Adjustable pulsed laser and plates launcher for shock generation Numerical tool to determine adequate shock parameters
Linear and non-linear ultrasound	Linear ultrasound uses bulk and guided waves to detect changes in impedance Non-linear ultrasound relies on non-linear phenomena to detect weak bonds	• Linear ultrasound showed limitations for the detection of weak bonds • Non-linear ultrasound is hard to generate and the obtained data are difficult to interpret	Considers various guided wave modes that might be more sensitive to a lack of adhesion Considered a more powerful source. Generated non-linear effect in the kissing bond area
Electromechanical impedance	Detection of weak bonds due to release agent using piezoelectric transducer was demonstrated	• Adhesively bonding the piezoelectric transducer is required	Analysing the influence of the transducer bonding; the investigation of the proper frequency range for assessing weak bonds; the development of multivariate analysis for the features extracted from the measured electromechanical spectra; Measurement of electrical parameters without transducer

1.3.4 *Impacts and Contributions*

The use of CFRP as an innovative material for aircraft design has increased immensely within the few last years [22, 29, 62] (the Airbus A350 XWB is composed of up to 53% CFRP), leading to an increasing demand for both high quality and quantity of composite material and composite joints, accompanied by lower costs. Adhesive bonding as a joining technology can enable the use of the full potential of CFRP as a lightweight material for aircraft design and is already used for joining many CFRP-based aircraft parts. Admittedly, adhesive bonding has thus far not been applied for load-critical CFRP primary aircraft structures. However, it is highly desirable to facilitate adhesive bonding as a joining technology for load-critical primary structures because adhesive bonding and adhesive joints possess numerous advantages over other processes: homogeneous stress distribution, full automation capability, lightweight design, strong and even complex structure design, capability of joining two distinct materials, interesting properties in electrolytic and corrosion protection and finally high fatigue resistance furnishing a longer service lifetime compared to mechanically bonded structures.

Most of the aircraft produced nowadays contain a high amount of CFRP material and therefore require the adhesive bonding of CFRP in manufacturing, maintenance and repair [29]. From our point of view, this further emphasises the need for complete and reliable quality assurance concepts based on appropriate and resilient ENDT techniques for quality assessment. This large number of affected industry sectors leads to the huge impact of the findings and achievements reported in this book. The provision of quality assurance concepts (using ENDT techniques) for load-critical primary structures will allow the increased and optimal use of CFRP and the replacement of metallic assembly concepts resulting in weight, time and cost savings for the aeronautic and all related industries. Moreover, it directly supports the certification and continued airworthiness of repaired CFRP structures. The R&D work presented in this book is clearly aligned with this objective through the maturation, improvement and adaptation of ENDT technologies for the characterisation of, on the one hand, the CFRP adherend surface state before bonding (pre-bond) and, on the other hand, the CFRP bonded structures (post-bond) in order to establish complete and process-comprehensive quality assurance concepts. We consider this the key to overcoming current limitations for the use of CFRP in aeronautical applications.

We expect that the innovations reported in this book will result in substantial socio-economic, technical and environmental impacts. Firstly, an impact will be achieved by reinforcing the competitiveness of Europe's aircraft industry and European aircraft operators by assisting the development of high technology SMEs and by accomplishing an increment of safety resulting from more reliable components and processes, which will strengthen and augment the reputation of the European aviation and aeronautics industry. Secondly, our contributions will bring about an integration of innovative materials in aeronautics as well as automated measuring processes in combination with strong signal evaluation processes that will provide surface and structural health data and contribute significantly to the reduction of human errors; therefore, an increment of safety will result from more reliable components and processes, weight savings of aircraft will be facilitated, and an improvement of the safety and operational capabilities of aircraft will ensue. Thirdly, a significant reduction of energy for the performance of the inspection, for the manufacturing of replacement parts and for the manufacturing of primary structures will be feasible, as will a reduction of scrapping rates during manufacture or repair and a weight reduction of aircraft.

Finally, we would like to highlight and acknowledge again the invaluable contributions resulting from the strong cooperation within the ComBoNDT consortium and promoted by the dedicated project management. The consortium consisted of eleven partners, including major European aerospace companies as well as high-level research organisations and universities experienced in aeronautics research and development. The project team consisted of aircraft manufacturers (Airbus Group; AERNNOVA Composites Illescas S.A.) in close collaboration with the research partners Fraunhofer Institute for Manufacturing Technology and Advanced Materials IFAM (IFAM), Instytut Maszyn Przepływowych im. Roberta Szewalskiego Polskiej Akademii Nauk (IMP-PAN), Centre National De La Recherche Scientifique (CNRS), Agenzia nazionale per le nuove tecnologie, l'energia e lo sviluppo economico sostenible (ENEA), University of Patras (LTSM-UPAT), and the Commissariat à l'énergie atomique et aux énergies alternatives (CEA), together with the small and medium-sized enterprises EASN Technology Innovation Services BVBA (EASN), GMI Aero SAS (GMI), and Automation W + R GmbH (AWR).

1.4 Synopsis

In this chapter, we presented a short introduction to ENDT and quality assessment in adhesive bonding processes relevant to the manufacture or repair of composite structures. We highlighted their relevance as field-level sensing systems for industrial automation and also for literally safeguarding quality in various steps of adhesive bonding processes. They will essentially contribute to quality assurance and optimisation within a manufacturing technology that we consider the most auspicious of the twenty-first century for innovations in joining functional or lightweight materials and components.

Based on a fast, firm and formal description of adhesive bonding processes and conferred to quality assurance, we determined that the analytical requirement for a monitoring process is far less complex than predicting any property of the ultimately manufactured adhesive joint, like the initial (or even final) strength or fracture toughness, which is often set as a fundamental design specification. Rather, indicating any deviations from a known, understood and qualified procedure will facilitate the timely, purposeful and precise amendments for achieving and maintaining the technologically relevant material and environment states following the requirements. All things considered, the documented compliance with documented procedures contributes to risk reduction and safety, while also safeguarding the economic value and social acceptance of the processes and products.

We have detailed that, especially in the past decade, the monitoring of quality-relevant operand features which are characteristic for the adherend and adhesive materials has caught up with the virtually established methods and instrumentation for controlling the acting operators, i.e. process factors which effectuate changes of the operand features. A major contribution to this advance in operand QA was traced back to recent developments of ENDT tools, e.g. in the European joint research projects ENCOMB and ComBoNDT. Moreover, we explicated the layout of a concept that was developed in these interdisciplinary and applied research projects. This is based on an interdisciplinary, comprehensive and forthright analysis and a description of the supposedly controlled production environment and monitoring of deviations or events which occur when quality-relevant operator-related factors are intentionally and quantitatively altered in technologically relevant increments.

We hope that we have intrigued our readers, and we will animate this concept for ENDT and quality assessment in adhesive bonding within the subsequent chapters of this book.

References

1. Regert M (2004) Investigating the history of prehistoric glues by gas chromatography-mass spectrometry. J Sep Sci 27:244–254. https://doi.org/10.1002/jssc.200301608
2. Conard NJ, Malina M, Münzel SC (2009) New flutes document the earliest musical tradition in southwestern Germany. Nature 460:737–740. https://doi.org/10.1038/nature08169
3. Gegner J (2008) Klebtechnik – multifunktionales Fügen für den nachhaltigen Werkstoffeinsatz im 21. Jahrhundert. Mat.-wiss u Werkstofftech 39:33–44. https://doi.org/10.1002/mawe.200700228
4. Regert M, Rodet-Belarbi I, Mazuy A et al (2019) Birch-bark tar in the Roman world: the persistence of an ancient craft tradition? Antiquity 93:1553–1568. https://doi.org/10.15184/aqy.2019.167
5. Trehub SE, Becker J, Morley I (2015) Cross-cultural perspectives on music and musicality. Philos Trans R Soc Lond B Biol Sci 370:20140096. https://doi.org/10.1098/rstb.2014.0096
6. Conard NJ, Malina M (2008) New evidence for the origins of music from the caves of the Swabian Jura. In Both AA, Eichmann R, Hickmann E, Koch L-Ch (Eds), Challenges and objectives in music archaeology. Studien zur Musikarchäologie VI, Orient-Archäologie 22:13-22, Verlag Marie Leidorf GmbH, Rahden, Germany, ISBN 978-3-896-46652-5
7. Osipowicz G (ed) (2014) Kowal 14: Sepulchral and ritual place of people representing the Globular Amphora Culture. Wydawnictwo Naukowe Uniwersytetu Mikołaja Kopernika, Toruń, ISBN 978-83-231-3328-5
8. Rao H, Yang Y, Abuduresule I et al (2015) Proteomic identification of adhesive on a bone sculpture-inlaid wooden artifact from the Xiaohe Cemetery, Xinjiang, China. J Archaeol Sci 53:148–155. https://doi.org/10.1016/j.jas.2014.10.010
9. Brockmann W, Geiß PL, Klingen J, Schröder KB (2008) Adhesive bonding: materials, applications and technology. Wiley-VCH, Weinheim, ISBN 978-3-527-62393-8
10. Kozowyk PRB, Soressi M, Pomstra D et al (2017) Experimental methods for the Palaeolithic dry distillation of birch bark: implications for the origin and development of Neandertal adhesive technology. Sci Rep 7:8033. https://doi.org/10.1038/s41598-017-08106-7

11. Gross A, Lohse H (2015) Quality assurance in adhesive bonding technology—new DIN 2304 standard and its use in practice. Adhes Adhes Sealants 12–17. https://doi.org/10.1007/s35784-015-0534-4
12. Reiling JM, Middendorf P, Sindel M (2015) Quality assurance of adhesive processes in the body shop. In: Bargende M, Reuss H-C, Wiedemann J (eds) 15. Internationales Stuttgarter Symposium. Springer Fachmedien Wiesbaden, Wiesbaden, pp 1421–1431
13. Paul F (2017) Adhesive bonding roadmap—increasing trust in adhesives. DECHEMA, Frankfurt am Main
14. Encinas N, Oakley BR, Belcher MA et al (2014) Surface modification of aircraft used composites for adhesive bonding. Int J Adhes Adhes 50:157–163. https://doi.org/10.1016/j.ijadhadh.2014.01.004
15. Ledesma RI, Palmieri FL, Yost WT et al (2017) Surface monitoring of CFRP structures for adhesive bonding, 40th Annual Meeting, The Adhesion Society, St. Petersburg, Florida, USA
16. Kraft A, Brune K (2017) Assured use of release agents and adhesives in the same production process. Lightweight des Worldw 10:40–43. https://doi.org/10.1007/s41777-017-0013-5
17. Konstantopoulos S, Fauster E, Schledjewski R (2014) Monitoring the production of FRP composites: a review of in-line sensing methods. Express Polym Lett 8:823–840. https://doi.org/10.3144/expresspolymlett.2014.84
18. Hübner M, Lang W, Dumstorff G (2017) Surface integrated printed interdigital structure for process monitoring the curing of an adhesive joint. Proceedings 1:631. https://doi.org/10.3390/proceedings1040631
19. Soldatos J, Gusmeroli S, Malo P, Di Orio D (2016) Internet of Things applications in future manufacturing. In: Vermesan O, Friess P (eds) Digitising the industry: Internet of Things connecting the physical, digital and virtual worlds. River Publishers, Gistrup, Denmark, ISBN 978-8-793-37981-7
20. de Baas AF (2017) Data and documentation terminology, classification and ontology. Common terminology and classification, efficient communication. https://emmc.info/wp-content/uploads/2017/12/EMMC_IntOp2017_Cambridge_deBaas_EC.pdf
21. Gellrich A, Lunkwitz D, Dennert A et al (2012) Rule-driven manufacturing control based on ontologies. In: Proceedings of 2012 IEEE 17th international conference on emerging technologies & factory automation (ETFA 2012). IEEE, pp 1–8. https://doi.org/10.1109/ETFA.2012.6489545
22. Ullmann T, Schmidt T, Hoffmann S et al (2010) In-line quality assurance for the manufacturing of carbon fiber reinforced aircraft structures. In: NDT.net Issue: 2011-03, 2nd Int. Symposium on NDT in aerospace 2010—Tu.1.A.4
23. Dahmen T, Trampert P, Boughorbel F et al (2019) Digital reality: a model-based approach to supervised learning from synthetic data. AI Perspect 1. https://doi.org/10.1186/s42467-019-0002-0
24. Gifford C, Daff D (2017) ISA-95 evolves to support smart manufacturing and IIoT—New challenges and opportunities for manufacturing technologies and standards across industries, InTech, November/December
25. Feng SC, Bernstein WZ, Hedberg T et al (2017) Towards knowledge management for smart manufacturing. J Comput Inf Sci Eng 17. https://doi.org/10.1115/1.4037178
26. Katnam KB, da Silva LFM, Young TM (2013) Bonded repair of composite aircraft structures: a review of scientific challenges and opportunities. Prog Aerosp Sci 61:26–42. https://doi.org/10.1016/j.paerosci.2013.03.003
27. ENCOMB "Extended Non-Destructive Testing of Composite Bonds" (2010–2014) Project funded from the European Union's Seventh Framework Programme under grant agreement No 266226
28. ComBoNDT "Quality assurance concepts for adhesive bonding of aircraft composite structures by advanced NDT" (2015–2018) Project funded from the European Union's Horizon 2020 research and innovation programme under grant agreement No 636494
29. Kanterakis G, Chemama R, Kitsianos K (2020) Bonded repair of composite structures. In: Pantelakis S, Tserpes K (eds) Revolutionizing aircraft materials and processes. Springer Nature, [S.l.], pp 359–392. https://doi.org/10.1007/978-3-030-35346-9_13

30. Light GM, Kwun H (1989) Nondestructive evaluation of adhesive bond quality. State-of-the-Art Review: SwRI Project 17-7958-838, Nondestructive Testing Information Analysis Center San Antonio, Texas, USA
31. Mattmann I (2017) Modellintegrierte Produkt- und Prozessentwicklung. Springer Fachmedien Wiesbaden GmbH, Wiesbaden, Germany, ISBN 978-3-658-19408-6
32. Espie AW (1998) Quality assurance in adhesive technology: EUREKA Project EU716. Abington Publishing, Cambridge
33. DIN EN ISO 9000:2015-11 (2015) Qualitätsmanagementsysteme - Grundlagen und Begriffe (ISO 9000:2015): Deutsche und Englische Fassung EN ISO 9000:2015(DIN EN ISO 9000:2015-11)
34. Markus S, Wilken R, Dieckhoff S (2006) Fehlervermeidung durch Inline-Monitoring des Oberflächenzustandes. Adhaes Kleb Dicht 50:20–22. https://doi.org/10.1007/BF03253336
35. DIN 2304-1:2016-03 (2016) Adhesive bonding technology—quality requirements for adhesive bonding processes: Part 1: Adhesive bonding process chain (DIN 2304-1:2016-03)
36. Allotrope Foundation (2020) Data Standard | Allotrope Foundation. https://www.allotrope.org/
37. Qin SJ, Zheng Y (2013) Quality-relevant and process-relevant fault monitoring with concurrent projection to latent structures. AIChE J 59:496–504. https://doi.org/10.1002/aic.13959
38. Peng K, Zhang K, You B et al (2016) A quality-based nonlinear fault diagnosis framework focusing on industrial multimode batch processes. IEEE Trans Ind Electron 1. https://doi.org/10.1109/TIE.2016.2520906
39. Huang J, Yan X (2017) Quality relevant and independent two block monitoring based on mutual information and KPCA. IEEE Trans Ind Electron 64:6518–6527. https://doi.org/10.1109/TIE.2017.2682012
40. Huang J, Yan X (2018) Relevant and independent multi-block approach for plant-wide process and quality-related monitoring based on KPCA and SVDD. ISA Trans 73:257–267. https://doi.org/10.1016/j.isatra.2018.01.003
41. Moyne J, Iskandar J (2017) Big data analytics for smart manufacturing: case studies in semiconductor manufacturing. Processes 5:39. https://doi.org/10.3390/pr5030039
42. Berger D, Zaiß M, Lanza G et al (2018) Predictive quality control of hybrid metal-CFRP components using information fusion. Prod Eng Res Devel 12:161–172. https://doi.org/10.1007/s11740-018-0816-1
43. Wassink CHP (2012) Innovation in non destructive testing. Dissertation, TU Delft
44. Custódio J, Cabral-Fonseca S (2013) Advanced fibre-reinforced polymer (FRP) composites for the rehabilitation of timber and concrete structures: assessing strength and durability. In: Bai J (ed) Advanced fibre-reinforced polymer (FRP) composites for structural applications. Elsevier Science, Burlington, pp 814–882. https://doi.org/10.1533/9780857098641.4.814
45. DIN 6701-2:2015-12 (2015) Adhesive bonding of railway vehicles and parts—Part 2: Qualification of manufacturer of adhesive bonded materials(DIN 6701-2:2015-12)
46. Michaloudaki M (2005) An approach to quality assurance of structural adhesive joints. PhD thesis, Technical University Munich
47. Davis M (2004) Best practice in adhesive bonding. https://www.niar.wichita.edu/niarworkshops/Portals/0/BestPracticesinAdhesiveBondingMaxDavis.pdf
48. Niermann D, Gross A, Brede M et al (2005) Quality assurance in adhesive bonding technology. Adhaes Kleb Dicht 30–32
49. Niermann D, Gross A, Brede M et al (2005) Quality assurance in adhesion (Part 1): Construction phase—which general valid quality assurance measures apply to be observed? Adhaes Kleb Dicht 49:36–38
50. Niermann D, Gross A, Brede M et al (2005) Quality assurance in adhesive technology (Part 2): Production phase. Adhaes Kleb Dicht 49:28–32
51. Pirolini A (2014) Quality assurance trends in the automotive industry: an interview with Alex Koprivc. AZO Materials. https://www.azom.com/article.aspx?ArticleID=11065
52. Flämmich M, Herdin H, Kunz U, Mannschreck K, Schulze L, Straub R (2020) Guidelines for Quality Assuring Process Management in Parts Cleaning. Fachverband Industrielle Teilereinigung e.V. (FiT)

53. Ebnesajjad S, Landrock AH (2015) Chapter 13—Quality control. In: Ebnesajjad S, Landrock AH (eds) Adhesives technology handbook (Third Edition) William Andrew Pub, Norwich, NY, pp 353–374. https://doi.org/10.1016/B978-0-323-35595-7.00013-9
54. Messler SW (1993) Joining of advanced materials. Elsevier Science, Burlington
55. Breuer F, Muckel P, Dieris B (2018) Reflexive Grounded Theory: Eine Einführung für die Forschungspraxis, 3., vollständig überarbeitete und erweiterte Auflage. Lehrbuch. Springer, Wiesbaden
56. Nielsen J (1994) 10 Usability heuristics for user interface design. https://www.nngroup.com/articles/ten-usability-heuristics/
57. Kleining G, Witt H (2000) The qualitative heuristic approach: a methodology for discovery in psychology and the social sciences. Rediscovering the method of introspection as an example. Forum: Qualitative Social Research 1(1), Art. 13. http://nbn-resolving.de/urn:nbn:de:0114-fqs0001136
58. Budhe S, Banea MD, de Barros S et al (2017) An updated review of adhesively bonded joints in composite materials. Int J Adhes Adhes 72:30–42. https://doi.org/10.1016/j.ijadhadh.2016.10.010
59. Kreling S, Blass D, Wilken R et al (2017) Sauber und prozesssicher vorbehandeln — Teil 1: [Pretreating cleanly and in a process-reliable manner - PART 1]. Adhaes Kleb Dicht 61:18–23
60. Kreling S, Blass D, Wilken R et al (2017) Sauber und prozesssicher vorbehandeln - PART 2: [Pretreating cleanly and in a process-reliable manner - PART 2]. Adhaes Kleb Dicht 24–28
61. Halliwell S, Repair of fibre reinforced polymer (FRP) structures. National composites network best practice guide
62. Tserpes K (2020) Adhesive bonding of aircraft structures. In: Pantelakis S, Tserpes K (eds) Revolutionizing aircraft materials and processes. Springer Nature, [S.l.], pp 337–357. https://doi.org/10.1007/978-3-030-35346-9_12
63. Tornow C, Brune K, Noeske M (2014) Verfahren und Vorrichtung zur Bestimmung einer Oberflächengüte (DE102014209862A1)
64. Jeenjitkaew C, Guild FJ (2017) The analysis of kissing bonds in adhesive joints. Int J Adhes Adhes 75:101–107. https://doi.org/10.1016/j.ijadhadh.2017.02.019
65. Acare (2020) Acare—Advisory Council for Aviation Research and Innovation in Europe. https://www.acare4europe.org/
66. EASA (2010) Composite Aircraft Structure: Annex II to ED Decision 2010/003/R of 19/07/2010(AMC 20-29). https://www.easa.europa.eu/sites/default/files/dfu/Annex%20II%20-%20AMC%2020-29.pdf
67. Clean Sky (2020) Welcome to the Clean Sky | Clean Sky. https://www.cleansky.eu/
68. BOPACS "Boltless assembling Of Primary Aerospace Composite Structures" (2012 – 2016) Project funded from the European Union's FP7-TRANSPORT programme under grant agreement No 314180. https://cordis.europa.eu/project/id/314180/de
69. ABiTAS "Advanced Bonding Technologies for Aircraft Structure" (2006 – 2010) Project funded from the European Union's FP6-AEROSPACE programme under grant agreement No 30996
70. Esquivel O, Seibold RW (2004) Capabilities and Limitations of Nondestructive Evaluation Methods for Inspecting Components Beneath Thermal Protection Systems. AEROSPACE REPORT NO. ATR-2004(5081)-1, Cambridge, MA 02142-1093
71. ASTM F22 - 13: Standard Test Method for Hydrophobic Surface Films by the Water-Break Test. In: ASTM Volume 15.03 Space Simulation; Aerospace and Aircraft; Composite Materials. ASTM International
72. Federal Aviation Administration (2017) AMT Airframe Handbook Volume 1 (Full Version) (Faa-H-8083-31). By: Federal Aviation Administration. Aviation Maintenance Technician Handbook - Airframe, vol 1. Createspace Independent Publishing Platform

2

Effect of Pre-Bond Contamination and Aging on Carbon Fiber Reinforced Plastic

Konstantinos Tserpes, Elli Moutsompegka, Mareike Schlag, Kai Brune, Christian Tornow, Ana Reguero Simón, and Romain Ecault

Abstract In this chapter, the pre-bond contamination and ageing effects on carbon fiber reinforced plastic (CFRP) adherends and CFRP bonded joints are characterized by means of reference laboratory non-destructive testing (NDT) methods, mechanical tests, and numerical simulation. Contaminations from two fields of application are considered, namely in aircraft manufacturing (i.e. production) and for in-service bonded repair. The production-related scenarios comprise release agent, moisture, and fingerprint, while the repair-related scenarios comprise fingerprint, thermal degradation, de-icing fluid, and a faulty curing of the adhesive. For each scenario, three different levels of contamination were pre-set and applied, namely low, medium and high level. Furthermore, two types of samples were tested, namely coupons and pilot samples (a stiffened panel and scarf repairs). The CFRP adherends were contaminated prior to bonding and the obtained surfaces were characterized using X-ray photoelectron spectroscopy. After bonding, the joints were tested by ultrasonic testing. To characterize the effects of each contamination on the strength of the bonded joints, mode-I and mode-II fracture toughness tests, and novel centrifuge tests were conducted on the coupons, while tensile tests were performed on the scarfed samples. Additionally, numerical simulation was performed on CFRP stiffened panels under compression using the LS-DYNA finite element (FE) platform.

K. Tserpes (✉) · E. Moutsompegka
Laboratory of Technology & Strength of Materials, Department of Mechanical Engineering & Aeronautics, University of Patras, 26500 Patras, Greece
e-mail: kitserpes@upatras.gr

M. Schlag · K. Brune · C. Tornow
Fraunhofer Institute for Manufacturing Technology and Advanced Materials IFAM, Wiener Str. 12, 28359 Bremen, Germany

A. Reguero Simón
Aernnova Composites Illescas Sau, Parque Industrial y Tecnológico, Avda Barajas, 3 Illescas, 45200 Toledo, Spain

R. Ecault
Airbus Operations S.A.S., 316, route de Bayonne, B.P. D4101, 31060 Toulouse Cedex 9, France

Keywords Pre-bond contamination · Hygrothermal aging · Fracture toughness · Centrifuge testing · Ultrasound testing · X-ray photoelectron spectroscopy

2.1 Introduction

The use of adhesive bonding in aircraft structures is increasing, both in the assembly of structural parts and when applying composite patch repairs due to the numerous advantages it offers over conventional joining techniques [1–4]. These include a more uniform stress distribution in the area of the joint, the ability to join dissimilar materials, the improved fatigue properties, and the attractive strength to weight ratio. However, the use of adhesive bonding technology is presently limited to the joining and patch repairing of secondary structures that are not load-critical. Amongst the reasons inhibiting the certification of adhesive bonding for primary structures is the sensitivity of the bondline integrity to the presence of defects, which might counteract the strength of the joints. These defects are not accessible to visual monitoring during the bonding process, and they usually are caused by a pre-bond contamination of the adherend surface during either the manufacture of the joints or their repairs. As conditions on the aircraft production line and in the maintenance/overhaul shed are different, defects are categorized as either production-related or repair-related. Table 2.1 lists the pre-bond contamination scenarios under consideration in the ComBoNDT project.

Amongst the defects that might arise during the manufacture of adhesive joints, the most critical ones are those that are not detectable by the available NDT methods. In addition to developing extended NDT (ENDT) methods capable of detecting such effects [6], evaluating their effect on the strength of adhesive joints is of equal importance. Early experimental studies conducted within the ENCOMB project [1, 7, 8] have shown that undetectable defects caused by pre-bond contamination may significantly degrade the mode-I fracture toughness of CFRP joints. In the ComBoNDT project, the experimental characterization has extended the current state of research by conducting mode-I fracture toughness tests for additional contamination scenarios, mode-II tests on the complete set of contamination scenarios (Table 2.1), and novel centrifuge tests. The centrifuge tests are fast and cost-effective tests that can be used to determine the adhesion strength of bonded joints. They serve as an alternative to the fracture toughness tests, which are more expensive and time consuming. In the test program of the ComBoNDT project, mode-II tests on aged

Table 2.1 The contamination scenarios studied in Chaps. 2 through 5 in the ComBoNDT project [5]

Production-related (P)	Repair-related (R)
Release agent (P-RA)	Thermal degradation (R-TD)
Moisture (P-MO)	De-icing fluid (R-DI)
Fingerprint (P-FP)	Fingerprint (R-FP)
	Faulty curing of the adhesive (R-FC)

joints have been also conducted with the aim of assessing the combined effect of pre-bond defects and after-bond hygrothermal aging. The material geometry tested comprises not only coupon-level samples, but also a series of samples that have more complex geometries derived from real geometries in the fields of application and are highly relevant for aerospace applications. These are the pilot samples and consist of scarfed samples and individual aerospace component parts (stiffened panels). These samples are used to evaluate the efficiency and check the applicability of ENDT methods on more complex/curved geometries with multiple contaminations in conjunction with a clean reference condition. With these samples, ENDT methods can be adapted to overcome the limitations arising from measurements on non-flat surfaces.

The debonding process is already complicated since it involves a three-material system (adherend, adhesive, and adherend/adhesive interface) as well as the geometric challenges of the parts themselves. Therefore, to determine the effects of the contaminations on the bonded joints of individual aerospace component parts, a numerical simulation was performed under compression loading using the LS-DYNA finite element (FE) platform.

The present chapter describes the respective contributions of the individual partners of the ComBoNDT consortium. The manufacturing of the CFRP adherends for the coupons and the pilot samples was performed by Aernnova Composites; the pre-bond contamination and the bonding of the samples was performed by Fraunhofer IFAM; the characterization of the adherends using X-ray photoelectron spectroscopy was performed by Fraunhofer IFAM; the ultrasound testing of the bonded plates and pilot samples was performed by Airbus; the mechanical testing of the coupons and the pilot samples as well as the aging of the coupons was performed by the University of Patras; and the numerical simulation of the stiffened panels was also conducted by the University of Patras. The datasets obtained from the mechanical testing represent design-relevant operand features. The subsequently detailed procedures offer a well-tried, step-by-step approach for compiling these features and, therefore, constitute an essential input into the framework of the applied concept for the quality assessment of adhesively bonded joints described in this book.

2.2 Materials and Sample Geometries

2.2.1 Basic Materials

Hexcel® M21E was used for the preparation of the test coupons. HexPly M21E/IMA, developed from Hexcel®'s M21 third-generation thermosetting epoxy resin system, uses an intermediate modulus fiber to balance superior strength and stiffness, and it was developed specifically for aircraft applications performed at Airbus. The matrix resin was developed to ensure an optimal translation of the carbon fiber properties whilst delivering outstanding fracture resistance. The sample plates were produced by Aernnova Composites using the liquid water-based silicone-containing release agent Frekote® C-600 in order to obtain smooth surfaces.

Regarding the structural layout, CFRP monolithic structures were manufactured according to the Airbus AIPS 03-02-019 standard for CFRP ("Manufacture of monolithic parts with thermoset prepreg materials"). For the fracture toughness testing, the adherends consisted of eight unidirectional plies and their layup sequence was $[0_2, \pm45]_s$ according to the AITM 1-0053 standard [9]. A release film—25 mm in length for the contaminated samples and 30 mm in length for the contaminated/aged samples—was inserted at one end of the sample prior to bonding to obtain an initial delamination for the fracture toughness tests.

The specimens used in the centrifuge tests had a stamp-to-plate configuration. The modular test stamps bonded to the CFRP adherends consisted of an aluminum (EN AW-2007) adherend screwed onto a body of mass made of copper. The CFRP adherends were manufactured from the M21E/IMA prepreg material. The layup sequence of the panels was $[(0/90/45/\text{-}45)_3]_s$.

For the adhesive bonding of the adherends, a film adhesive was used instead of a paste in order to standardize the thickness and increase the reliability of the results. Specifically, the film adhesives FM 300 K and FM 300-2 from Cytec® (0.20 and 0.25 mm thickness, respectively) were used for the production and the repair scenario, respectively.

2.2.2 *Sample Geometries*

2.2.2.1 Coupons

For the mode-I fracture toughness tests, a double cantilever beam (DCB) specimen was used to mirror the relevant geometry and the loading of the adhesive joint. A DCB specimen consists of rectangular adherends bonded along their length, incorporating a region of non-adhesive release film at one end for the introduction of the initial crack in the bondline during testing (Fig. 2.1a). For the mode-II fracture toughness tests, an end notched flexure (ENF) specimen was used. Similar to the DCB specimens, ENF specimens also consist of rectangular adherends, but with a longer pre-crack that is embedded through the width at the end of the specimen to accommodate the sliding deformation of the adherends that result from the flexural loading (Fig. 2.1b). In order to provide crack growth stability, the initial crack length was considered to be equal to 70% of $L/2$ [10, 11]. The test specimens were cut from the residual part of the mode-I specimens.

Finally, the specimen for the centrifuge tests comprised a rectangular composite plate (adherend) bonded to a metallic cylindrical stamp using a technically relevant adhesive system. The test stamps had a diameter of 10 mm on the bonding face. The samples were cut to the desired size by dry diamond cutting. The final dimensions of the CFRP adherends were 25 mm × 25 mm × 4.4 mm.

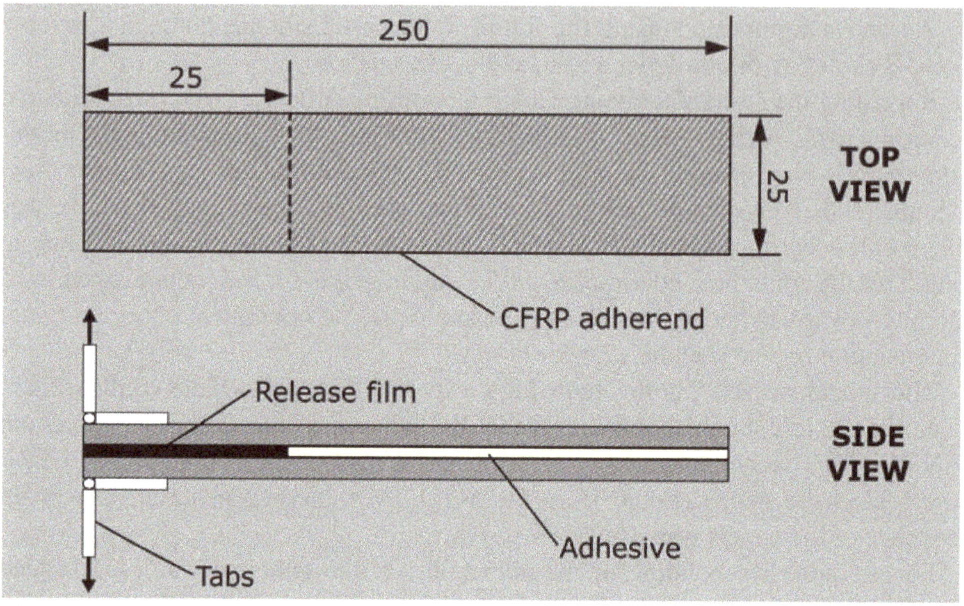

a)

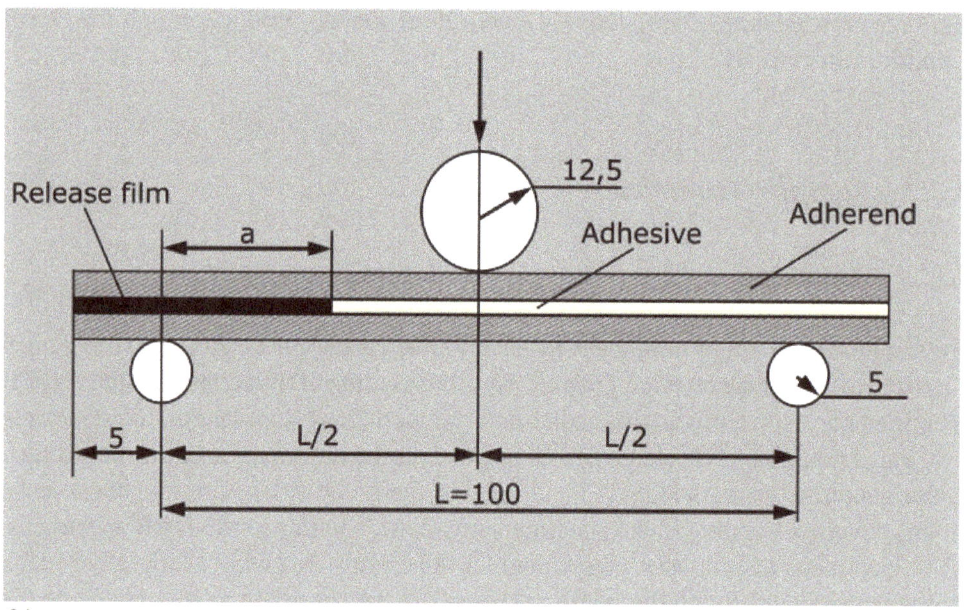

b)

Fig. 2.1 Geometry, dimensions and boundary conditions of **a** DCB coupon and **b** the ENF coupon

2.2.2.2 Scarfed Samples

As already stated, the two main relevant scenarios for CFRP adhesive joint quality assurance are manufacturing, when the joints are first bonded during aircraft construction, and maintenance, when aerostructures are repaired. Scarfed joints were used for the second scenario. When damage is detected, the aircraft part is locally scarfed

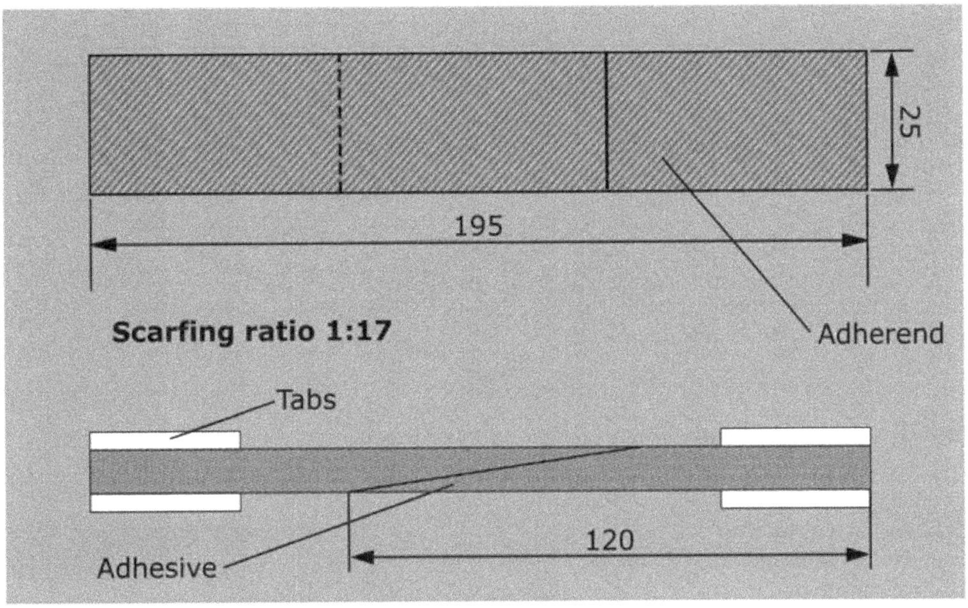

Fig. 2.2 Geometry and dimensions of the scarfed sample

to remove the damaged outer layers, which are then substituted with a patch that is bonded over the scarfed area in order to restore the load-carrying capacity.

The scarfed samples used in this work were rectangular and consisted of two CFRP plates scarfed by milling with a ratio of 1:17 (Fig. 2.2).

2.2.2.3 Panels

The individual aerospace component parts considered here were two stiffened panels: A flat panel with a laminated skin and two laminated T-stringers for the production scenario, and a curved panel with a laminated skin and two laminated Ω-stringers for the repair scenario. The stringers were spaced at equal distances on the panels.

The dimensions of the panel and the T-stringers are given in Fig. 2.3. The dimensions of the Ω-stringers are 800 mm × 100 mm × 33 mm (Fig. 2.4).

2.3 Manufacturing

2.3.1 Adherend Manufacturing

The CFRP laminates for the coupons were manufactured using the automated tape laying (ATL) technique (Fig. 2.5a). The diagram showing the applied vacuum bag

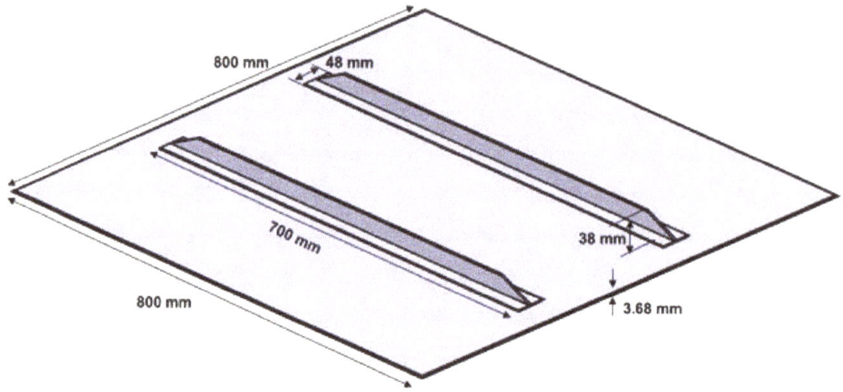

Fig. 2.3 Geometry and dimensions of the flat stiffened panel

Fig. 2.4 Curved stiffened panel with a laminated skin and two laminated Ω-stringers

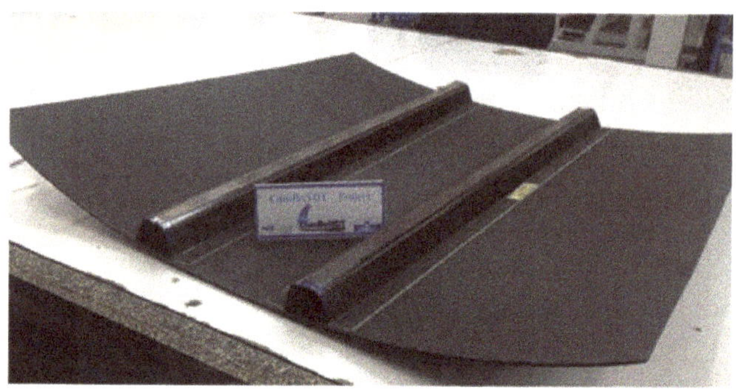

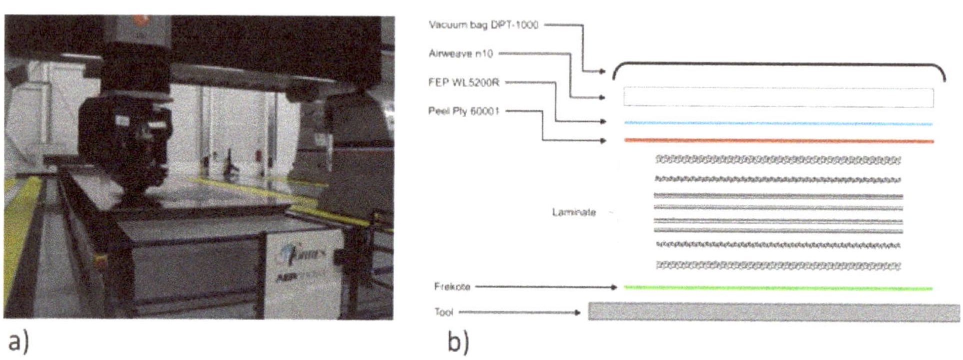

Fig. 2.5 **a** Photograph showing the automated tape laying process of the laminated panels; **b** diagram of the vacuum bag

is given in Fig. 2.5b, while Fig. 2.6 shows the panels placed inside the autoclave as well as the cutting of the panels.

The CFRP material for the scarfed samples was also manufactured by Aernnova Composites, as described previously, and was then delivered to Fraunhofer IFAM. The CFRP plates were scarfed with a ratio of 1:17 by milling and were then manually ground and cleaned with methyl ethyl ketone (MEK) to remove any handling

Fig. 2.6 Photographs showing **a** the panels inside the autoclave and **b** the cutting of the panels

contamination from thermoplastic residues resulting from the milling process. Lastly, the samples were cut to the final size using dry diamond cutting (Fig. 2.7).

The laminated skin for the flat panels (Fig. 2.8a) was manufactured using the same process as was applied for the laminates of the coupons. For the manufacturing of the T-stringers, the web was shaped by joining two C-shaped preforms that had been manufactured using the ATL technology. First, a flat panel was laminated and placed on a hot-forming tool to obtain the C-shape. After that, both C-shaped preforms were joined. Finally, the stringer was trimmed to obtain the six T-stringers with the required dimensions. The T-stringers are shown in Fig. 2.8b.

In the curved stiffened panels, the Ω-stringers were first manufactured using a specific shaping tool (Fig. 2.9). As the skin was curved, a difference in the orientation

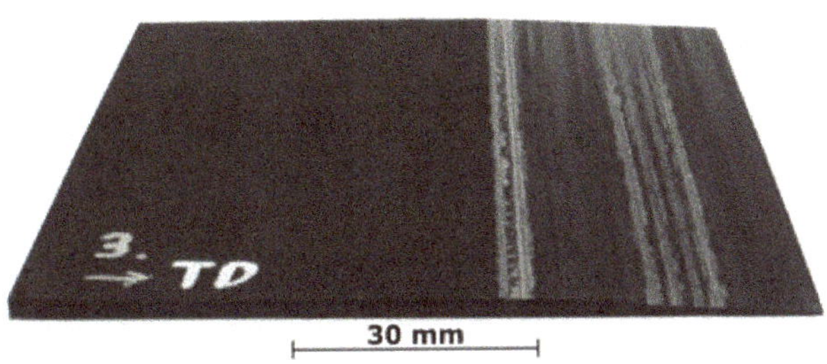

Fig. 2.7 Overview of scarfed sample after milling

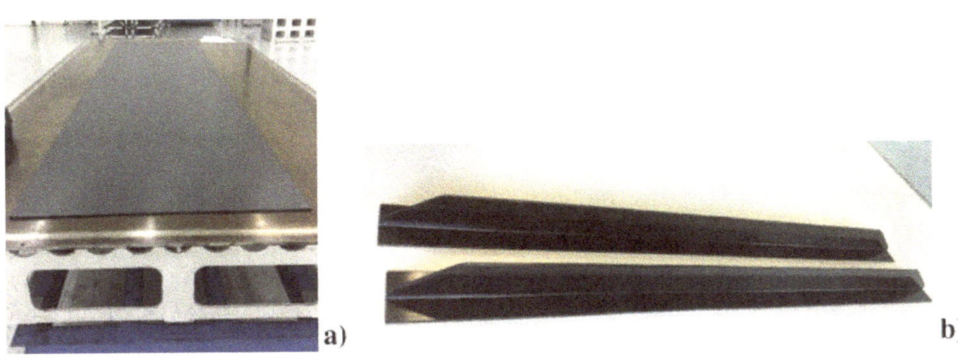

Fig. 2.8 Photographs showing **a** the flat laminated skin before being placed into the autoclave and **b** the T-stringers

Fig. 2.9 Photograph showing the Ω-stringers inside the shaping tool

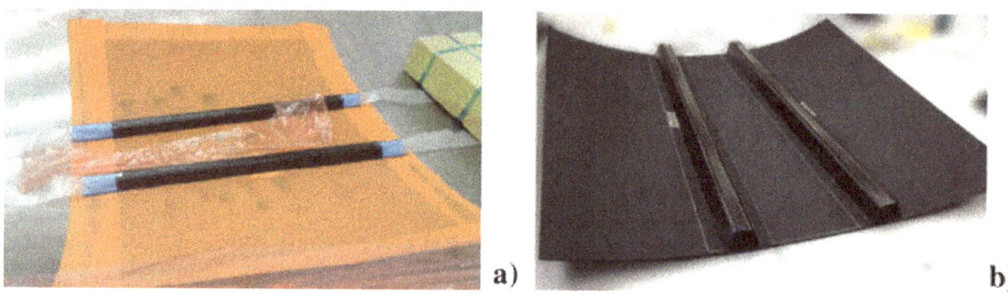

Fig. 2.10 Photographs showing **a** the preparation of the final curing for the curved panel and the Ω-stringers and **b** the curved and stiffened panel

of the Ω-stringers was considered. The bonding of the Ω-stringers with the skin was done immediately after the manufacturing of the skins. Once the stringers had been placed in their respective positions, the preparation for the final curing cycle started. In addition to the added tubular bags, the panel needed to be protected before entering the curing cycle (Fig. 2.10a). Special attention was paid to the adaptation of the complicated geometry parts to the vacuum bag during the pre-compaction. The curved and stiffened panel is shown in Fig. 2.10b.

2.3.2 *Adherend Pre-bond Contamination*

Implementing the concept for quality assessment during joining processes, e.g. as applied in the ComBoNDT project, requires the development of ENDT technologies for their integration into adhesive bonding process chains in aircraft production as well as for in-field repair. Such implementation is based on the identification and definition of all test scenarios to be considered and investigated; therefore, possible quality-relevant contaminants with high relevance for all, or at least for the majority, of aerospace applications must be identified. Following this decision, a detailed description of the sample preparation, including the surface contamination, was prepared in order to guarantee a precise and reproducible sample preparation process, which is an important key element in the success of the entire approach, as highlighted during the ComBoNDT project. Table 2.2 presents more details on quality-relevant scenarios affecting CFRP adherends or adhesive layers in adhesive joints investigated during the ENCOMB [8] or ComBoNDT [5] projects; a more detailed description is provided in the following sections.

2.3.2.1 Production Scenarios

Most aircraft produced nowadays contain a significant number of components made of CFRP, and therefore require the adhesive bonding of CFRP in manufacturing, which further emphasizes the need for complete and reliable quality assurance concepts using ENDT techniques. Three different contaminations were investigated in detail, namely release agent (scenario RA), moisture (scenario MO) and (human) fingerprint (scenario FP).

During the molding process of the composite panels, silicone (Si)-based release agents are used to facilitate the easy removal of the component from the mold. A Si-containing contamination on the CFRP surface caused by release agent residue hinders the adhesion of the adhesive to the substrate [1, 4, 7].

The release agent used was Frekote® 700NC. This is an Si-based liquid that needs to be removed from the CFRP surface before the bonding process because it prevents wetting and adhesion. Therefore, it is necessary to detect possible residues on surfaces prior to adhesive bonding. The release agent was applied to the CFRP surfaces by dip-coating with fixed immersion times and fixed withdrawal speeds. Different concentrations of Frekote® 700NC in heptane were used to yield different degrees of contamination. After the dip-coating of the adherend, the polymerization of the release agent was allowed to occur by drying for 30 min under ambient conditions followed by a heat treatment for 60 min in an oven at 80 °C. The first tests used to yield the desired amounts of Si-containing contaminations on the surface were conducted with solutions of Frekote® 700NC in heptane with the following concentrations (vol%): 1, 2, 3, 4, 5 and 8%. Based on the results of the preliminary tests (X-ray photoelectron spectroscopy measurements and mechanical tests with lap shear specimens), one of the two CFRP adherends per joint was intentionally exposed using dip-coating solutions with the following volumetric concentrations: 1, 2, and 4% of Frekote® 700NC in heptane.

Pre-bond moisture penetration into a composite adherend can occur via either air humidity or direct contact with liquid water. CFRP panels often undergo several pre-treatment procedures, such as wet abrasion and the water brake test, to ensure the effectiveness of the cleaning procedure [4, 7]. Although precautionary measures are implemented, such as using large autoclaves to remove moisture by heat-drying, the problem persists due to the ubiquity of water, e.g. in the surrounding atmosphere, and the limitation of water removal from the bulk of the CFRP thermoset resin through

Table 2.2 Quality-relevant scenarios affecting a CFRP adherend or an adhesive layer in an adhesive joint as assessed in the European joint research projects ENCOMB [8] and ComBoNDT [5]. In each scenario, the formation and properties of at least one interphase region (close to one of the adherends) is impaired as compared to a joint prepared following the qualified joining process

Quality-relevant scenario	Technological implementation and denotation		Affected joint region	Comment
	ENCOMB [8]	ComBoNDT [5]		
Reference (during production of the joint)	**X ("RE")** grinded down to fibers	**X ("RE")** grinded	Following the qualified bonding process	ComBoNDT [5]: P-RE (slightly grinded) and R-RE (grinded down to fibers) for production and repair scenarios, respectively
Release agent (during production of the joint)	**X ("RA")** higher amount	**X ("RA")** lower amount	CFRP surface covered by nanoscale film	Same silicone-containing agent used in ENCOMB and in ComBoNDT
Moisture (during production of the joint)	**X ("MO")**	**X ("MO")**	CFRP surface covered by nanoscale water film; moist CFRP bulk	
Fingerprint (during manufacture of the joint)	–	**X ("P-FP")** (following DIN ISO 9022-12)	CFRP surface covered by (dried) aqueous film	Artificial hand perspiration solution, according to DIN ISO 9022-12 [12]
Thermal impact (during joint application; repair scenario)	**X ("TD")** (thermo-oxidative)	**X ("TD")** (thermal)	CFRP surface thermo-oxidatively affected during application; CFRP bulk thermally affected	Removal of oxidatively affected surface region by grinding only in ComBoNDT
Exposure to components of hydraulic oil (during joint application; repair scenario)	**X ("HF")** (immersion in aqueous extract of oil)	**X ("R-FP")** (fingerprinting of hydraulic oil)	CFRP surface covered by a film	Different liquids used in ENCOMB and ComBoNDT
De-icer (during repair of the joint)	–	**X ("DI")**	CFRP surface covered with salt particles	De-icer liquid based on potassium formate

(continued)

Table 2.2 (continued)

Quality-relevant scenario	Technological implementation and denotation		Affected joint region	Comment
	ENCOMB [8]	ComBoNDT [5]		
Faulty curing of adhesive (during repair of the joint)	–	**X ("FC")**	Adhesive layer; interphases to adherends	Initiated by selective pre-curing of the adhesive

diffusion as it is intercepted by the fiber layers. Moisture uptake mainly affects the properties of the matrix, resulting in swelling and the development of stresses large enough to pull the matrix away from the fiber [4, 7, 13]. Moreover, moisture also affects the adhesion properties [13]. CFRP can absorb moisture by up to 1.5–2.0 wt%. The range for concern at production sites often goes up to 0.5 wt%. A higher moisture uptake needs to be avoided because it negatively influences the adhesion properties and leads to a loss of performance of both the CFRP as well as the adhesive bond.

The preparation of moist CFRP samples was performed following two different procedures with different environmental conditions: One for use in the development of ENDT monitoring technologies for the quality assurance of adherend surfaces and the second one for the measurement of the mechanical properties of the bonds and for the further development and adaptation of ENDT technologies for the quality assurance of adhesive bondlines. Regarding the first category, the samples were contaminated in a defined climate that was established in small boxes in an oven at 70 °C. The humidity in the boxes was adjusted using beakers of demineralized water (MO-3) and saturated salt solutions (MO-1 and MO-2), which were placed in the boxes together with the samples until a constant weight of the samples had been achieved. The beakers in the boxes contained the following aqueous liquids in terms of saturated salt solutions: $MgCl_2 * 6H_2O$ saturated solution for MO-1, resulting in an approximately 30% relative humidity (RH); NaCl saturated solution for MO-2, resulting in an approximately 75% RH; and pure demineralized water for MO-3, resulting in an approximately 99.5% RH.

Clean CFRP samples were dried at 80 °C until they had achieved a mass constancy resulting in the dry weight. Afterward, they were stored in the respective boxes with moist atmospheres until the weight was constant (at least 40 days) and then taken out directly prior to the measurement with the respective surface inspection method. With this method, the following mass uptake of water was achieved:

- 0.4 (±0.2) mass% water for MO-1
- 0.8 (±0.1) mass% water for MO-2
- 1.4 (±0.2) mass% water for MO-3

For the second MO conditioning, the samples were prepared using a different procedure. These adherends were dried in an oven at 80 °C until mass constancy.

They were then stored in a climate chamber (70 °C and with the respective and well-defined relative humidity) for two weeks prior to bonding. After the removal from the climate chamber, the samples were directly bonded. The following RH conditions were adjusted in the chamber:

- 30% RH for MO-1
- 75% RH for MO-2
- 98% RH for MO-3

All the results presented in this chapter refer to the second MO conditioning.

Contamination by fingerprints can occur due to inadequate cleaning of a bonding surface or inappropriate handling after the cleaning process [14, 15]. Fingerprint contamination leads to the formation of thin contaminant films on the bonding surfaces and, ultimately, to a lower adhesion quality [15]. This may occur during both production and repair processes. Even though the occurrence of fingerprints seems to be easily avoidable, they are often responsible for adhesion failures, and therefore the detection of fingerprints is an essential requirement for an appropriate quality assurance approach.

Concerning the samples with fingerprint contamination to be investigated for the production scenario, the preparation was performed using a standardized salty fingerprint solution (artificial hand perspiration solution) according to DIN ISO 9022-12 [12]. This liquid formulation contains sodium chloride, urea, ammonium chloride, lactic acid, acetic acid, pyruvic acid, and butyric acid dissolved in demineralized water. Samples were prepared by manually applying this solution onto a surface area of the samples that correspond to the size and extent of a wet fingerprint. Different degrees of contamination were achieved by using different dilutions (with demineralized water) of the FP solution:

- 10% FP solution for P-FP-1
- 50% FP solution for P-FP-2
- pure FP solution for P-FP-3

Finally, in addition to the single contamination cases described above, the occurrence of a combined contamination case was also considered. Combined contaminations for the production scenario included the combination of release agent and fingerprint contaminations (RA+FP). Two levels of contamination were investigated:

- Low-level contamination (RA1+FP3): level RA-1 of release agent followed by the application of level FP-3 salt-based fingerprint solution.
- Medium-level contamination (RA2+FP3): level RA-2 of release agent followed by the application of level FP-3 salt-based fingerprint solution.

2.3.2.2 Repair Scenarios

In the second field of feasible application scenarios, distinct composite "repair" cases were defined, implemented, and examined. Hereby, the effects of contacting adherend surfaces with either of the two contaminant materials de-icing fluid or hydraulic oil (applied in a fingerprinting process) were examined in detail. The third scenario dealt with thermally degraded CFRPs and aimed to account for CFRP parts that may have been exposed to heat (fuselage parts or alighting gear, for example, or aircraft structures that affected by lightening impact) and are then subjected to a mechanically abrasive surface pretreatment process. A fourth scenario that was investigated comprised a faulty curing of an adhesive that is then applied to distinct pretreated adherend surfaces. In the following, these technologically relevant contamination issues will be assessed in more detail.

In winter, airports use a de-icer to maximize runway friction during plane taxiing. Runway de-icing fluid is one of the most commonly encountered fluids to which aircraft structures may be exposed, as it can be swirled up from the runway and onto the outer parts of the aircraft [16]. During the patch repair of composite parts, inadequate cleaning can result in residues or the transfer of de-icing fluid onto adherend surfaces. After drying, potassium formate, which is present in the de-icing fluid, forms a thin layer on the CFRP part, thus affecting the bonding quality.

The de-icer used (DI scenario) was SAFEWAY® KF from CLARIANT, which contains potassium formate (KF) as the freezing point depressant. It was diluted with demineralized water to obtain solutions with the following concentrations in vol%: 2, 5, 7, 10, 30, and 50%. It was applied to the surfaces by dip-coating in the respective aqueous solution; finally, drying was performed in an oven for 2 h at 40 °C in air. Subsequently, acclimatization at room temperature was allowed for at least 24 h.

With the aim of narrowing the applied range of de-icer solution concentrations, three lap shear specimens comprising one contaminated adherend each were manufactured for each of these DI concentrations, and these were then used for adhesive bonding and subsequent mechanical testing. A significant loss in bond strength was observed for contamination levels characterized by surface concentrations of approximately 4 at.% potassium as measured by XPS. The fracture pattern also showed an impact when potassium surface concentrations of approximately 4 at.% were present. Based on these preliminary tests, it was decided to assess such samples in more detail; these were obtained using de-icer dip-coating solutions with the following concentrations for the ComBoNDT final coupon level samples: 2, 7, and 10% de-icer in demineralized water.

Moreover, CFRP aircraft parts may be exposed to high temperatures during service, for example, when fuselage parts are exposed to lightning [17], which causes local overheating and damage to the matrix or the wing parts situated close to the engines. Damage can also be caused by an overheating of an aircraft part by an external source of heat (gas, liquid, beam, etc.) that has inadvertently been placed near the aircraft. Besides affecting the mechanical properties of the structural parts, the thermal impact on and resulting degradation of the CFRP parts might also affect bonding in a repair situation.

For the sample preparation, all thermal impact treatments were carried out in an oven with air circulation. The samples (subsequently denoted as TD samples) were placed inside the oven and then underwent the heating phase at different temperatures. Once the indicated temperature was reached, the samples remained inside the oven for 2 h. Prior to both the surface inspection and the subsequent steps of the bonding process, all samples were grinded down to the fibers (using Si-free sandpaper, grit size 120). A significant loss in bond strength (lap shear specimens) was observed for the samples treated with 280 °C. In a comparison with joints prepared from adherends that had not suffered such a thermal impact, the fracture pattern demonstrated an impact from a heat treatment of 260 °C. Based on these results, the following three different temperatures were used to realize three different levels of thermal impact (TD):

- 220 °C for TD-1
- 260 °C for TD-2
- 280 °C for TD-3

Concerning a further feasible contamination scenario, oily fingerprints can accidentally be applied to CFRP bonding surfaces when, for example, a worker wears gloves while working in an area where hydraulic oil is used and afterwards touches a bonding surface. Even though this contamination scenario seems easily avoidable, it is nevertheless of great importance in the field of aircraft repair because such a transfer of oil is unlikely to be detected during a visual inspection. Concerning the sample preparation for the repair scenario, fingerprints containing Skydrol 500B-4 hydraulic oil from Eastman were applied to the surfaces using a plastic finger. The oil was diluted in heptane to obtain formulations with the following contamination concentrations in vol%:

- 20% for a low level (denoted as R-FP-1)
- 50% for a medium level (R-FP-2)
- 100% for a high level (R-FP-3)

In order to achieve adequate adhesive bonding processes, it is important that all adhesive-related parameters, like pot life and curing times, comply with regulations as well as the specifications of the qualified bonding process. If the adhesive is out of specification with respect to its pot life due to, e.g., too high temperatures in the working area, the result can be weak or kissing bonds in the resulting joint. In this scenario, the bonded joint does not contain any foreign materials or contaminants that have erroneously remained after the cleaning, pretreatment, or conditioning steps. Instead, the loss of performance of the bonded joint is due to irregularities affecting the adhesive material that was used in the manufacturing process of a limited number of joints.

Regarding the sample preparation, a faulty curing of the adhesive was initiated through a selective pre-curing of an adhesive that was subsequently introduced into the bonding process. In the resulting selected and pre-cured areas, the bond strength may be reduced drastically, possibly due to its impeding any force transfer. This scenario will represent the cases of weak and kissing bonds. Three levels of pre-curing resulting in a faulty curing (FC) of the adhesive were realized:

- a slight pre-curing for FC-1
- a medium pre-curing for FC-2
- a strong pre-curing for FC-3

Finally, besides the single contamination cases described above, a combined contamination case was also considered. Combined contaminations in the repair scenario include the combination of thermal degradation and de-icing fluid (TD+DI), whereby two levels of contamination were investigated:

- Low level of contamination (TD1+DI1): Thermal degradation at 220 °C for 2 h followed by dip-coating in the DI1 concentration of the de-icing fluid solution.
- Medium level of contamination (TD1+DI2): Thermal degradation at 220 °C for 2 h followed by dip-coating in the DI2 concentration of the de-icing fluid solution.

2.3.3 *Bonding*

For the production scenarios, the samples described in this book were bonded in an autoclave using the adhesive FM® 300 K (0.2 mm) from Cytec® following the curing cycle shown in Fig. 2.11. The heating rate (starting from room temperature) was 3 K/min up to 175 °C. The pressure was 3 bars and the final temperature of 175 °C was held for 1 h.

Plates with the dimensions 30 cm × 30 cm and 30 cm × 15 cm were bonded and afterwards cut into the desired sizes for measurements with the respective measuring techniques as well as for mechanical testing in the specified geometries. The cutting was performed dry (diamond cutting) to prevent any contamination of the cleaned surfaces as might be the case when using cooling liquids. After cutting, the surfaces were cleaned again with isopropanol (IPA) soaked tissues. Figure 2.12 shows the preparation of the samples for bonding in the autoclave (at Fraunhofer IFAM facilities).

For the repair scenarios, all the samples were bonded in the autoclave using the adhesive FM® 300-2 (0.25 mm), which is specially designed for bonded repair. The

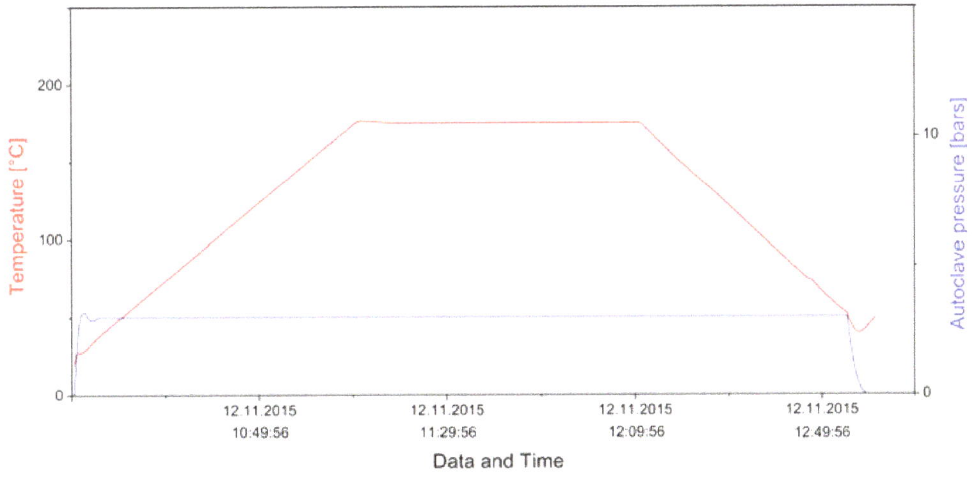

Fig. 2.11 Autoclave thermal and pressure cycle for bonding production samples with adhesive FM® 300 K

a) b)

Fig. 2.12 Photographs showing the preparation of samples for bonding **a** in the autoclave and **b** using a vacuum bag

respective curing cycle is shown in Fig. 2.13. The CFRP plate sizes and the cutting of the samples into the final sizes after the bonding were as described for the production of the samples.

For the bonding of the centrifuge samples (Fig. 2.14a), the film adhesive was hole-punched to a diameter of 10 mm and then deposited onto the test stamp (cleaned by sonication for 5 min in isopropanol), which was then placed onto the CFRP sample (Fig. 2.14b). The production samples bonded with the FM 300 K adhesive were cured in an autoclave using a custom-made curing device at 3 bars and 175 °C for 60 min (heating up to 175 °C in 60 min, cooling down to room temperature in 60 min). The repair samples bonded with the FM 300-2 adhesive were cured at 2 bars and 121 °C for 90 min (heating up to 121 °C in 30 min, cooling down to room temperature in 60 min).

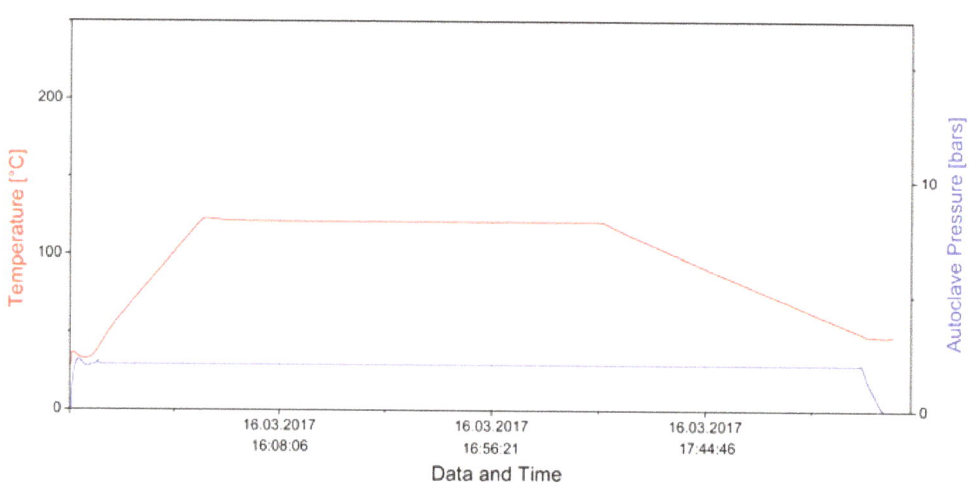

Fig. 2.13 Autoclave thermal and pressure cycle for bonding repair samples with the adhesive FM® 300-2

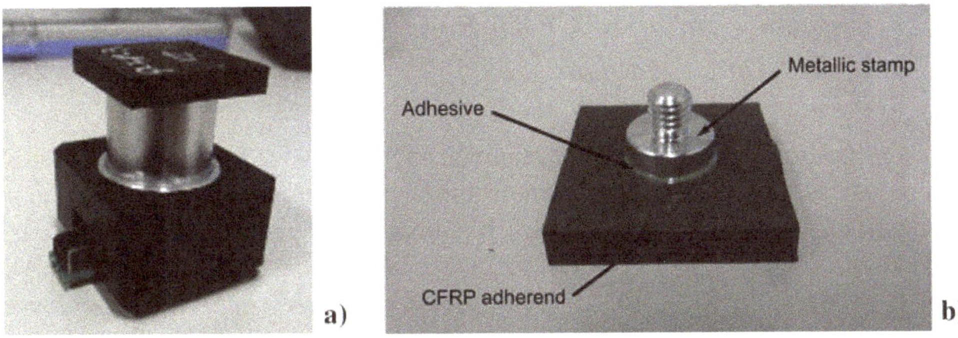

Fig. 2.14 Photographs showing the stamp-to-plate specimen used in the centrifuge tests; **a** full specimen configuration and **b** metallic stamp bonded to the CFRP adherend

2.4 Experimental Procedure

2.4.1 Characterization of CFRP Adherend Surfaces by Reference Methods

Spectroscopic surface characterization was performed on the CFRP adherends before the adhesive bonding to verify the contamination level obtained by the contamination procedure. X-ray photoelectron spectroscopy (XPS) was used as a spectroscopic reference method, and XPS measurements were performed on detached and cut plates in the state "as delivered" on three different surface positions.

X-ray photoelectron spectroscopy (XPS) is a widely used surface analysis method for the characterization of the elemental and chemical composition of a sample surface, which is positioned inside a vacuum system. XPS is based on the photoelectric effect and enables the study of the energy distribution of the photoelectrons emitted by X-ray irradiated compounds [18]. Monochromatic soft X-rays irradiate the sample surface, and upon interaction with the sample material electrons are emitted, mainly from the atomic core levels. These ejected electrons have discrete kinetic energies, and the portion of electrons passing the electron energy analyzer is detected within the photoelectron spectrometer. Signal intensities are given by the number of emitted photoelectrons as a function of the photoelectron kinetic energy. A high vacuum environment is required to enable the emitted photoelectrons to be analyzed without interference from gas-phase collisions, and in cases of electrically non-conducting surfaces special care is taken to control electrostatic surface charging.

2.4.2 Characterization of CFRP Bonded Samples by Reference Methods

Ultrasonic testing is considered a conventional NDT technique for the quality control of components. This technology is widely used for composite material inspection in aeronautics as well as in other domains. It is thus considered as a standard method by end users.

Within the ComBoNDT project, the aim of the ultrasonic inspection was to check the integrity and quality of the produced bonded samples. Indeed, according to the literature and also with regards to the context of the project, a "weak bond" is not expected to be detectable by conventional NDT methods. Otherwise, a common interpretation of the obtained data is that such a bond is considered a bond with defects (such as voids, porosity, gap-like disbonding) rather than as a possible weak bond. Therefore, within the framework of the characterization using ultrasonic testing as a reference method for the bonded samples, one issue was proving that the contamination and bonding processes do not lead to such defects. This information, in combination with the results of the terminal destructive mechanical testing (i.e. strength and fracture pattern), is required in order to consider the respective joints as samples with weak bonds.

Three categories of samples were investigated:

(a) Coupons were widely used by all the project partners in order to develop their ENDT technology, and thus it is of great importance that the quality of these samples is known. The obtained dataset comprises results from more than 360 samples.
(b) Multi-contaminated samples were used to increase the maturity of the ENDT technologies.
(c) Curved specimens were also tested. In this case, the curvature of the samples led to challenges in the signal reception, which rendered the inspection of these samples more prospective than quantifying.

The ultrasound-based inspections were performed in the Airbus laboratory using an M2M ultrasonic generator and a 6-axis mechanism (Fig. 2.15b, c). The immersion configuration was selected to maximize the signal quality. Samples were placed in consistent groups in the water tank on metal beams (Fig. 2.15d). The water path for passing the oscillation from the sonotrodes to the sample, i.e. the distance between the probe and the samples, was set to 40 mm on average. Two different phased array probes were used for the inspection (Fig. 2.15a). The characteristics are given below, with the trajectory parameters for each:

- 5 MHz linear probe, 64 elements, 1.0 mm pitch, 64 mm of aperture, 10 mm elevation, flat focusing; linear scanning; scanning step: 2 mm (standards), increment 30 mm
- 10 MHz linear probe, 64 elements, 0.5 mm pitch, 32 mm of aperture, cylindrical focusing (R = 40 mm); scanning step: 1 mm, increment 20 mm (i.e. 33% overlap)

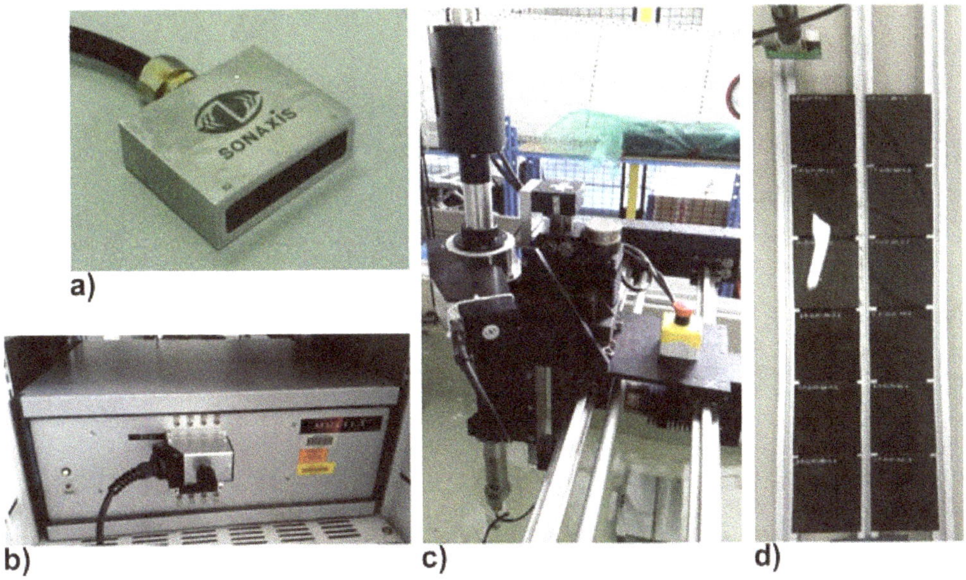

Fig. 2.15 Overview of the ultrasonic inspection setup; **a** example of a phased array probe; **b** ultrasound generator for emission and reception (M2M MultiX); **c** water tank and the 6-axis mechanism used to scan the parts; and **d** typical placement of the samples in the water tank for the inspections

The 5 MHz probe corresponds to Airbus standards. It was important to check that it is not possible to highlight any contamination-induced defects using current production tools. The 10 MHz probe was chosen because it is more accurate due to its higher central frequency and the smaller element size.

For each probe, different settings were used for the production samples and the repair samples. The global gain of the signal and the time correction gain (TCG) are slightly different between scenarios. This is mainly due to the bond material, which is different in each case and thus induces different ultrasonic responses.

In order to display ultrasonic cartographies, software settings (or gates) are necessary. A typical B-scan and A-scan are given in Fig. 2.16a with the gate display. Three main echoes can be observed, namely the front wall echo (with the gate or, respectively, obtained signal dataset denoted as "FWE"), bond echo ("Bond") and back-wall echo ("BWE"). All gates are synchronized using a synchronization gate tracking the entry echo. These gates can subsequently be used for the analysis. The respective details are

- "FWE" is the maximum of the front wall echo. It can be used to check the acquisition quality and to highlight surface defects.
- "g+" records the highest echo after FWE. In this case, it is typically the bond or the back-wall echo. The signal from this gate is particularly useful to compare the echoes.
- "Bond" is centered on the bond echo and tracks its maximum.
- "BWE" is centered on the BWE and tracks its maximum.

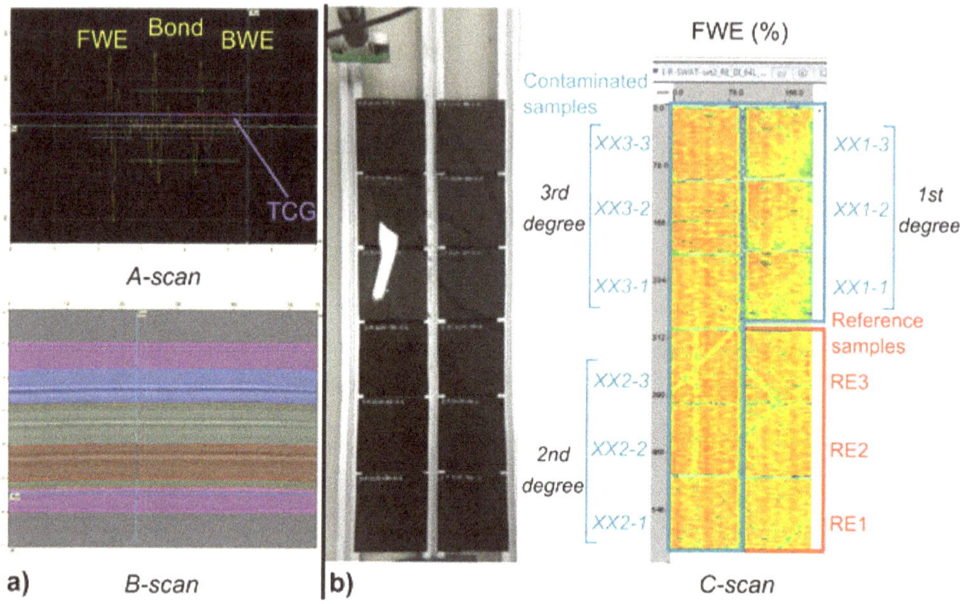

Fig. 2.16 Use and display of the ultrasound results from gates related to the front wall echo (FWE), bond echo (Bond), and back-wall echo (BWE); **a** typical A-scan and B-scan with gates setting display and **b** example of cartographies in amplitude for bonded coupon coupons with the corresponding sample positions

These gates are used to generate cartographies, also called C-scans, where the amplitude (given in %) or the time-of-flight (TOF, given in μs) of the recorded echo is displayed on the inspection plan (for an example of amplitude cartography see Fig. 2.16b). Samples are always placed in the same way with the references situated on the bottom right and the contaminated samples in the remaining spaces. For each set of samples, the origin is taken at the top left corner for the defect positioning.

In order to be complete, there must be precision on the phased array acquisition mode. Two different pulse-echo modes were used:

- The linear scanning (LS) mode, which consists of emitting a group of elements (E10 typically) and then receiving the same group of elements (R10). This configuration increases the scan accuracy. A single point focusing (SPF) can be added to direct the ultrasonic beam along the bondline, for example.
- The PaintBrush (PB) mode with the additional surface adaptative ultrasonic laws (SAUL) option, which consists of emitting with all the elements (E64) and then summing the responses by groups of elements (R10 for the 5 MHz probe and R16 for the 10 MHz probe). Such an investigation is faster but can lead to "strip-like" marks within the cartographies. The SAUL algorithm was also used in some specific cases. This option is particularly interesting for curved parts or to achieve a higher tolerance to a misalignment between the probe and the coupons.

2.5 Mechanical Testing

2.5.1 Fracture Toughness Testing

2.5.1.1 Mode-I Testing

Mode-I fracture toughness tests were conducted with double cantilever beam specimens according to the standard AITM 1-0053 [9] using a Tinius Olsen H5KT universal testing machine with a load cell of 5 kN under ambient conditions (25 °C, 55% RH) and under constant displacement control. Loading was applied to the DCB specimen via metallic piano hinges bonded to the adherends at one end. In order to avoid any influence of the incorporated release film, the specimen was preloaded until an initial crack length of 10–15 mm was achieved. The pre-cracked specimens were then loaded continuously by opening forces until a total propagated crack length of 100 mm was reached. After that, the test was stopped, and the specimen was unloaded. Six specimens per scenario were tested in this manner. During the crack propagation, the load and crosshead displacement of the test machine was recorded continuously. A traveling microscope was used to facilitate the visual measurement of the crack length. Figure 2.17 illustrates the mounting of a specimen onto the tensile testing machine during the mode-I test.

Fig. 2.17 Photograph showing a DCB specimen under mode-I loading

The AITM 1-0053 standard specification specifies the area method to determine the mode-I fracture toughness energy G_{IC} of CFRP bonded joints [9]. The crack extension is related directly to the area enclosed between the loading and unloading curves:

$$G_{IC} = \frac{A}{a \times w} \times 10^6 \left(J/m^2\right) \tag{2.1}$$

where

- A is the energy required to achieve the total propagated crack length (J) (integration of the area of the load-crosshead displacement diagram)
- a is the propagated crack length (mm) ($a = a_{final} - a_{initial}$)
- w is the width of the specimen (mm)

The most popular approach to investigating delamination mechanisms in mode-I tests is the examination of fracture surfaces. Therefore, in order to accurately assess the causes of bondline failure, the fracture patterns were examined after the tests based on a visual inspection supported by photography. The classification, identification, and characterization of the failure mode of the CFRP bonded joints were conducted according to the ASTM D5573 standard [19]. For increased accuracy, a grid drawn on a clear film placed over the failure surface was used and the square areas showing a certain type of fracture pattern were counted, providing input to calculate the area percentage attributed to each failure mode. The main failure modes that were observed for the tested CFRP adhesive joints are schematically described in Fig. 2.18:

a. Adhesive (ADH) failure, which occurs when a separation takes place at the adhesive/adherend interface (respectively, within the three-dimensional adhesive/adherend interphase).
b. Cohesive (CO) failure, which results when a separation takes place within the adhesive.
c. Fiber tear (FT) failure, which is perceived when a failure occurs exclusively within the matrix of the CFRP adherend, resulting in the appearance of fibers on both fracture surfaces.
d. Light fiber tear (LFT) failure, which follows when a failure occurs within the adherend, near the interface characterized by a thin layer of the matrix on the fracture surface with few or no fibers transferred from the substrate to the adhesive.
e. Thin layer cohesive (TLC) failure, which is observed when the separation takes place within the adhesive in proximity to one adherend and not around the mid-thickness area of the adhesive layer.

Usually, a mixed failure occurs and symptoms of several failure modes are observed simultaneously for each tested specimen.

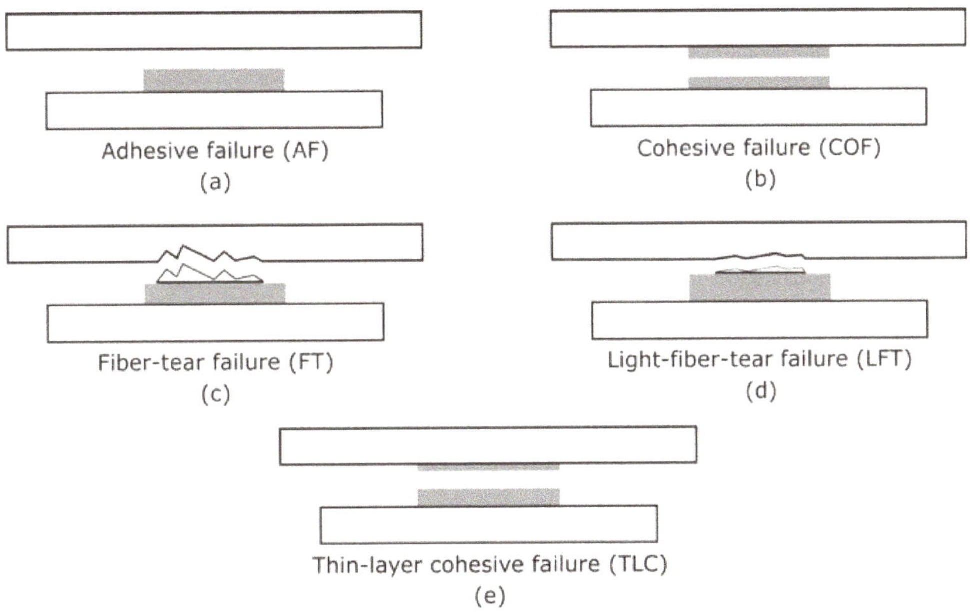

Fig. 2.18 Schematic representation of the main failure modes of CFRP bonded joints observed after destructive mechanical testing

2.5.1.2 Mode-II Testing

Since there is currently no standardized mechanical test to measure the fracture toughness energy of bonded joints under pure mode-II loading, we decided to use the ENF test, which we have identified as the most convenient mode-II fracture toughness test [16]. Figure 2.19 provides a schematic representation of the ENF test,

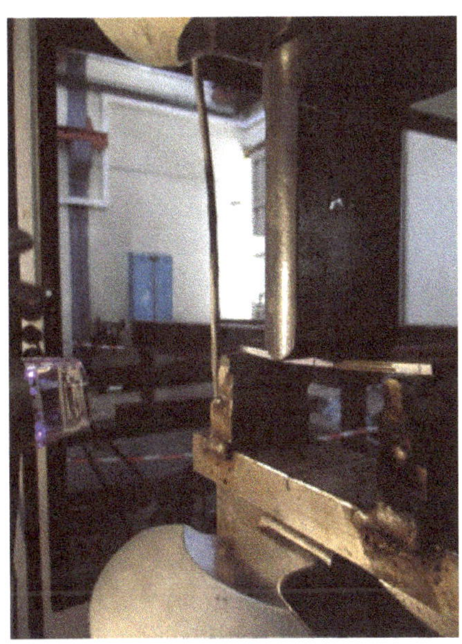

Fig. 2.19 Photograph showing an ENF specimen under mode-II loading

wherein a pre-cracked specimen is loaded into a three-point bending fixture until the crack propagation onset occurs.

Mode-II tests were conducted according to the AITM 1-0006 standard [20] under a constant displacement rate of 1 mm/min using an MTS universal testing machine with a load capacity of 100 kN. The test specimens were cut from the residual parts of mode-I specimens so that a pre-crack of 35 mm was achieved. Three specimens were tested for each condition within the considered scenarios. In order to facilitate the optical observation of the crack tip and the detection of the crack propagation onset, a digital microscope was used, and a thin layer of white ink was applied to the longitudinal side faces of the specimen.

Both the load applied to the specimen and the crosshead displacement of the test machine were continuously recorded during the test. To calculate the G_{IIC} fracture toughness energy, the following formula was used [20]:

$$G_{IIC} = \frac{9 \times P \times a^2 \times d \times 1000}{2 \times w \times \left(1/4 \times L^3 + 3 \times a^3\right)} \ (\text{J/m}^2) \tag{2.2}$$

where

- d is the crosshead displacement at onset of the crack propagation (mm)
- P is the critical load to start the crack propagation (N)
- a is the initial crack length (mm)
- w is the width of the specimen (mm)
- L is the span length (mm)

2.5.1.3 Centrifuge Testing

Standardized methods only allow the testing of specific bond strength parameters, and the achieved statistics are often limited due to the high cost and work effort required. Additionally, the respective measurements are time and cost-consuming due to complex sample and fixture preparation, single-sample testing, and manual evaluation of the mechanical load and fracture surfaces. Against this background, the novel centrifuge test is introduced in order to overcome these limitations.

Up to eight samples can be measured within 5 min and the measured mechanical properties have a defined accuracy with very good precision and reproducibility. The novel testing is cost-efficient, fast, and reliable. In the ComBoNDT project [5], the information value of mechanical testing was increased as compared to the results obtained from the abovementioned standardized mechanical tests.

The centrifuge testing principle for bonded joints is illustrated in a schematic diagram in Fig. 2.20 [21]. The centrifuge test is based on the physical law of inertia of a body [22]. Due to rotation, a progressively increasing radial centrifugal force is applied synchronously to each of the specimens being tested. The load increase is adjusted through a variation of the rotor's rotational speed.

Across the bondline, the axial centrifugal force acts as a normal tensile force. If the applied load exceeds the tensile strength of the joint, a rupture occurs, and the

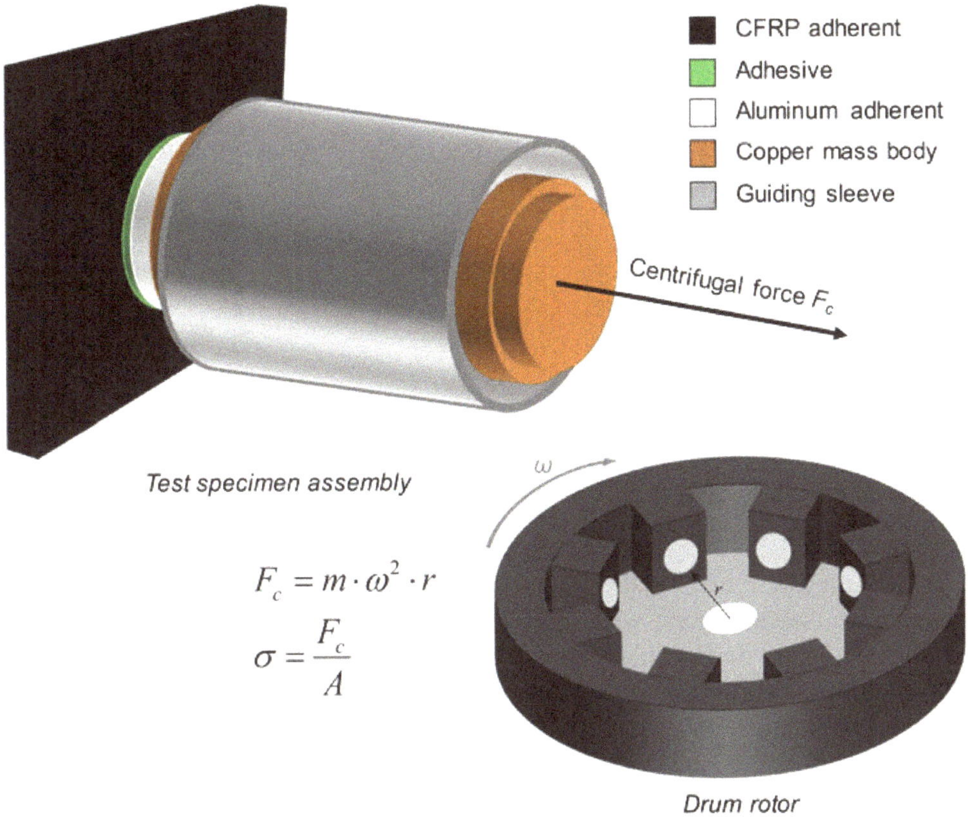

Fig. 2.20 Diagram highlighting the measurement principle of the centrifuge test

test stamp changes its position within the guiding sleeve. The detachment of the test stamp from the CFRP adherend at the moment of rupture is automatically detected and a position-coded infrared signal is sent from the turning rotor, transmitting the current rotor speed as well as the rupture time [23].

The centrifugal force F_c (N) is derived from

$$F_c = m \cdot \omega^2 \cdot r \quad (2.3)$$

where m (kg) is the mass of the stamp, r (m) is the distance of the test stamp to the rotational axis, and ω (rad/s) is the angular velocity related to frequency v by

$$\omega = 2 \cdot \pi \cdot v \quad (2.4)$$

Dividing the centrifugal force F_c (which is effective at the time of the adhesive fracture) by the area of the bondline A (mm^2), the tensile adhesion strength σ (MPa) is derived:

$$\sigma = \frac{F_c}{A} \quad (2.5)$$

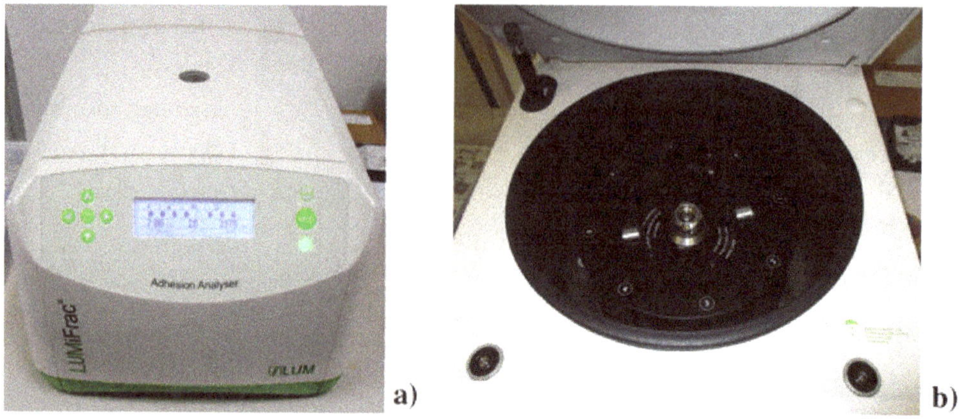

Fig. 2.21 Photographs displaying the setup for the centrifuge test; **a** the LUMiFrac desktop analyzer and **b** stamp-to-plate test specimens inside the drum rotor

For the preparation of the thus investigated joints, the composite substrates were subjected to contamination with the release agent, moisture, fingerprint, thermal degradation, or de-icing fluid before being bonded to the metallic stamp. The centrifuge tests were carried out using a LUMiFrac desktop adhesion analyzer equipped with an LSFR-ST: 200.42 drum rotor with up to eight testing units (Fig. 2.21). The fully loaded rotor allows for a maximum rotational speed ω of 13,000 rpm, corresponding to a centrifugal acceleration of 13,715 g [23].

By means of the SEPView software, the desired load-controlled testing sequence was realized. To achieve compatibility with conventional testing machines (in load-controlled mode), the increase in the rotational speed of the rotor was designed to be quadratic. According to Eq. (2.3), a square root-like increase in the rotational speed is accompanied by a linear increase in the centrifugal force [24]. Subsequently, the rotor and centrifuge lid were closed, and the testing procedure was initiated. The duration of each test lasted from 6 to 20 s on average, depending on the contamination scenario. The rupture event was detected online outside the centrifuge using a position-coded and rpm-correlated infrared data transmission from the inside of the testing units mounted in the drum rotor. After testing, high-resolution microscopy images of the failure surfaces of both the CFRP adherend and the test stamp's side were taken and examined with the aim of characterizing the failure patterns.

2.5.1.4 Tensile Testing

Scarfed samples were loaded under ambient conditions (25 °C/48%RH) by tensile stress using an MTS universal testing machine with a load capacity of 100 kN under a constant crosshead speed of 0.5 mm/min (Fig. 2.22) until a final failure (separation) of the two scarfed adherends occurred. Aluminum end tabs (30 mm × 25 mm × 2 mm) were bonded to the ends of the specimens using a two-part adhesive (PM Mega Cryl) in order to achieve a successful and smooth introduction of the load into

Fig. 2.22 Photograph showing a scarfed specimen under tensile loading

the specimen. Moreover, the end tabs prevented gripping damage to the adherends or premature failure as a result of a significant discontinuity. The load and crosshead displacement were recorded using a computerized data logging system. A total of four tests were performed for each contamination scenario and the failure load was the mechanical feature used for comparing the tested specimens.

Additionally, after the tensile tests, the failure surfaces were examined in order to accurately assess the causes of adhesive joint failure. The ASTM D5573 [19] standard was followed.

2.5.1.5 Environmental Aging

The procedure given in the EN 2823 [25] standard was used to determine the effects of after-bond exposure of the joints to a humid atmosphere on the mechanical characteristics of the contaminated joints. The specimens were exposed without mechanical loading to conditions of 70 °C and 85% RH until the moisture saturation point, which was reached after approximately 65 days of aging.

The specimens to be aged were placed inside an environmental chamber with an embedded pre-crack (Fig. 2.23), which was created a priori through mode-I tests conducted according to the AITM 1-006 standard [20]. Reference and contaminated specimens were subjected to hygrothermal aging using an ESPEC SH-641 environmental chamber for a period of 64–74 days, ensuring that the saturation point was reached.

Fig. 2.23 Photograph showing CFRP joint specimens inside the chamber during environmental aging

During the hygrothermal aging period, the weight of the specimens was measured at weekly intervals. After the hygrothermal aging, the specimens were stored in sealed containers and tested under mode-II loading conditions within 72 h according to the DIN EN 2823 standard [25].

As a measure of the absorbed moisture, the percental normalized weight gain $M(t)$ was used:

$$M(t) = \left(\frac{w_t - w_0}{w_0}\right) \times 100 \tag{2.6}$$

where

w_0 is the initial weight (g)
w_t is the weight at exposure time t (g)

The weight gain achieved with these hygrothermal aging conditions was 0.49–0.71%. Fick's law was used to define the equilibrium conditions in composite materials [25]. The diffusion coefficient D of water is derived from the slope of the linear part of $M(t)$ curve as

$$D = \pi \times \left(\frac{h}{4 \times M_\infty}\right)^2 \times S^2 \left(\frac{\text{mm}^2}{2}\right) \tag{2.7}$$

where

M_∞ is the water uptake at saturation (wt%)
h is the specimen thickness (mm)
S is the slope of the $M(t)$ curve $\left(1/s^{0.5}\right)$.

2.6 Experimental Results

2.6.1 Spectroscopic Surface Characterization

To prepare the clean reference samples from the delivered CFRP plates, a thorough cleaning procedure was followed to remove any contaminations or residues remaining from the manufacturing process (Fig. 2.24). Each step was monitored using X-ray photoelectron spectroscopy (XPS) analyses to measure the amount of release agent on the plate surface:

1. Pre-cleaning of the plates with isopropanol (IPA) soaked tissues to remove part of the release agent and any other soluble contaminations, e.g. fingerprints, remaining from the manufacturing process. XPS measurements performed on the "as delivered" plates on three different positions showed an inhomogeneous distribution of Si-containing release agent on the CFRP surface (Table 2.3). The XPS results for the cleaned plates showed that pre-cleaning with IPA is effective in that the amount of release agent on the CFRP surface can be reduced to 0.5–1.4 at.% within the information volume of the investigation. This indicates the

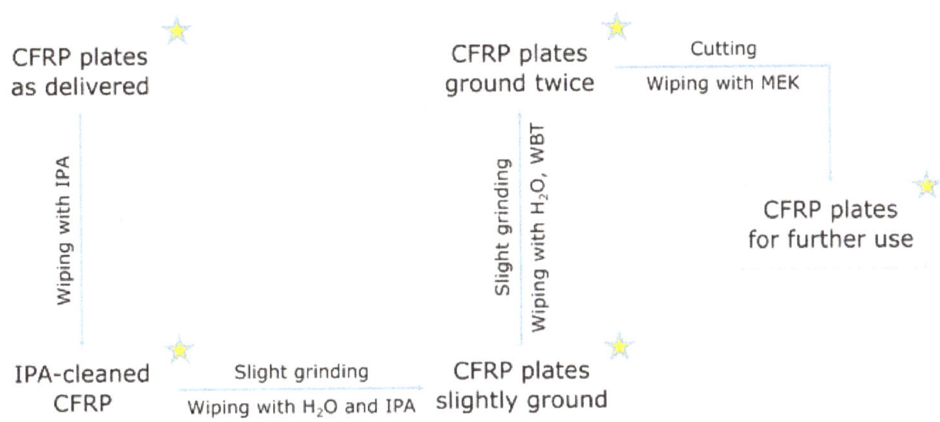

Fig. 2.24 Process of the cleaning of as-received CFRP plates by moist wiping with water, isopropanol or methylethylketone soaked tissues and grinding with sandpaper

Table 2.3 Surface Si concentrations (in at.%) from XPS investigations performed for the CFRP sample plates following the given cleaning steps with isopropanol (IPA)

CFRP plates	Si (at.%)
CFRP "as delivered" sample plates	5.3 ± 1.3
CFRP sample plates after IPA cleaning	0.9 ± 0.5
CFRP sample plates after IPA cleaning and slight grinding	0.3 ± 0.2
CFRP sample plates after IPA cleaning and two slight grinding steps with cleaning in between	0.1 ± 0.04

amount of dehesive material that can easily be removed by a subsequent abrasive grinding step without significant smudging.

2. Slight grinding of the surfaces to remove residual release agent that had penetrated or was incorporated into the topmost resin layers, and afterward wiping with demineralized H_2O and IPA to remove the dust from the grinding, which comprised the inhering residual silicone. Further XPS measurements showed that a small amount of silicone remained on the surface (Table 2.3).
3. A second slight grinding step followed by wiping off the dust with demineralized water and IPA. On these samples, XPS measurements showed a very clean surface Table 2.3.

After the cleaning steps, the sample plates were wiped with methylethylketone (MEK) soaked tissues prior to performing the intentional contamination (corresponding to the respective scenario) and the subsequent adhesive bonding.

For the release agent contamination scenario (RA), the first and last dip-coating samples for each dip-coating solution (labeled RA-1, RA-2, or RA-3) were used for the XPS measurements; these were conducted on three positions (top, middle, bottom) on each sample. The results are shown in Table 2.4.

Based on these results, samples with the final concentrations of Si on the CFRP surfaces were obtained, as shown in Table 2.5.

For the production fingerprint contamination, three samples with each FP concentration were prepared for the XPS measurements. The results are shown in Table 2.6.

Based on these results, the concentrations of Na and Cl on the CFRP surfaces for each contamination level could be determined, as described in Table 2.7.

For the repair fingerprint scenario (FP), the locally applied Skydrol has the tendency to spread over the surrounding surface, thus no clear fingerprints can be observed after some time. Systematic XPS measurements were not conducted for this contamination scenario since most of the oil evaporates in the vacuum of the analysis chamber.

Since the de-icer contains potassium formate, the potassium content on the surface is taken as a measure for the degree of de-icer contamination. Dip-coating of the final de-icer contaminated samples was performed using solutions of 2, 7, and 10% de-icer in demineralized water. The samples for XPS control were dip-coated together with these samples. The XPS results are shown in Table 2.8.

Based on these results, the final concentrations of potassium on the CFRP surfaces for each contamination scenario were determined (Table 2.9).

Table 2.4 XPS results (indicating the surface concentrations of the main elements [in at.%] as obtained) for CFRP plates after dip-coating in Frekote solutions (following RA-1, RA-2, RA-3) with different concentrations

Samples	C (at%)	O (at%)	N (at%)	Si (at%)	S (at%)	Na (at%)
RA-1 start, top	69.7	19.6	6.8	2.6	1.2	<0.1
RA-1 start, middle	67.1	22.2	5.2	3.6	1.4	<0.1
RA-1 start, bottom	66.6	20.5	6.6	5.1	0.8	<0.1
RA-1 end, top	70.0	19.2	6.9	2.5	0.9	0.2
RA-1 end, middle	70.4	18.6	7.1	2.6	1.1	0.1
RA-1 end, bottom	69.9	19.1	7.1	2.6	1.0	0.2
RA-2 start, top	67.3	20.9	6.5	4.3	1.1	<0.1
RA-2 start, middle	64.9	22.1	6.1	5.6	1.0	<0.1
RA-2 start, bottom	66.1	20.8	6.9	5.2	0.7	<0.1
RA-2 end, top	68.4	19.2	7.2	4.2	0.8	<0.1
RA-2 end, middle	65.4	21.3	6.6	5.2	0.9	0.1
RA-2 end, bottom	65.2	21.1	6.5	5.9	0.7	0.3
RA-3 start, top	66.0	20.4	6.9	5.8	0.7	<0.1
RA-3 start, middle	63.2	23.1	5.7	6.4	0.9	0.3
RA-3 start, bottom	64.8	21.0	6.6	6.6	0.6	<0.1
RA-3 end, top	65.3	22.2	5.0	6.3	1.1	<0.1
RA-3 end, middle	65.4	21.0	6.5	6.1	0.7	<0.1
RA-3 end, bottom	64.5	23.0	4.8	6.2	1.1	<0.1

Table 2.5 Average concentration of Si on CFRP surfaces (obtained by XPS) for the RA contamination scenario

Scenario	Si (at.%)
RA-1	3.2 ± 1.0
RA-2	5.1 ± 0.7
RA-3	6.2 ± 0.3

Table 2.6 XPS results indicating the surface concentrations (main elements) of three CFRP plates contaminated with different (salty) solutions (FP-1, FP-2, FP-3) applied as a fingerprint

Samples	C (at%)	O (at%)	N (at%)	Si (at%)	Zn (at%)	Cl (at%)	S (at%)	Na (at%)
FP-1, sample 1	74.6	15.3	8.5	0.2	0.1	0.4	0.6	0.2
FP-1, sample 2	74.5	16.6	6.3	0.2	0.2	0.8	1.0	0.3
FP-1, sample 3	74.7	16.6	6.8	0.2	<0.1	0.3	1.0	0.2
FP-2, sample 1	74.0	15.0	8.8	<0.1	<0.1	0.9	0.7	0.5
FP-2, sample 2	73.2	16.8	7.3	<0.1	<0.1	0.9	1.0	0.6
FP-2, sample 3	74.5	17.2	5.2	0.2	<0.1	0.9	1.4	0.5
FP-3, sample 1	73.8	16.0	7.5	0.1	-	0.9	1.1	0.6
FP-3, sample 2	73.3	15.3	8.5	0.1	<0.1	1.2	0.8	0.8
FP-3, sample 3	73.3	16.1	7.4	0.1	-	1.2	1.1	0.8

Table 2.7 Selected average surface concentrations (from the XPS results) of CFRP samples treated with differently concentrated solutions (FP-1, FP-2, and FP-3) applied as a fingerprint

Scenario	Na (at.%)	Cl (at.%)
P-FP-1	0.2 ± 0.1	0.5 ± 0.3
P-FP-2	0.5 ± 0.1	0.9 ± 0.0
P-FP-3	0.7 ± 0.1	1.1 ± 0.2

2.6.2 Ultrasound Results

2.6.2.1 Coupons

In this section, the results from the two different phased array probes used are presented, depending on the highlighted feature. Various types of defects were observed over the complete set of samples; these can be grouped into the three categories detailed below. The first one contains observations of a minor defect due to manufacturing; the second one comprises slight deviations from the reference;

Table 2.8 Final XPS results indicating the surface concentrations at distinct positions (pos.) for CFRP samples contacted with different de-icer dip-coating concentrations (DI1, DI2, DI3)

Samples	C (at%)	O (at%)	K (at%)	S (at%)	Si (at%)	Cl (at%)	N (at%)	Na (at%)
DI-1 sample 1, pos. 1	58.4	28.5	8.9	0.8	0.3	0.3	2.6	0.3
DI-1 sample 1, pos. 2	62.5	25.9	6.2	0.9	0.3	0.3	3.7	0.2
DI-1 sample 1, pos. 3	64.1	25.7	4.3	0.9	0.3	0.3	4.3	<0.1
DI-1 sample 2, pos. 1	61.9	26.2	8.4	1.0	<0.1	0.5	1.8	0.1
DI-1 sample 2, pos. 2	65.0	24.3	5.2	0.9	<0.1	0.2	4.2	<0.1
DI-1 sample 2, pos. 3	63.5	25.0	5.6	0.6	<0.1	0.2	4.7	<0.1
DI-2 sample 1, pos. 1	55.7	30.5	11.0	0.6	1.0	0.2	0.8	0.2
DI-2 sample 1, pos. 2	51.8	32.9	12.8	0.5	0.6	0.2	0.8	0.3
DI-2 sample 1, pos. 3	51.5	31.8	13.8	0.6	1.0	0.2	0.9	0.2
DI-2 sample 2, pos. 1	63.8	24.9	7.4	1.0	<0.1	0.2	2.6	0.1
DI-2 sample 2, pos. 2	62.6	24.4	10.2	0.6	0.3	0.2	1.5	0.1
DI-2 sample 2, pos. 3	63.9	23.3	10.0	0.6	0.4	0.2	1.5	0.2
DI-3 sample 1, pos. 1	52.3	31.9	13.0	0.4	0.6	0.2	1.1	0.2
DI-3 sample 1, pos. 2	55.2	30.5	11.2	0.6	1.1	0.1	1.1	<0.1
DI-3 sample 1, pos. 3	55.7	29.2	11.8	0.6	0.9	0.2	1.3	0.3
DI-3 sample 2, pos. 1	57.5	28.7	11.9	0.5	0.4	0.2	0.4	0.3
DI-3 sample 2, pos. 2	57.3	29.6	9.9	0.9	0.2	0.1	1.7	0.2
DI-3 sample 2, pos. 3	53.8	29.3	14.0	0.6	0.6	0.1	0.8	0.2

and the third one is the category of obvious defects with possible consequences for the ENDT measurements. Illustrations are given in the following sub-sections.

Table 2.9 Average concentration of potassium on CFRP surfaces for the DI contamination scenario

Scenario	K (at.%)
DI1	6.4 ± 1.8
DI2	10.9 ± 2.3
DI3	12.0 ± 1.4

First category of defects: Minor defects due to sample manufacturing

The defect types within this category can be itemized into three groups. These are most likely due to the sample manufacture and are expected to have no impact or a low impact on the ENDT measurements:

- Bending of the bonded specimens (Fig. 2.25a). This defect type was particularly observed for the repair reference samples as well as some of the samples within the TD scenario. The observed curvature is evidenced on the FWE amplitude cartography with the 5 MHz probe, as shown in Fig. 2.25a. In fact, a phased array measurement was only possible using the SAUL option to partially compensate for the curvature and the induced misalignment of the probe with the sample along its

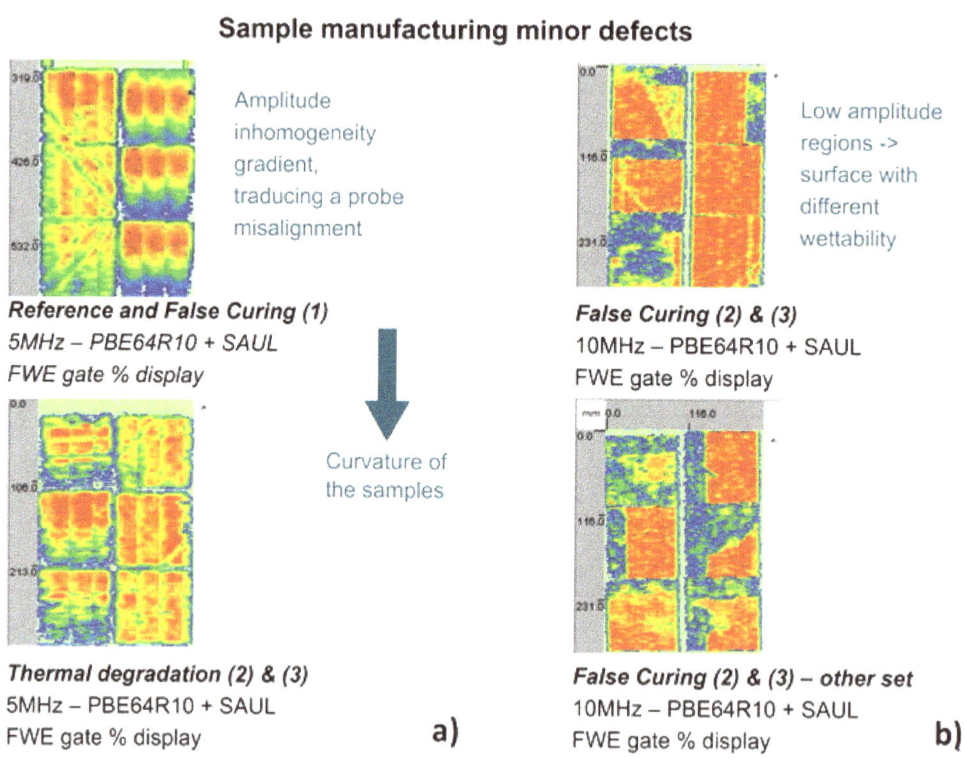

Fig. 2.25 Overview of the minor manufacturing defects observed during the ultrasonic inspection of CFRP specimens, with typical examples of **a** the bending of samples and **b** the surface quality problem

length. However, even if this defect has an effect on the ultrasonic measurements, it should have no consequences for the ENDT investigations.

- Bad quality of the composite surface (Fig. 2.25b). This defect type was evidenced using the 10 MHz probe on the FWE amplitude cartographies. Indeed, the cylindrical focusing of the probe increases the sensitivity to such surface defects. These were generally located on edges and were probably due to marks left by the adhesive tape used during the manufacture. In these areas, wettability with the ultrasound coupling medium is probably different, thus leading to low amplitude regions. In other rare cases, the wall surface was covered, probably due to resin leakage.

Second category of defects: Minor deviation from the reference, potentially due to contamination

Other sample inspection results presented a significant deviation from the reference measurements, albeit without clear evidence of a defect (Fig. 2.26). Specifically, in case of CFRP samples from the P-MO and P-FP scenarios, the bond echo amplitude is of approximately the same order of magnitude as the BWE, whereby this is not the case for the reference and other contaminated samples (as shown by the release agent samples in Fig. 2.26). This observation could be due to higher bond impedance or lower back composite skin impedance. In other words, it could be a sign of bond alteration or composite alteration (especially in the case of moisture). This was evidenced using g+gate TOF cartography, which displays the position of

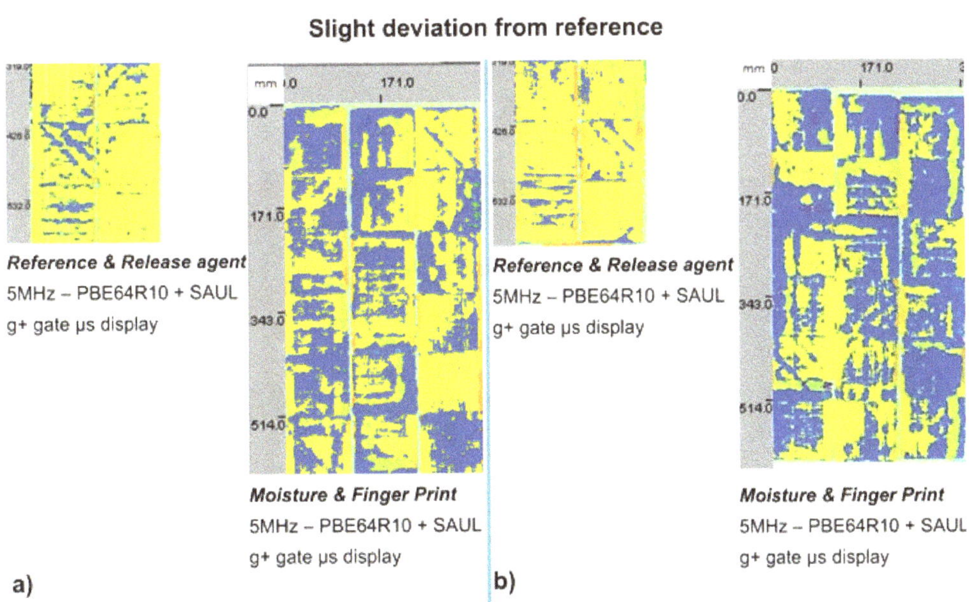

Fig. 2.26 Example of ultrasonic inspection results for CFRP specimens showing a slight deviation from the reference, taking the "g+" gate display for the reference and release agent scenario or for the moisture and fingerprint combined contamination scenario; **a** first sample set as an example and **b** second sample set as an example

the maximum echo in the sample thickness. Referring to the color coding used for displaying the results of the ultrasound investigations, blue corresponds to the bond echo, and yellow to the BWE. A mix between blue and yellow means that the echoes are of approximately the same order of magnitude, while only yellow means that BWE dominates. Two examples are given in Fig. 2.26a for the first set of samples, while Fig. 2.26b shows the second set. This effect is so far neither well defined nor explained, but it might be due to contamination. However, this does not prevent these samples from being considered as weak bonds.

Third category of defects: Obvious defects due to "contamination"

Finally, more sizeable and obvious defects with a potentially highly detrimental effect on the future ENDT measurements were observed, namely delamination and disbonding. CFRP specimens having undergone faulty curing or thermally degraded samples are the main concerns here. In these cases, the state of the joint of the samples cannot be considered as presenting as a weak bond but rather as a bond with defects. All the observed defects are presented in Figs. 2.27 and 2.28.

In the FC-1 samples (Fig. 2.27), the regions showing the effects of debonding are evidenced in the middle of each plate. A defect signature is visible in the bond amplitude and TOF cartographies. The highest amplitude and a small shift in TOF, respectively, show debonding with a high contrast. Consequently, defects are also visible on the BWE amplitude gate, with a low amplitude region, and in the "g+" TOF cartography. The sizes of the debonding regions vary depending on the samples but are usually above 1 cm and up to 5 cm. These debondings are located in the middle of the bonded coupons and, thus, they could have a significant impact on the ENDT measurements. Note that some small debonding spots (smaller than 1 cm) are also observed in some of the FC-2 samples. These defects were probably induced by the way the samples were manufactured.

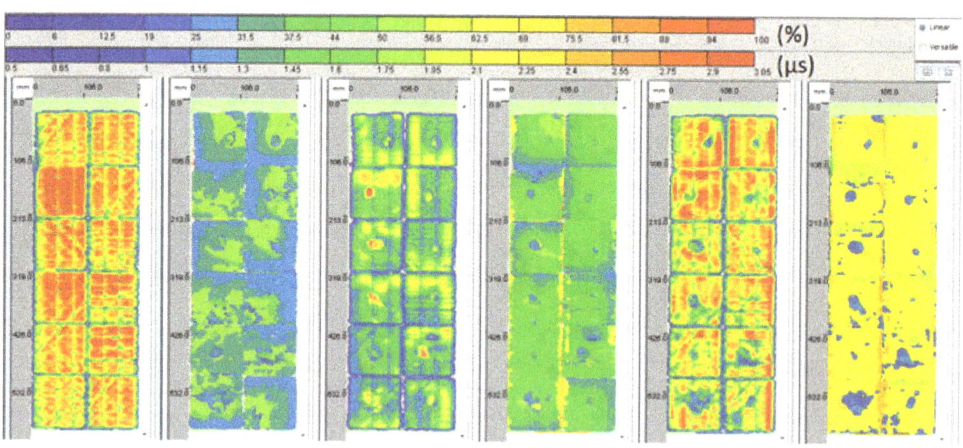

Fig. 2.27 Results of the ultrasound investigations for samples within the FC-1 scenario—observation of disbonding in all the samples with defects showing a size range of 1–5 cm

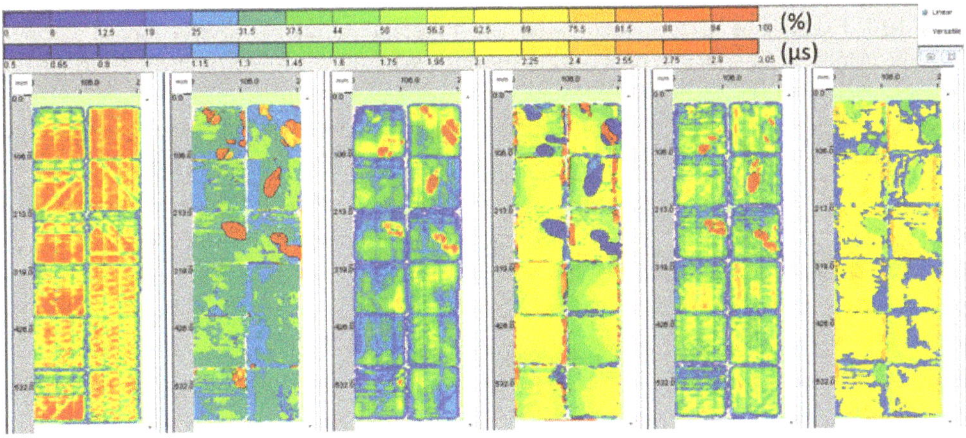

Fig. 2.28 Results of the ultrasound investigations for all samples within the TD-3 scenario—observation of delamination, in some samples with a size in the range of 2–5 cm

For the TD-3 samples, important delaminations were highlighted by ultrasonic inspection (Fig. 2.28), however not for all samples. The delaminations are evidenced by all the gates because they are located on the first composite skin, i.e. the one that has been thermally affected. This is confirmed by the "g+" TOF cartography (cf. light green color coding in the figure of Table 2.10). Therefore, the defect signature enters the bond gate and has consequences on all the other gates. Because of their considerable size (several centimeters) these delaminations could have a strong influence on results obtained using ENDT techniques.

The most profound investigations were achieved using a post-treatment tool. In this work, the NDTKit analysis software was used to obtain the ultrasound data for the CFRP samples (the commercial version of this software is named Ultis), whereby "defect detection" was enabled to size some of the defects. An example of a case of faulty curing is presented in Table 2.10, together with the main characteristics typical for disbonding and described according to the parameters of position on the plate, surface, outline surface, length, and mean value (for TOF). We note that the origin of the defect is at the top left-hand corner of each plate.

2.6.2.2 Combined Contamination Coupons

Multi-contaminated samples were also investigated using the same methodology and the same instrument settings as described above. Essentially, the same kinds of defects were observed for these samples, which is as expected since their manufacturing and contamination procedures were the same. Some examples are given in Fig. 2.29, and are these are further detailed below.

Minor defects were also observed; in this case, they are attributed to clamp marks, as shown in Fig. 2.29a. These are located on the edges and occur during the sample manufacture (probably while stabilizing the plate for bonding). On the ultrasound data acquisitions, they are visible on the bond amplitude C-scan as well as on the

Table 2.10 Example of an ultrasound-based defect detection result for CFRP coupons prepared within the faulty curing scenario, indicating the position in the plate, surface, outline surface, length, and mean value (for TOF)

Defect detection	Names (centers)	Surface (mm^2)	Outline (mm^2)	Length (mm)	Mean (μs)
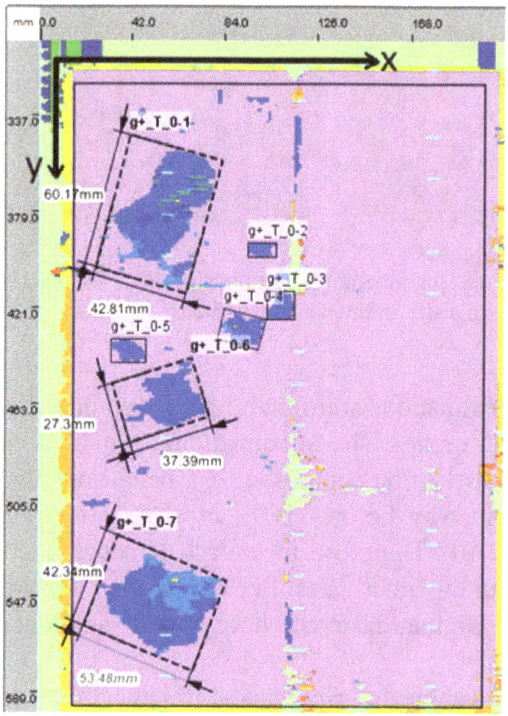	g+_T_0-1 (X = 43, Y = 62)	1328.0	2575.6	60.2	1.08
	g+_T_0-2 (X = 92, Y = 81)	57.0	78.0	13.0	1.08
	g+_T_0-3 (X = 98, Y = 7)	115.0	144.0	12.0	1.09
	g+_T_0-4 (X = 80, Y = 20)	174.0	260.2	19.2	1.12
	g+_T_0-5 (X = 28, Y = 27)	108.0	160.0	16.0	1.08
	g+_T_0-6 (X = 49, Y = 49)	517.0	1020.9	37.4	1.07
	g+_T_0-7 (X = 44, Y = 44)	1350.0	2264.3	53.5	1.12

FWE gate. Indeed, this type of defect is often associated with a bond material leakage on the composite surface (due to bond creep). Therefore, the composite surface is also modified. Note that these areas are also often associated with a bond thickness reduction. Moreover, slight deviations from the reference can also be evidenced in the case of the "release agent+fingerprint" scenario. The effect is in this case probably the same as the one explained in the previous section, namely that a higher bond impedance may have been induced by the contamination. Since this effect was not observed when considering the release agent contamination alone, it might mean that additionally applying the fingerprint is causal for this signature in the combined contamination samples.

2.6.2.3 Curved Specimens

Finally, the curved specimens were tested. Compared to inspecting the flat samples using ultrasound, these inspections are more challenging because of the geometry of the part. The echo entry face reverberates almost all the incoming energy in a divergent beam. Therefore, it is very hard to inspect such a type of geometry with

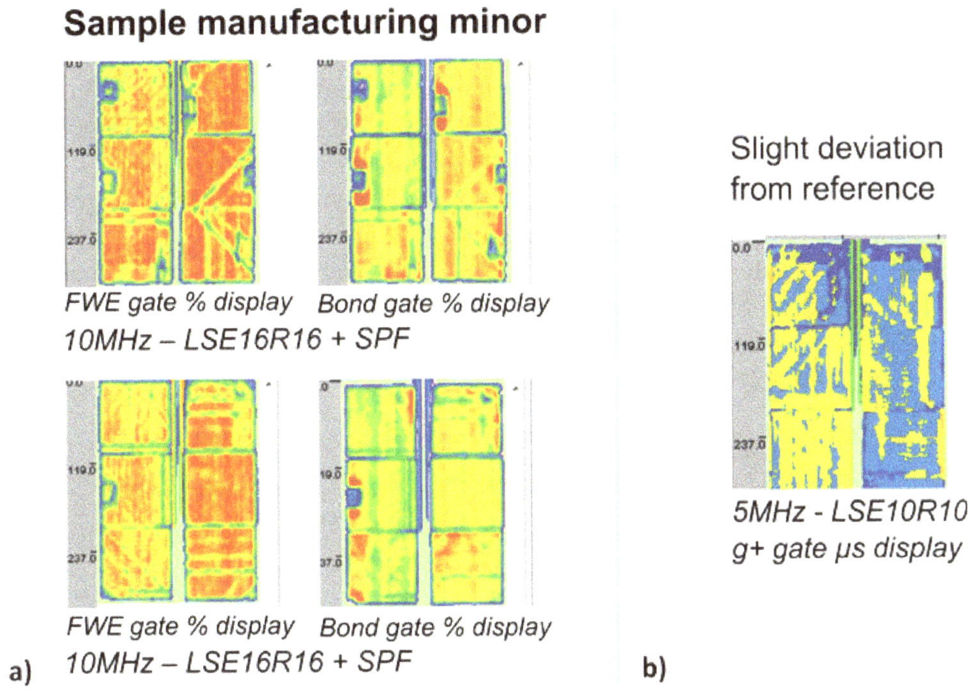

Fig. 2.29 Overview of the defects observed by ultrasound investigations of CFRP samples from the combined contamination scenarios, especially the scenario "release agent+finger print"; **a** minor manufacturing defect and **b** minor deviation from reference

a regular phased array setting. An example of an inspection result when using a conventional linear scanning configuration is shown in Fig. 2.30. Only one small part of the sample appears in the cartography for the curved specimen. The thus inspected region corresponds to the place where the sample surface is oriented closest to parallel to the probe, plausibly in the middle of the curved specimen. Everywhere else, no echo was measured, which is attributed to the fact that all the waves were reflected away from the probe.

To solve this issue, the SAUL algorithm was used. The idea is to first describe the surface geometry by using the ultrasonic phased array like a radar, whereby the

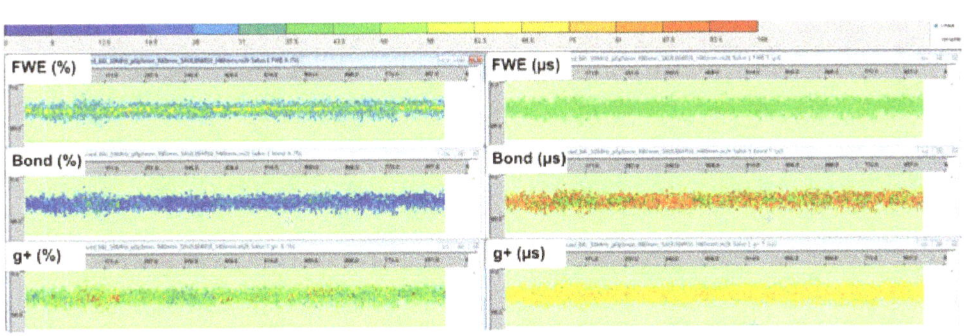

Fig. 2.30 Inspection results for a curved CFRP when using a linear scanning configuration

distance to each element of the probe is detected. Subsequently, the delay laws are calculated to generate an ultrasonic wavefront that will fit the sample geometry. In Fig. 2.31, an example based on a CIVA numerical simulation is presented in order to explain the principle. The elements on the edges are shot first, while those in the middle are shot in the last position.

The curved CFRP sample inspections when applying SAUL are presented in Fig. 2.32. The results show that the specific algorithm helps to complete the inspection. We did not notice any lack of data acquisition in any region of the specimen, and the echo-wall "ew" amplitude was more or less homogeneous. Inspections were thus possible thanks to SAUL, and the obtained signal could be used to make a statement

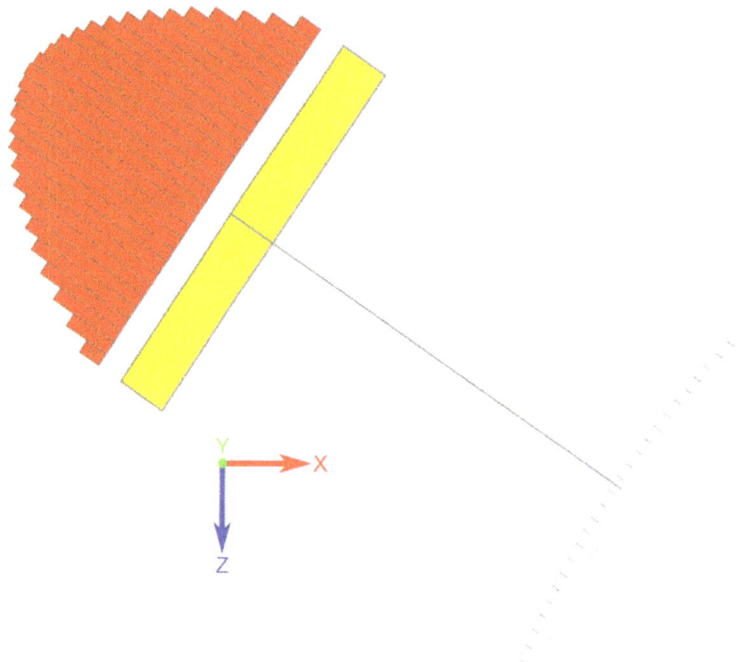

Fig. 2.31 Sketch showing an example of a SAUL (surface adaptative ultrasonic laws) calculation applying CIVA software

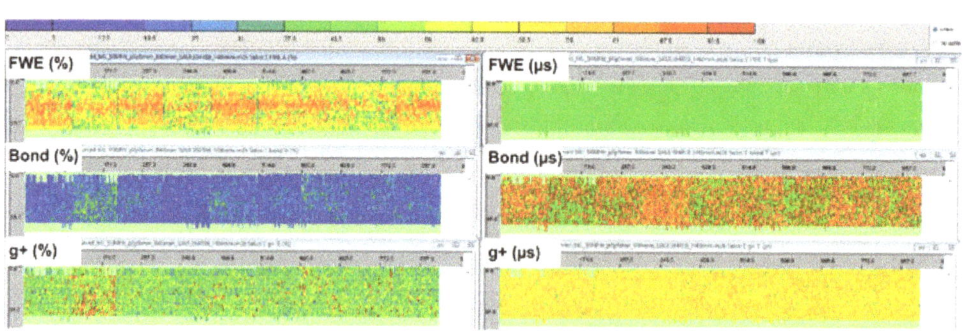

Fig. 2.32 Ultrasound inspection results for a curved CFRP specimen using a SAUL configuration

about the material quality, which appears to be very poor. Indeed, the amplitude of the BWE turned out to be very low. The TCG had to be significantly increased in order to receive a signal from the back wall. This necessary increase in the gain is an indication of strong attenuation, probably due to a high porosity content. Thus, it is difficult to give a clear statement on the bond quality, since the quality of the composite adherends might be the main issue.

2.6.3 *Fracture Toughness Results*

As presented in the previous sections, pre-bond contamination was systematically arranged and intentionally performed on CFRP adherends, and the adherend surfaces as well as the resulting CFRP bonded joints were then characterized by means of reference laboratory non-destructive testing (NDT) methods. Subsequently, the results from the destructive testing regarding the respective joint strength are reported and the effects of the carefully adjusted deviation from the qualified bonding process are evaluated based on the observed joint strength and fracture pattern, which are a common design quantity constituting a joint specification. Using the terminology introduced in Chap. 1 for the concept of quality assessment in adhesive bonding suggested in this book, in the aircraft production this strength is considered a design-relevant operands feature.

2.6.3.1 Mode-I Testing

The average G_{IC} values of the specimens manufactured within the production and repair scenarios are presented and compared in the compendious histograms in Fig. 2.33 and Fig. 2.34, respectively. The reference category samples denoted as

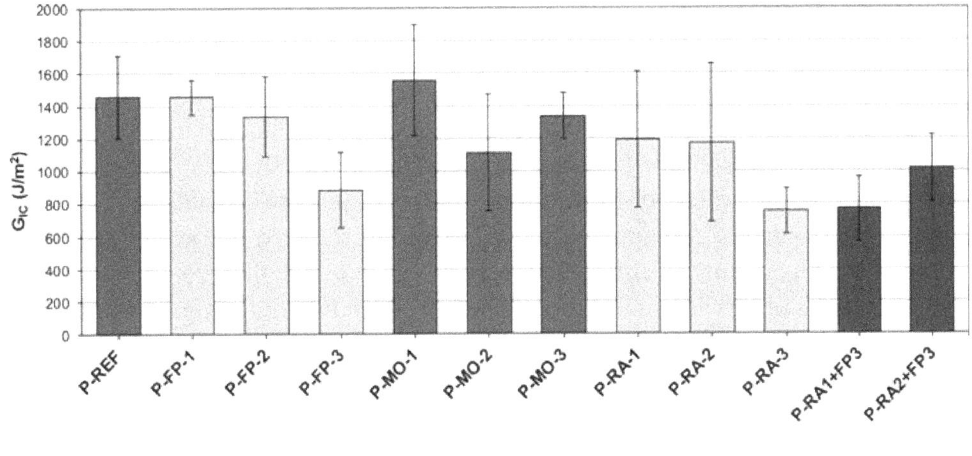

Fig. 2.33 The average G_{IC} values for bonded CFRP joints in a comparison of the production (P) scenarios

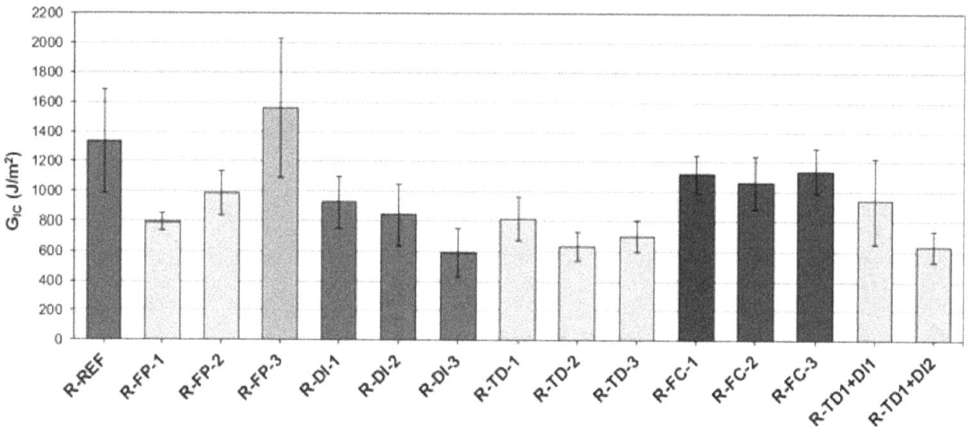

Fig. 2.34 The average G_{IC} values for bonded CFRP joints in a comparison of the repair (R) scenarios

P-REF and R-REF exhibited the highest fracture toughness values while the experimental results indicate a mainly negative effect of the contamination introduced in the respective scenario.

The characterization of the failure mode observed for the respective fracture surfaces showed that adhesive failure was the dominant failure for all the production-related scenarios regardless of the contamination case or level. The adhesive fracture occurred for both the intentionally contaminated substrates in the production and the repair scenarios, but in a different way for each specimen (the pattern and amount of separated adhesive differ), contributing to the large scatter effect observed in the G_{IC} values. In the repair reference samples, a mixed-mode failure was observed (Fig. 2.35), with the dominant failure being a light fiber tear failure, at 50%.

When investigating the joints manufactured following the production fingerprint contamination scenario (P-FP), the respective observed G_{IC} values decreased as the level of contamination by the artificial hand perspiration solution increased. Specifically, for the joints prepared following the P-FP-1 deviation from the qualified production process used for the set of reference joints, the average G_{IC} value was the same as obtained for the reference joints. This indicates that a low concentration level of the FP contamination does not affect the performance of the bond. For P-FP-2, the average G_{IC} decreased by 8%, while for the high contamination level P-FP-3 the G_{IC} fracture toughness of the joints decreased significantly, by 39%. These findings show the detrimental effect of FP contamination on bond performance.

Regarding moisture contamination, for the joints produced following MO-1, the average G_{IC} values show an almost 7% increase compared to the reference category. Considering the observed mixed-fracture pattern, such a finding might be attributed to a modification of the CFRP material by a moderate water uptake, which causes plasticization of the polymeric matrix due to dispersing water molecules. However, in the MO-2 production scenario, the average G_{IC} value was reduced by 24%, and in the MO-3 by 8% as compared to the reference value. Considering the large standard deviation observed, especially for the MO-2 samples, extrapolating a straightforward

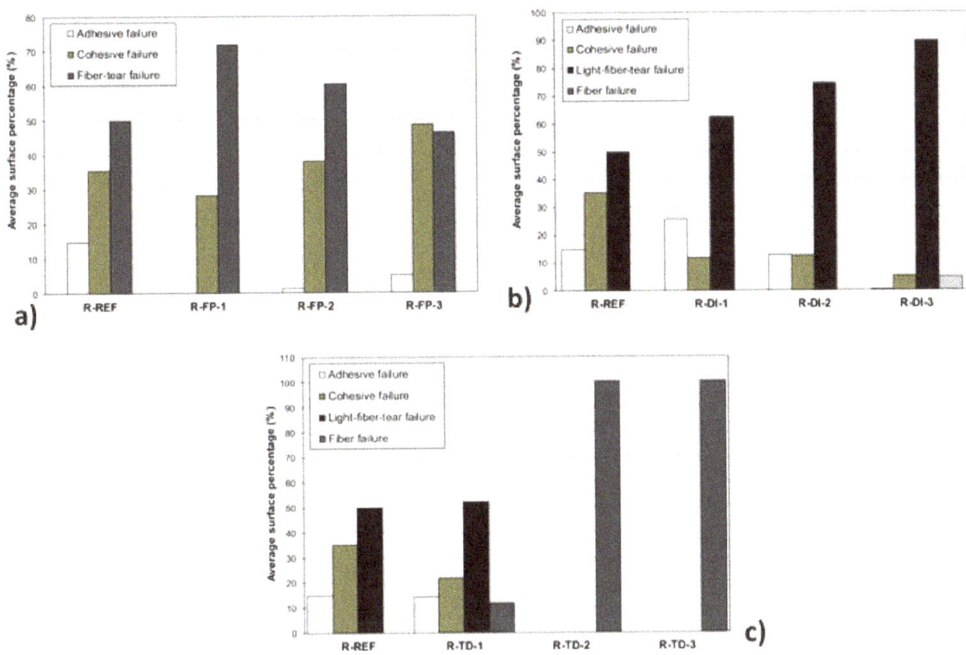

Fig. 2.35 Average percentages of fracture modes obtained after mode-I testing of the joints manufactured within the repair scenarios **a** R-FP, **b** R-DI, and **c** R-TD

structure-property relationship regarding the effect of a medium or high level of moisture contamination on bondline integrity is hindered. In any case, observing a changed average value or a higher standard deviation for the findings related to the design quantity fracture toughness indicates that moist CFRP adherends should be considered an issue for the quality assessment of the resulting bonded joints.

Evaluating the effects of applying even small amounts of release agent to CFRP adherend surfaces within the RA scenario indicates that there are substantial effects on the observed fracture toughness for the thus produced adhesive joints. When the production follows the low-level contamination RA-1 scenario, the average G_{IC} values show an almost 18% reduction compared to the reference category, and for RA-2 the average G_{IC} is reduced comparatively clearly by 20%. The large scatter of the G_{IC} values of the RA-1 and RA-2 samples denotes that there is no statistically significant difference between the effects of the RA-1 and RA-2 contaminations; however, the negative effect of the release agent contamination on the fracture toughness of the bonded joint is evident when compared to the joints produced following the qualified process. For RA-3 the fracture toughness of the joints degrades significantly, by 43%, demonstrating the detrimental effect of the release agent on bond performance. Moreover, the mixed-fracture pattern indicating a weak bond is observed in this case, in clear contrast to the fracture pattern of the joints produced following the P-REF scenario.

Finally, the effects of following a combined contamination scenario with release agent and fingerprint during the production process were investigated with respect to the fracture toughness. The results indicate a significant reduction of the G_{IC} value of the bonded joints that is greater than the reduction caused by each contaminant separately, indicating that the combination of contaminations may be more detrimental to the composite bonded joints' performance. Specifically, the G_{IC} values in the RA1+FP3 and RA2+FP3 scenarios decreased by 48% and 30%, respectively, compared to the reference category values. It is worth mentioning that a consecutive combination of the contamination RA2+FP3, i.e. first a contamination as for RA2 and then as for FP3, led to a smaller G_{IC} reduction than the nominally lower contamination level of RA1+FP3. This finding may indicate that the interaction of a high RA level with FP3 affects the fracture toughness of the bondline less than the interaction of a low RA level with FP3. With both the release agent and the artificial hand perspiration solution resulting in filmy deposits on the CFRP surface upon drying, the supposed interaction between the contaminations is attributed to the effect that the hydrophobic and barely water-soluble release agent film exercises on the transfer and/or film formation of the aqueous solution, which finally dries on top of it.

Subsequently, we discuss the G_{IC} findings displayed in Fig. 2.34, meaning that the focus will be on the mechanical characteristics of joints manufactured according to a repair scenario. The evaluation of the, respectively, obtained fracture pattern is presented in Fig. 2.35. In contrast to the deleterious effect of films from the artificial hand perspiration solution observed for joints prepared within the P-FP category, for the R-FP scenario the contamination with a Skydrol fingerprint seems to have a different scaling effect on the mechanical performance of the joint. We refer to the finding that although R-FP contamination degrades the mode-I fracture toughness of the joint, a reduced decrease in the G_{IC} values was surprisingly observed for samples provided with higher contamination levels. This phenomenon was also supported by the failure mode presented in these samples, whereby an increasing cohesive failure mode was observed (Fig. 2.35a). While a discussion based on additional findings from the surface characterization will not be initiated here, intuitively such a trend would hardly be expected if—in an analogy to the RA scenario—an increasingly thicker inert film with a low cohesion were formed on CFRP surfaces upon contact with increasingly concentrated Skydrol formulations.

Returning to Fig. 2.34 and moving on to the contamination scenario based on depositing residues from a drying aqueous solution of de-icing agents onto CFRP adherends, the fracture toughness results indicate a detrimental effect of dried de-icer fluid on the bond performance as G_{IC} is reduced for all three contamination levels (up to 56% for the DI3 contamination level). The large scatter of G_{IC} values is attributed to the complexity of the adhesion mechanisms and the failure mechanisms (unstable crack propagation, varying failure modes) and possibly to a non-uniformity of the contamination [1, 17]. Considering the respective fracture patterns, it was observed that when increasing the contamination level there is an increase in the percentage of areas exhibiting LFT failure (Fig. 2.35b), which is a clear sign that contact with de-icing fluid degrades the tensile strength of the matrix.

Concerning the TD scenarios, thermal impact and degradation constitute an external influence on a well-characterized material rather than a contamination, e.g. by deposited substances. High temperatures can cause local overheating, damage the CFRP resin and even affect the fiber/matrix interaction due to the differences in thermal elongation between matrix and fiber. The average G_{IC} was reduced by 39%, 53%, and 58% for the joints prepared following the TD-1, TD-2, and TD-3 cases, respectively, as compared to the reference category. However, the G_{IC} value obtained following an exposure of the adherends to the higher degradation temperature (TD-3) was not significantly lower than that observed after an exposure to the temperature of the TD-2 scenario. Although the opposite might be expected, it should be noted here that in some cases [17, 26, 27] it has been reported that high temperatures can cause oxidation, especially at the surface of the resin, which, due to the formation of carbonyl surface groups, may even improve adhesion. This aspect is expected to be less relevant when comparing the effects of the TD-2 and TD-3 scenarios because the thermo-oxidatively affected CFRP surface region had been removed in a grinding process prior to the bonding step. In any case, thermally affected CFRP adherends are clearly an issue for the quality assessment of adhesively bonded joints. Based on the fracture pattern evaluation, the dominant failure was the LTF failure and its percentage increased as the temperature to which the CFRP adherends were exposed increased, with LTF portions reaching up to 100% of the fracture surface area for joints prepared following the TD-2 and TD-3 scenarios (Fig. 2.35c). This indicates that considerable damage was caused to the CFRP adherends due to thermal degradation.

When inspecting the G_{IC} values of the joints with the faulty curing of the adhesive and comparing them with those obtained for the reference joints, it is evident that there was a degradation of 15–21%. Evidently, the non-proper curing of the adhesive in a joint can be very detrimental. Considering the rather large scatter between the fracture toughness values of the joints with the faulty curing of the adhesive, a significant distinction between the effects of the three contamination levels within the TD scenario was not found.

Finally, a contamination scenario that combined the thermal impact on the CFRP adherends and a deposit of dried de-icer was studied and a loss of the bond quality—mirrored by a lower fracture toughness—was observed. Specifically, the G_{IC} values for samples of the combined scenarios R-TD1+DI1 and R-TD1+DI2 decreased by 30% and 52%, respectively, compared to the values observed for joints from the reference category. Especially for the higher level of combined contamination, i.e. TD1+DI2, the G_{IC} reduction was greater than the reduction that each contamination scenario induced separately.

2.6.3.2 Mode-II Testing

The average G_{IIC} values of the samples prepared either following the qualified production and repair processes or after intentionally introducing process deviations during production or repair are presented and compared in the histograms displayed in Fig. 2.36 and Fig. 2.37, respectively. Subsequently, we will present and discuss our findings beginning with the tested as-bonded specimens and then regarding the effects of environmental aging for specimens that underwent a hygrothermal exposure prior to testing the mode-II fracture toughness.

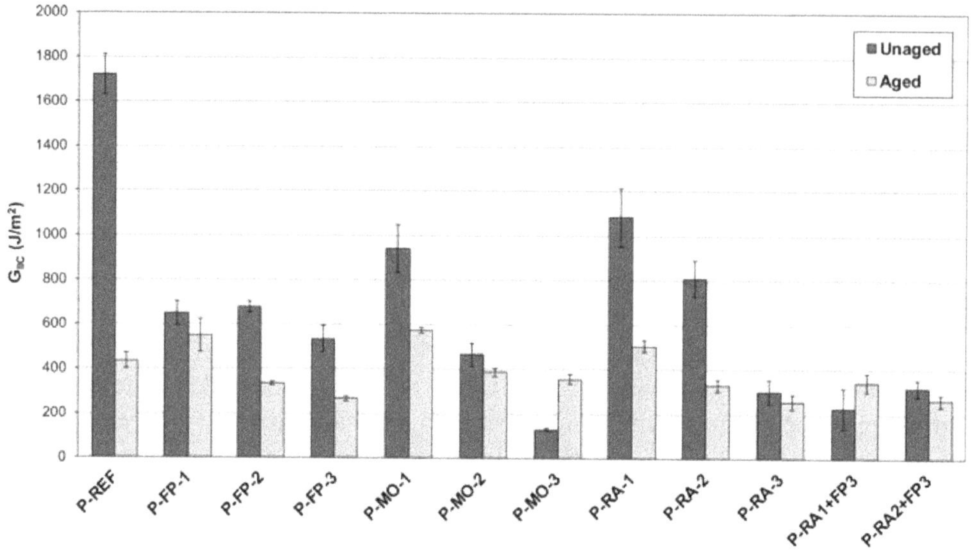

Fig. 2.36 Comparison of the average G_{IIC} values for bonded CFRP joints in the production scenarios before and after hygrothermal environmental aging

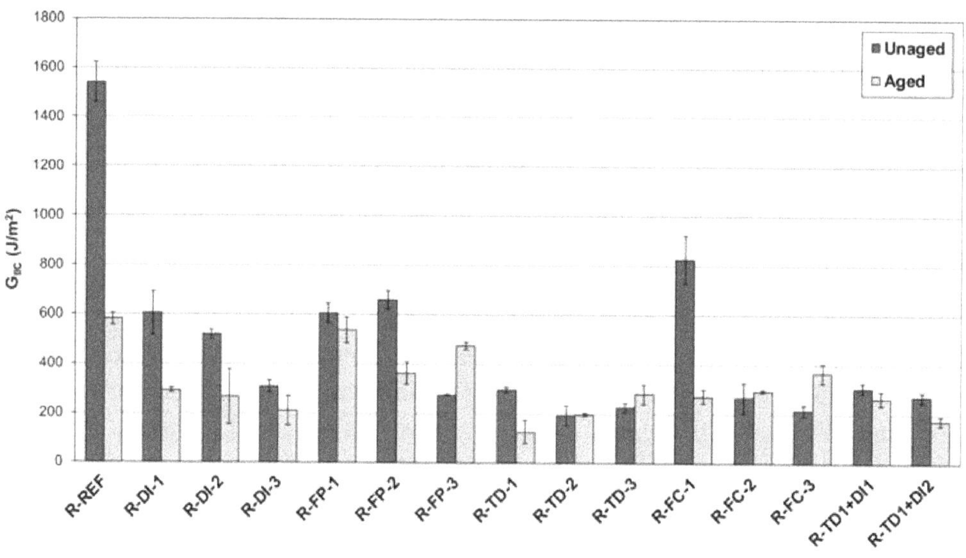

Fig. 2.37 Comparison of the average G_{IIC} values for bonded CFRP joints in the repair scenarios before and after hygrothermal environmental aging

As was observed for the average G_{IC} values of the unaged specimens, the reference category samples exhibited the highest fracture toughness values, also with mode-II characterization. Implementing any of the previously described contamination scenarios during production or repair cases caused a decrease in the G_{IIC} fracture toughness. The observed reduction, as compared to the values found for specimens from the P-REF and R-REF scenarios, respectively, was always greater than the decrease of the G_{IC} values that was observed for the correspondingly prepared sample sets. This finding indicates that the composite bond is more sensitive to contamination when loaded in mode-II (shear-induced crack propagation).

Starting with the G_{IIC} tests of the unaged specimens, a detrimental impact of fingerprinting the CFRP adherend surface prior to bonding in the P-FP scenario was observed, also for the mode-II fracture toughness. When intentionally applying increased contamination levels for further sample sets within this scenario, this highly significant adverse effect was confirmed and a further reduction of the value of G_{IIC} was found. Specifically, for both the P-FP-1 and the P-FP-2 cases a reduction of 61% was observed regarding the reference values, while for P-FP-3 the reduction reached 69%.

Clearly exceeding the effects observed the in mode-I test results, a profound impact of implementing the moisture contamination scenario for the adherends before bonding was revealed in the mode-II tests. An increase of the moisture concentration in the atmosphere applied during the storage of the adherends even caused a further G_{IIC} reduction. Specifically, a reduction by 45% and 73% compared to the reference values was observed for the MO-1 and MO-2 cases, respectively, while for MO-3 the reduction reached 93%. These findings clearly reveal the detrimental effect of moisture absorption on CFRP adherends in the mode-II fracture toughness of the composite bonds. Moisture significantly lowers the quality of adhesion, and it also leads to a loss of performance in the CFRP material itself and, by extension, causes a loss of performance of the adhesive bond [1].

Concerning the mode-II investigations of specimens prepared from adherends intentionally contaminated by release agent, a detrimental effect on the fracture toughness was observed, which corresponds to the findings of the mode-I tests. Increasing the release agent concentration causes an even stronger G_{IIC} reduction. Specifically, for RA-1 a reduction of the G_{IIC} value by 37% was observed with regard to the reference values, while for RA-2 the corresponding value was 53% and for RA-3 the reduction reached 82%.

Finally, it was also observed that the combined contamination with release agent and fingerprint resulted in a pronounced reduction of the G_{IIC} values for the sets of bonded joints. The decrease in fracture toughness was greater than the reduction caused by each contaminant separately, indicating that the effect of successively implementing two deposit-forming combination scenarios of contaminations may prove even more deleterious to the performance of bonded composite joints. Specifically, the G_{IIC} values which were found after having applied the combined contamination RA1+FP3 and RA2+FP3 on the CFRP adherends during the manufacture of the joints were decreased by 87% and 82%, respectively, compared to the reference category values. As in the mode-I tests, the combination RA2+FP3 did not lead to a more distinct reduction of the G_{IIC} value than the RA1+FP3 combination, in which a fingerprint was applied using the same diluted artificial hand perspiration solution but on top of a thinner release agent film.

The following will cover the results of G_{IIC} tests performed with adhesive joints prepared following the distinct repair scenarios and presented in Fig. 2.37.

Intentionally applying runway de-icing fluid to the CFRP adherend surface before bonding has a similar impact on the mode-II fracture toughness of the resulting joints as was observed in the mode-I fracture toughness testing. An increase of the de-icing fluid concentration used for the intentional CFRP surface contamination caused a further G_{IIC} reduction. Specifically, when following the DI1 scenario a reduction of 56% was observed with regard to the reference values, while for DI2 the corresponding value was 62% and for DI3 the reduction reached 80%.

With respect to specimens from the R-FP contamination scenario, it was observed that the mode-II fracture toughness of the joints was drastically reduced. Specifically, applying R-FP-1 and R-FP-2 contamination levels caused a reduction of approximately 61%, while introducing R-FP-3 reduced the G_{IIC} even further to 82% compared to the R-REF category. For all the tested joints a mixed-fracture image was found. Basically, this reduction in bond strength could be attributed to the fact that the FP contamination, with the hydraulic fluid transferred by fingerprinting, led to poor adhesion between the adhesive and the adherend, whereby kissing bonds were formed. However, the observed decrease in the fracture toughness as compared to the R-REF specimens and the obtained adhesive fracture image contrast with the findings for the joints based on correspondingly contaminated CFRP surfaces that were subjected to mode-1 testing and which, in the case of the R-FP-3 scenario, yielded increased G_{IC} values as compared to the R-REF scenario as well as an adhesive fracture image. Therefore, we essentially highlight once again that, under mode-II loading, the composite bond is strikingly more sensitive with respect to the applied contamination than under mode-I loading.

Concerning joints prepared from adherends that had experienced thermal impact before being bonded within the TD scenario, an increase in the exposure temperature caused a further G_{IIC} reduction. The average G_{IIC} was reduced by 81%, 88%, and 86% for the TD-1, TD-2, and TD-3 cases, respectively, compared to the reference category. Again, the reduction for the TD-3 case was lower than for the TD-2 thermal degradation. The scatter hinders any clear distinctions to be made between the effects of applying the TD-2 and TD-3 scenarios on the adhesive composite bond integrity.

As in the mode-I tests, the mode-II tests revealed that the intentionally applied faulty curing of the adhesive within the repair scenario of CFRP joints can be detrimental to their resulting properties. Specifically, for the FC-1 scenario a reduction of 46% was observed with regard to the reference values, while for FC-2 and FC-3 the reduction was even more drastic, amounting to 83% and 86%, respectively.

Finally, applying the more complex repair scenario, which comprises a combination of thermal impact and degradation with an application of de-icing fluid contamination on the CFRP adherends before being bonded, results in a reduction of the G_{IIC} value of the joints by 80% and 83% for the R-TD1+DI1 and R-TD1+DI2 cases, respectively. These effects are greater than the reduction of the G_{IIC} value caused by each contaminant separately. These findings indicate that monitoring effects of successively applied contaminations is a task in the quality assessment of adherend surfaces as well as the performance of the resulting joints.

Aging Effect

Exposure to a hygrothermal environment is reported to be a critical issue regarding the durability of adhesively bonded joints whenever the applied demands are characterized by a combination of elevated temperature, moisture, exposure time, and mechanical loading [28]. For example, elevated parameter settings are applied in scenarios for highly accelerated life tests (HALT), which are used in the development of a joint design to quickly achieve indications of weak points [29] or to identify the functional operating limits exceeding the operational area defined by the product specifications [30]. Against this background and aiming at a comprehension of feasible deviations from the qualified bonding process in the development phase of an adhesive joint, we have decided to include such elevated hygrothermal parameter settings in the concept of the quality assessment for the manufacturing processes of CFRP composite adhesive joints. The effect of the externally applied environment in terms of moisture and temperature has been thoroughly investigated [31–33], essentially indicating a significant loss of the bond strength of joints subjected to aging, especially after reaching a moisture saturation in adhesive joints that were exposed to a high relative humidity or water immersion.

The combination of a pre-bond contamination of adhesive composite joints and after-bond exposure to hygrothermal aging leads to a drastic reduction of the G_{IIC} (Figs. 2.36 and 2.37). In the majority of the investigated scenarios, the effect of applying the combination of pre-bond contamination and after-bond long-term hygrothermal aging to a set of bonded joints is more severe than the effect of applying either of the two scenarios separately.

The drastic reduction of the G_{IIC} values ascertained when comparing the findings for sets of aged and unaged CFRP joints coincides with the fact that the onset of crack propagation was observed at the side of the specimen where the moisture concentration and thermal effect are higher [34]. Additionally, the crack propagation occurred near the adhesive region, and diffusion through the adhesive is regarded as the primary access route for moisture to enter a joint [35].

For example, the aging procedure resulted in a reduction in the mode-II fracture toughness of the R-REF samples by 62% compared to the unaged repair reference scenario (Fig. 2.37). Additionally, for all of the aged specimens of any of the applied contamination scenarios, the observed G_{IIC} values of the aged samples were drastically reduced for all contamination levels compared to the respective unaged reference REF category. Surprisingly, it was observed that there were some sets of joint specimens for which the G_{IIC} values after the hygrothermal aging were higher than for the respective set that was destructively tested before aging, namely for the sample sets prepared following the MO-3, RA1+FP3, R-FP-3, FC-2, and FC-3 scenarios. We attribute this finding of increased fracture toughness after hygrothermal aging to a plasticization caused by the swelling of the adhesive layer and the CFRP matrix due to a specific water uptake. Additionally, that the Skydrol contained in the R-FP series is known to react with the water from the environmental chamber, producing phosphoric acid [36]. Discriminating between such effects or deriving more detailed insights into effects caused by reversible water uptake, e.g. a plasticization of polymeric material, or by irreversible chemical reactions with moisture, e.g. the hydrolysis of phosphoric acid esters, might be possible if re-dried aged specimens were investigated in addition to unaged and aged specimens.

2.6.4 Tensile Testing

In this section we present and discuss the load–displacement curves and the respective fracture patterns observed when performing a tensile testing of adhesive joints prepared from scarfed CFRP adherends. The results obtained for specimens prepared following the reference scenario or by implementing distinct scenarios characterized by introducing contaminations during the manufacturing process are presented in Fig. 2.38. In all the contamination cases, one of the adherends was contaminated while the other was intentionally left in the respective reference state, a setup intended to replicate the real-life application of repair patches.

During tensile testing, it was observed that prior to the final failure (i.e. the separation of the two adherends), all specimens presented an initial failure revealed by a first load drop in the region of the load–displacement curves, which corresponds to plastic deformation. This failure was localized at the edges of the scarfed area and is attributed to stress concentrations at this point as well as to edge effects. Having overcome this marginal fracture, the initial failure propagated along the scarfed area and led to the final separation of the two adherends.

For the specimens prepared following one of the contamination scenarios, we observed that all samples of the II-R-REF+TD1 scenario presented a higher failure

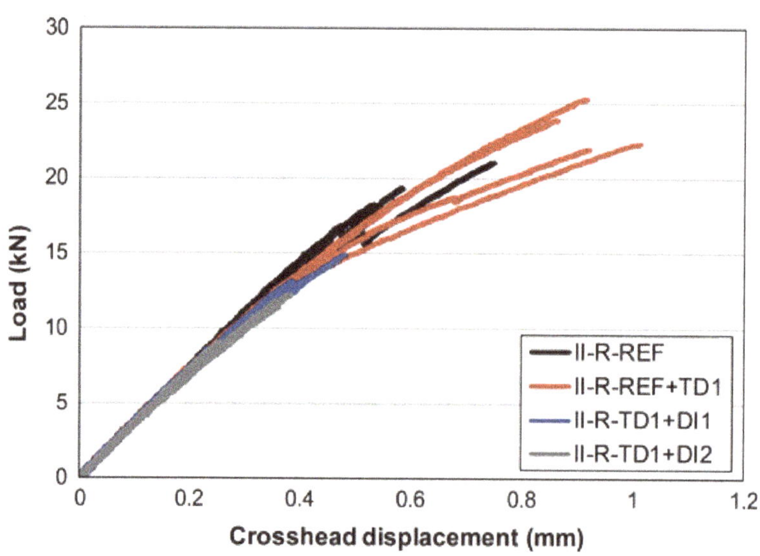

Fig. 2.38 Load–displacement curves for adhesive joints manufactured from scarfed CFRP adherends that were prepared following the reference and contamination scenarios

Fig. 2.39 Failure load comparison for the scarfed samples

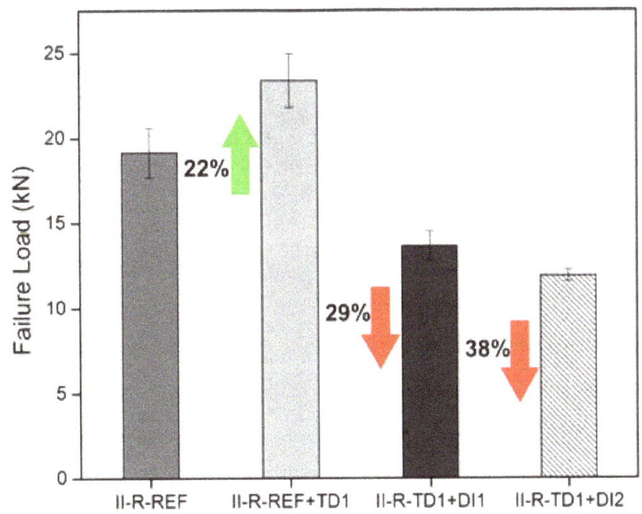

load, by 22%, than the reference samples (Fig. 2.39). Although the exposure of flat CFRP adherends to an elevated temperature following the TD-1 scenario decreased the fracture toughness tested in mode-I and mode-II loading, such a decrease was not prominently observed in the tensile testing of the joints prepared from scarfed CFRP adherends. Even though heat usually damages the CFRP structure or causes chemical changes in the matrix, there have been reports that high temperatures can cause oxidation of the resin which may improve adhesion due to the formation of carbonyl groups at the surface [17, 26, 27].

When implementing scenarios comprising the successive application of two contamination cases, the obtained results revealed that the contamination combining thermal impact and deposits of dried de-icing fluid had a negative effect on the mechanical performance of the scarfed repair joints, reducing the failure load as compared to specimens prepared following the reference scenario. Specifically, for samples from the TD1+DI1 scenario, the reduction of the failure load was 29%, while for the TD1+DI2 scenario the reduction reached 38% (Fig. 2.39).

Additionally, the failure surfaces of the joints were examined after the tensile tests in order to characterize the failure modes and correlate them with the tensile test results. Figure 2.40 depicts the representative fracture surfaces of each contamination scenario studied, showing the main failure modes observed in the tensile specimens.

The percentages of the different failure modes are compared for the different sample sets in Fig. 2.41. For the reference samples, a mixed-mode failure was observed, with the dominant failure being the FT failure, at 63% of the surface area, while adhesive (ADH) failure was observed for 37% of the surface area. In contrast, the tested samples prepared following the II-R-REF+TD1 case presented a higher amount of CO failure (30%), while FT failure and the ADH failure modes showed a reduction (43 and 27%) as compared to the reference samples. This change of the fracture pattern coincides with the increase in the failure load observed for the specimens prepared from scarfed adherends that had been exposed to elevated

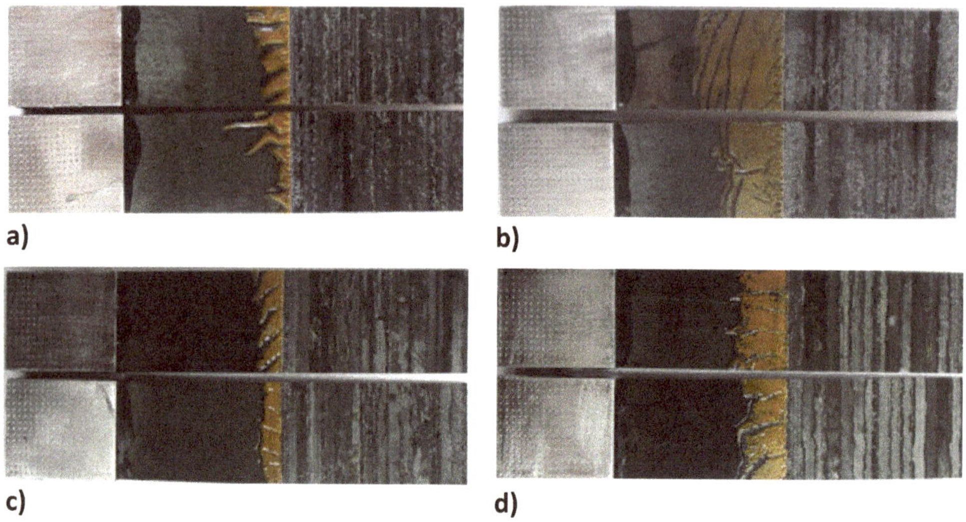

Fig. 2.40 Images showing the representative fracture surfaces of the joints prepared from scarfed CFRP adherends obtained after loading in tension, sorted according to the contamination scenario; **a** II-R-REF; **b** II-R-REF+TD1; **c** II-R-TD1+DI1; and **d** II-R-TD1+DI2

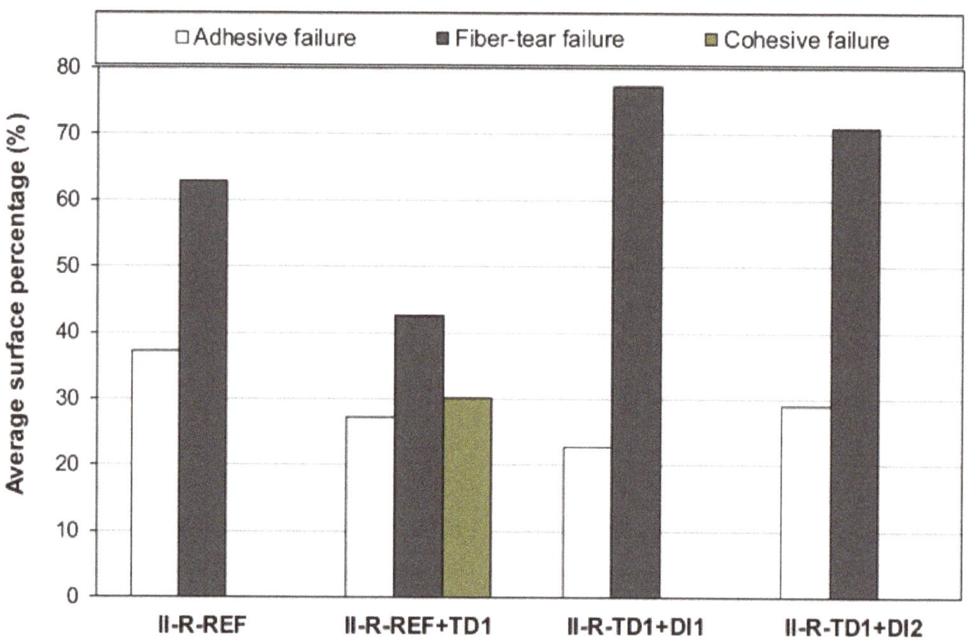

Fig. 2.41 Average percentages of the failure modes presented in the tensile tested scarfed joints, sorted according to the contamination scenario

temperatures following the II-R-REF+TD1 scenario. When discussing these observations, the effects should be considered in relation to the loading geometry of the scarfed specimens or the performed scarfing and subsequent cleaning processes. It is plausible, according to the occurrence of a CO failure, to infer a greater amount of energy required for a crack to develop and propagate than that required to cause an ADH or FT failure. As a consequence, an increase in the failure load is observed.

Inspecting the fracture patterns obtained for the II-R-TD1+DI1 samples, we found that the FT remained the dominant failure. The observed increase of the area percentage by 77%, as compared to samples prepared following the REF scenario, indicates that the combined successive contamination with TD1 and DI1 had a deleterious impact mainly on the behavior of CFRP adherends under tensile loading. Finally, after testing the II-R-TD1+DI2 set, the FT failure showed a slight reduction compared to the findings for II-R-TD1+DI1. Specifically, 71% of the surface area with an FT failure was observed, while ADH failure increased to 29%. These findings indicate that the TD1+DI2 combined contamination affects mainly the bondline performance (the interphase between the adherends and adhesive).

2.6.5 Centrifuge Test Results

In this section, the effect of pre-bond contamination scenarios related to production and repair processes on adhesion strength between the intentionally contaminated CFRP surface and the adhesive layer is assessed based on investigations by means of the novel centrifuge testing technology. The plots shown in Fig. 2.43 and Fig. 2.45 display the average adhesion strength values, as derived from Eq. (2.5) for the applied geometry used in the tests, for the production-related and repair-related samples, respectively.

Figure 2.42 depicts representative microscope imagery revealing the different failure modes observed for the various samples. Meanwhile, Figs. 2.44 and 2.46 display the evaluated average surface area percentages of the failure modes for specimens prepared following the different production and repair-related contamination scenarios.

For all the intentionally implemented deviations from the qualified manufacture process for CFRP adherend surface preparation by applying contaminations defined for production scenarios, the results showed a decrease in the adhesion strength. This decrease was small for the low and medium contamination levels; however, for the high contamination level a more profound decrease was found. The lowest adhesion strength values were obtained for the specimens prepared following the P-MO-3 case (98% RH). The standard deviation is also considerable, but within acceptable limits for revealing the described trends. Remarkably, for all the sets of samples prepared with contaminated adherends, a significantly higher standard deviation of adhesion strength values was observed than for the P-REF specimens (Fig. 2.43).

Considering the specimens for the production-related scenarios, all the investigated fracture patterns revealed ADH and LFT failure modes (Fig. 2.44). The adhesive mainly remained on the metallic stamp, which is an indication of a stronger bond between the metallic stamp and the adhesive as compared to between the adhesive and the CFRP adherend. This stands for all cases of ADH failure here. The P-REF samples showed a much higher adhesion strength (almost double) than that of the

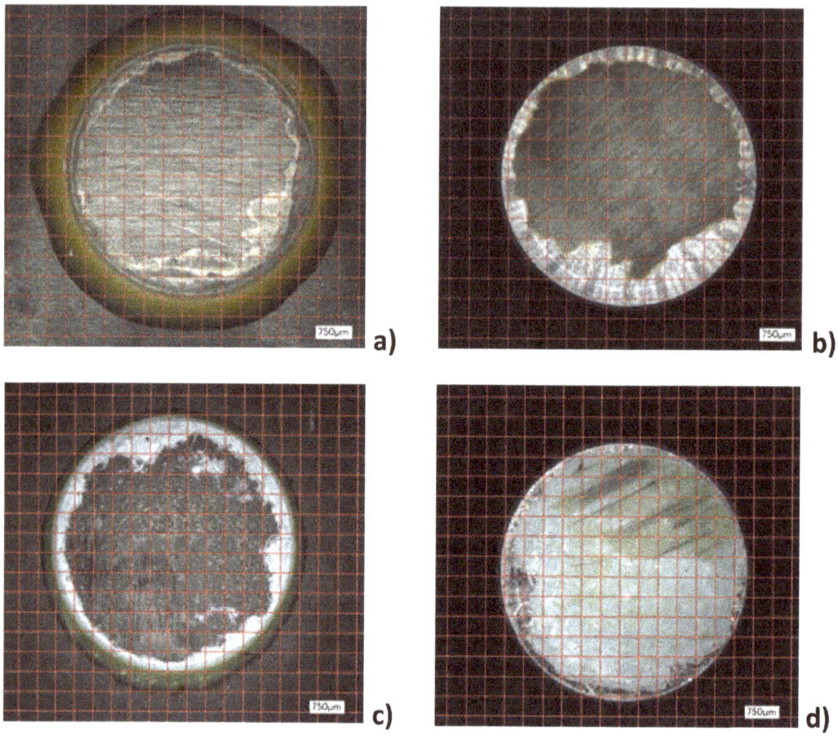

Fig. 2.42 Representative microscopy images revealing the failure modes of centrifuge-tested CFRP samples; **a** ADH+FT for R-TD sample (CFRP side); **b** ADH+FT for R-TD sample (stamp side); **c** ADH+LFT for P-MO sample (CFRP side); and **d** ADH+LFT for P-MO sample (stamp side)

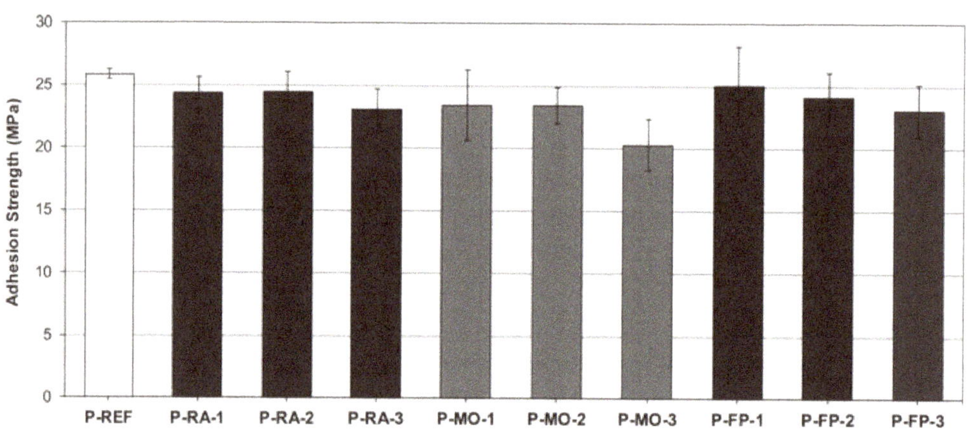

Fig. 2.43 Adhesion strength values obtained upon centrifuge testing specimens of the production-related scenarios

R-REF samples (Fig. 2.45), which is due to the different type of adhesive used and the different curing conditions applied.

Regarding the specimen sets prepared and tested within the repair scenario, the R-REF samples (Fig. 2.45) showed a much lower adhesion strength (almost half) than

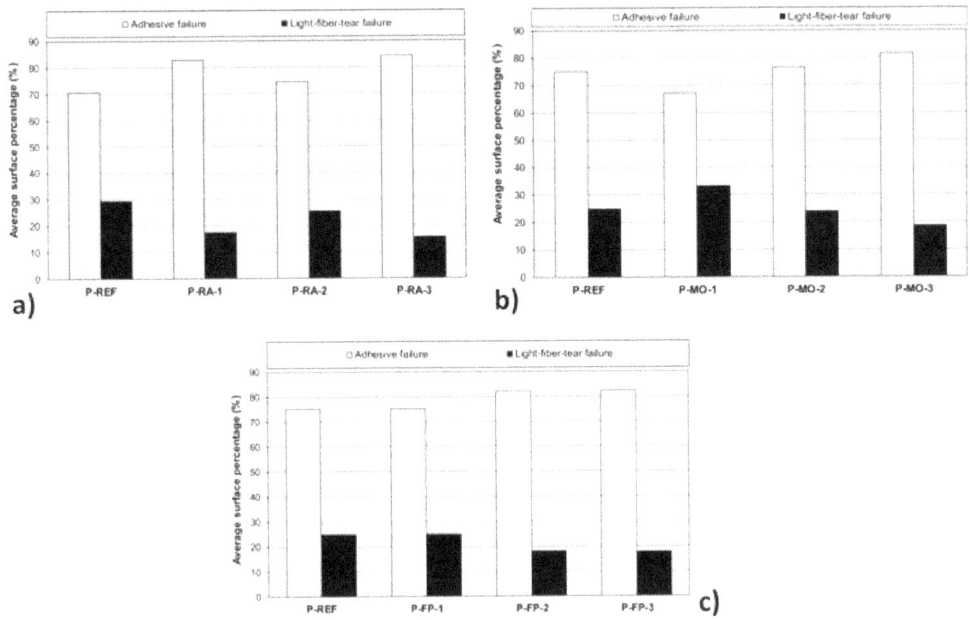

Fig. 2.44 Average surface percentage for the different failure modes of the centrifuge-tested samples **a** P-RA, **b** P-MO, and **c** P-FP

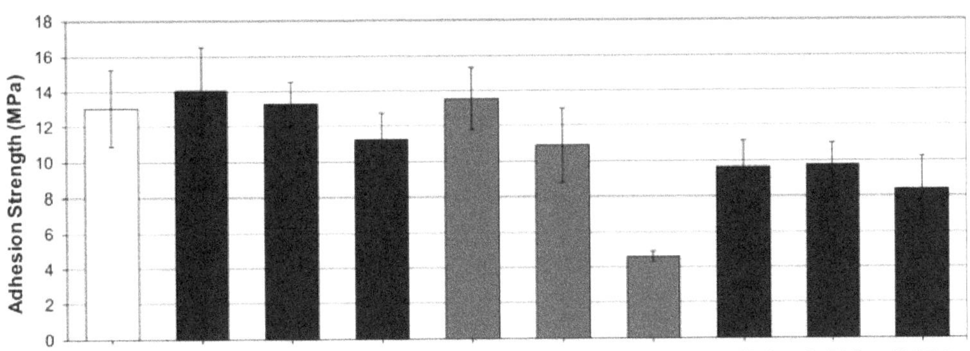

Fig. 2.45 Adhesion strength values for the repair-related sample categories

the P-REF samples. A significantly higher relative standard deviation of the measured strength values was found for the reference scenario R-REF as compared to in the P-REF scenario. These findings are attributed to the different types of adhesive used and the different curing conditions applied.

Against the background, quite insignificant effects of the implemented deviations from the reference joining process were found. Concerning the specimens from repair-related contamination scenarios, (except for R-FP-1 and R-TD-1), a decrease of the adhesion strength was generally observed (Fig. 2.45) as compared to the specimens prepared following the R-REF scenario. For the Skydrol-based R-FP scenario, there seemed to be a slight increase in the adhesion strength for the R-FP-1 case, while an insignificant variation in the adhesion strength was observed for the R-FP-2 case and a decrease of the adhesion strength for the R-FP-3 case. However, a robust conclusion cannot be drawn for the R-FP-1 and R-FP-2 cases due to the very high standard deviation of the results for these two cases and the R-REF case. Only for the R-FP-3 case was there a considerable decrease in the adhesion strength.

The percentages of the failure modes in Fig. 2.46a reveal a similar failure behavior for the R-FP-1 case and the R-FP-2 and R-FP-3 cases, namely a decrease of the LFT failure and an increase of the ADH—which is an indication of a weak bond—and TLC failure modes.

Bearing in mind that the TD scenario mimics effects of a thermal impact on the CFRP adherends, we may assume from the finding that for specimens of the R-TD-1 case a similar behavior in terms of adhesion and failure modes is observed as for the R-REF case (Fig. 2.46b). In contrast, the R-TD-2 and R-TD-3 cases present a lower adhesion strength, and this is attributed to the degradation of the polymer matrix, which becomes effective in the first layer of the CFRP adherend because of the increased temperature, which causes the FT failure mode. The lowest adhesion strength for this set of scenarios was measured for the R-TD-3 case (corresponding to one of the CFRP adherends experiencing a pre-bond exposure of 280 °C).

Finally, for the R-DI scenario, a detrimental effect of the presence of dried de-icer was revealed for the adhesion strength of the joint. The failure mode percentages (Fig. 2.46c) show that an increase in the DI concentration causes an increase of the ADH failure mode, as the deposition of a thin layer by the de-icer acts as a barrier during the bonding of the adhesive and the CFRP adherend. However, in contrast to

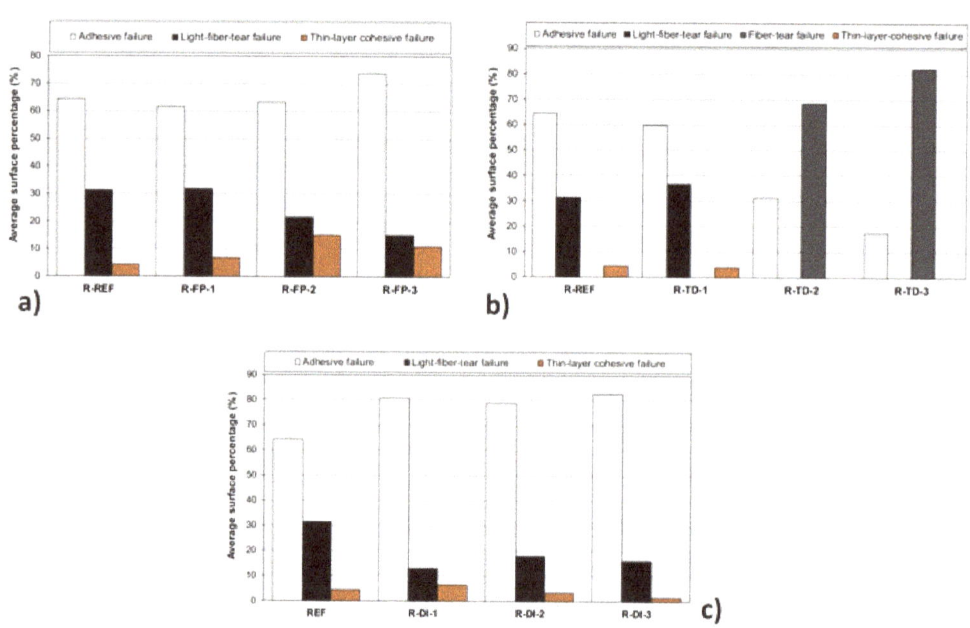

Fig. 2.46 Average surface percentages of the different failure modes for the **a** R-FP **b** R-TD and **c** R-DI centrifuge samples

the adhesion strength, there is not a clear differentiation regarding the failure mode percentages between the different concentrations of the dried de-icer.

2.7 Numerical Simulation

2.7.1 FE Model

A composite panel, stiffened with two stringers, was simulated under compression using the LS-DYNA FE platform. Besides the reference panel, all contamination scenarios were simulated, and their maximum load-bearing capacity was compared to the reference panel. Debonding growth was simulated using the cohesive zone model (CZM) method. This method has been widely used in the last decade to simulate the delamination progression in composite materials and the debonding progression in bonded joints, mainly due to its ease of use as it has been implemented in many commercial FE codes.

For the analysis, the linear elastic/linear softening (bilinear) traction-separation law was adopted. The constitutive law described in Fig. 2.47a is for the tension loading and separation of the adherends in the normal direction (mode-I). The mixed-mode behavior is described by the mixed-mode bilinear traction shown in Fig. 2.47b. The first region (until point 1) corresponds to the elastic part of the material's response. In this region, the material remains undamaged and the unloading at point 1 follows the elastic line. The region from point 1 to point 2 represents the material softening (damage growth) area. Once the loading has progressed beyond point 2, the material has suffered some damage (damage parameter is greater than zero but less than one), however, there is no adherend separation yet. This occurs at point 2, where the adherends separate permanently (damage parameter has reached unity). The total area under the triangle represents the energy required to debond the adherends and is known as the fracture energy. In LS-DYNA, the fracture energy

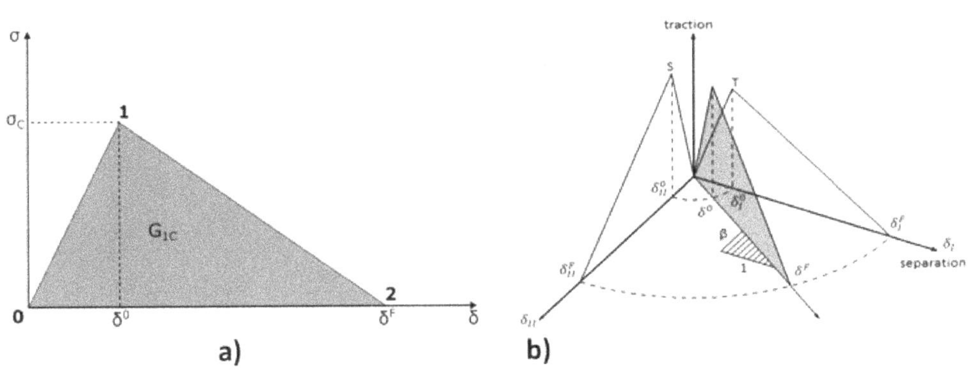

Fig. 2.47 Schematic representation of **a** the bilinear traction-separation law for the mode-I load case and **b** the bilinear traction-separation law for the mixed-mode load case [37]

is an input parameter. It has units of energy/area. In addition, the elastic stiffness (slope) and the peak stress (point 1) are required to completely define the bilinear law.

The progressive damage modeling method was adopted to simulate the damage in the composite adherends. To this end, the material model MAT_162 of the LS-DYNA was used, which has the capacity to predict several modes of damage to the composite material. The specific material model automatically implements the progressive damage modeling method by combining a set of strain-based Hashin-type failure criteria for predicting several failure modes like tension/shear fiber failure, compression fiber failure, perpendicular matrix failure, and delamination [38].

To this end, skin, cap, and web components were modeled using standard eight-node solid elements with three degrees of freedom per node (ELFORM = 1) and MAT_162_COMPOSITE_MSC_ DMG. In addition, MAT_162 automatically applied a property degradation module to simulate the damage effects.

FE mesh was created using ANSYS Workbench and imported into the LS-DYNA FE platform. All analyses were performed using LS-DYNA. The imported mesh is depicted in Fig. 2.48.

The adhesive layer was modeled using eight-node cohesive elements with three degrees of freedom per node (ELFORM = 19) and MAT_138_COHESIVE_MIXED_MODE, which applied a mixed-mode CZM with bilinear traction-separation law to the cohesive elements. Furthermore, the debonding growth was predicted using the B-K power law.

The FE model was loaded in compression by applying displacement and the nodes were fully supported at the end of panel, as can be seen in Fig. 2.48. In addition, in order to reduce the extensive out-of-plane deformation that would cause a buckling to the panel, the nodes that are depicted in Fig. 2.48 were also supported. These nodes represent a possible anti-buckling device that could be used during mechanical tests.

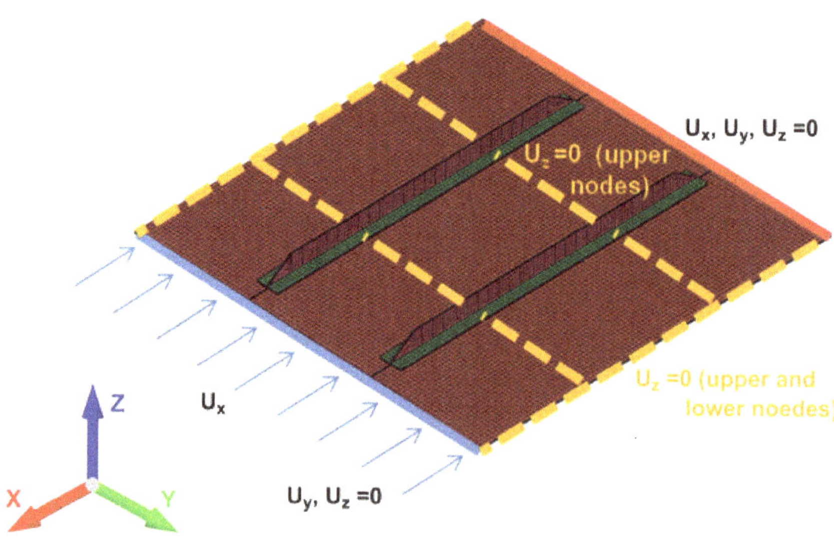

Fig. 2.48 Applied compression and boundary conditions on a flat stiffened panel

2.7.2 Numerical Results

The resulting load–displacement curves for all contamination scenarios are depicted in Fig. 2.49.

The comparison concerning the maximum load for all contamination scenarios is presented in Fig. 2.50, wherein it can be observed that all contamination scenarios have a negative influence on the load-bearing capacity of the stiffened panel.

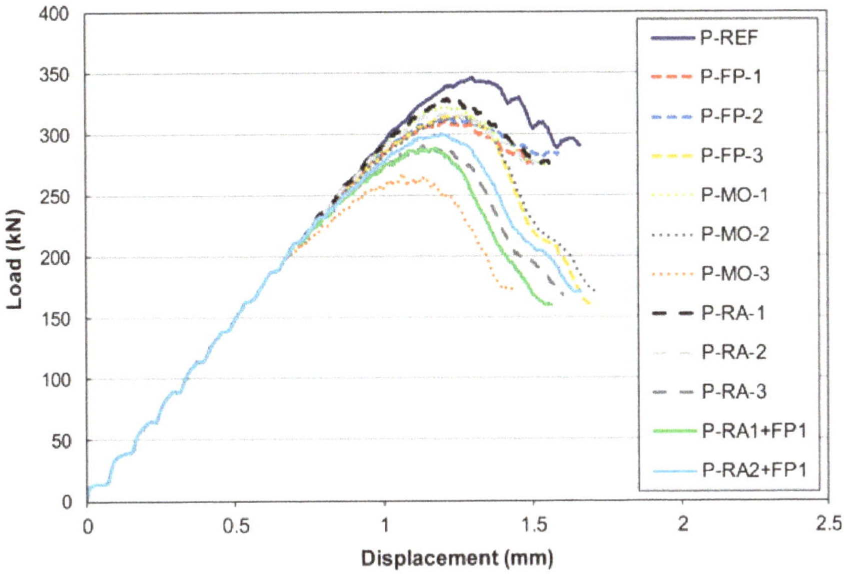

Fig. 2.49 Load–displacement curves under compression for all contamination scenarios

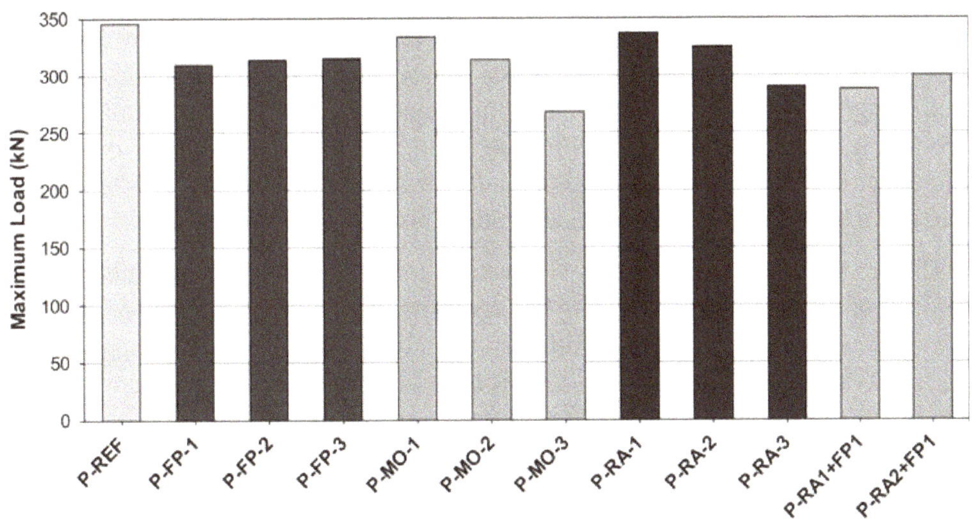

Fig. 2.50 Predicted maximum compressive load for all contamination scenarios

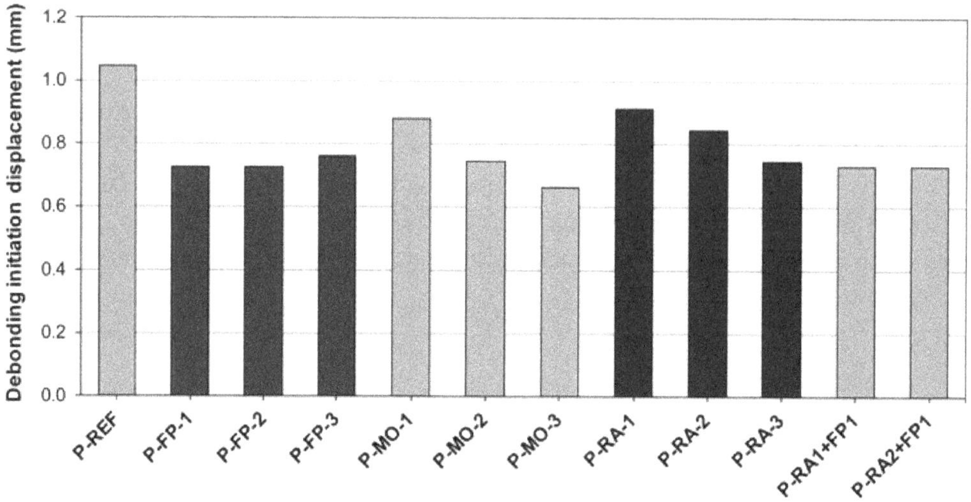

Fig. 2.51 Predicted debonding initiation for all contamination scenarios

In Fig. 2.51, the debonding initiation displacement is presented and compared for all contamination scenarios. It is clear that all contamination scenarios have a negative impact on debonding initiation, as in all scenarios the debonding initiates earlier than in the reference case.

2.8 Conclusions/Synopsis

With our objective in mind of providing the reader with a feasible concept for quality assessment in adhesive bonding technology that complies with the ten heuristics and systematics described in Chap. 1, in this chapter we detail procedures to introduce disturbances from one or even several operator-related process features in scheduled ways and to test quantitative and design-relevant joint features by applying pre-process or post-process methodologies. Hereby, the exemplification is based on identifying, defining, and intentionally implementing pre-bond contamination on carbon fiber reinforced plastic (CFRP) adherends in gradational levels quantified with spectroscopic laboratory methods and identified with respect to a reference state given by the respective qualified bonding process.

The contaminants investigated within this project have a high relevance for the majority of aerospace applications. The test scenarios cover two fields of application for the adhesive bonding of primary structures, namely aircraft manufacture and in-service bonded repair. The identification of all feasible (or, pragmatically, all imaginable) disturbances of process features resulted in the definition of production and repair scenarios, yielding distinct reference surface states differing in the depth of abrasion accomplished by the CFRP grinding process. For the relevant adherends, three sample geometries were defined, namely smooth coupon samples, scarfed samples, and curved panels. The production-related disturbances comprised the impact of release agent, moisture, or saliferous fingerprints and, with respect to repair-related disturbances, thermal impact, dried de-icing fluid, or a fingerprint with hydraulic fluid on the adherend surface; a faulty curing of the adhesive was also considered. For each of these identified and technologically crucial scenarios, we implemented discrete levels differing in the dimension of the applied contamination. Moreover, we accounted for the effects of hygrothermal aging before determining the joint strength using mode-I or mode-II testing of the fracture toughness as well as through a novel centrifuge test. In particular, specifications defined by the users of CFRP adhesive joints are often based on safeguarding adequate G_{IC} values. Exceeding the respective standards, we show that joint quality is supportively and sensitively mirrored by mode-II testing of the fracture toughness as well as by the novel centrifuge test.

The present chapter describes in detail the manufacturing of the adherends for all the sample geometries (Aernnova Composites), the pre-bond single and multiple contamination and bonding of the samples (Fraunhofer IFAM), the characterization of the adherends and the joints using the XPS method (Fraunhofer IFAM), the ultrasound testing of the bonded samples (Airbus), the mechanical testing and the after-bond contamination of the samples (University of Patras), and finally the numerical simulation of the stiffened panels (University of Patras).

The manufacture of the CFRP adherends for the coupons, scarfed samples, and stiffened panels was performed under consideration of the specifications and surface quality requirements determined by the internal procedures of the manufacturing company and end users. In order to obtain a high level of repetitiveness and quality, the process of the sample preparation was carried out under aeronautical specifications and in a controlled environment.

Following the identification of the relevant three production-related and four repair-related contamination scenarios, for each scenario three levels of contamination concentration were applied, namely a low level, a medium level, and a high level. In addition, a combined contamination case for each process field was realized. The contamination of the adherends was realized by Fraunhofer IFAM and investigated using XPS analyses to measure the amount of contamination on the adherend surface and to determine the exact contaminant concentration. After the contamination procedure, the adherends were bonded in the autoclave.

The resulting joints manufactured from intentionally contaminated adherends were inspected using conventional NDT. The objective was to make a statement on the sample quality as well as on the weak bond status. All the samples were investigated using two different probes (5 and 10 MHz). For the contaminated coupons and the multi-contaminated flat samples, three different kinds of defects were observed:

1. Manufacturing defects with a marginal impact on the use of the sample for ENDT evaluation, e.g. bending of the bonded specimens, adherend surface quality issues.

2. Minor deviation from the ultrasound reference signal in the case of moist CFRP samples or sample surfaces contaminated by fingerprints (including multi-contaminated specimens). This effect could be due to contamination if only compared to the reference signal, but no further proof has been found so far.
3. Contamination-induced defects such as disbonding (faulty curing) or delamination (thermal degradation). They might have a detrimental effect on future ENDT measurements.

Conjointly with the mechanical testing results, the status of weak bonds could then be confirmed for most of the samples. Finally, curved samples were successfully tested thanks to SAUL (surface adaptive ultrasonic laws) configurations. The inspection results revealed a very low-quality sample material, probably due to the CFRP composite adherend itself. Results obtained from such samples should be evaluated and interpreted carefully.

In order to evaluate the influence of the surface state (clean, single, or multiple contaminations) of one adherend on the mechanical properties on adhesively bonded joints, established mechanical tests like mode-I and mode-II fracture toughness tests as well as tensile and centrifuge tests were conducted by the University of Patras. Mechanical testing demonstrates the contamination level that affects the mechanical strength of a bond; the results can be correlated with the results from the reference analysis methods and ENDT methods.

We shortly highlight that for each of the investigated disturbances we found significant effects on the resulting CFRP composite joint strength for at least one level of contamination applied during the bonding process. Moreover, in many cases, the lowest applied level only caused a decrease in the joint strength as compared to production or repair procedures performed following the respectively qualified process. In this way, the prepared sets of specimens encompass two challenges for the aspired process monitoring procedures: On the one hand, (the effects of) the contaminations will need to be detected, and on the other hand, the measured values will need to facilitate the discrimination between more and less relevant levels of contamination.

Specifically, the reference category exhibited the highest fracture toughness values, while for almost all cases, except from R-FP-3, the presence of the contaminant proved to be detrimental for the fracture toughness of the joints. The higher the contamination level, the higher the decrease of the joints' performance. A combined contamination results in a reduction of the fracture toughness of the bonded jointed that is greater than the reduction caused by each contaminant separately, indicating that a combination of contaminations may be more detrimental to the composite bonded joints' performance.

Additionally, a novel test was used that is both time and cost-efficient, namely the centrifuge test, whereby the adhesion strengths of all the bonded joints were measured. Besides the rather large scatter presented in some scenarios, in almost all contamination scenarios, except for R-FP-1 and R-TD-1, there was a decrease of the adhesion strength. By evaluating the centrifuge test's experimental process and results, it can be concluded that the centrifuge testing technology has great potential to be established as a test method for the characterization of bonded joints as it is a fast testing process that generates repeatable tests capable of describing the strength of the joints.

In order to evaluate the combined effect of the pre-bond contamination and after-bond exposure to hygrothermal environment on the mode-II fracture toughness of CFRP bonded joints, the contaminated samples underwent aging inside an environmental chamber. Mostly, there was a negative effect of the contamination. After-bond hygrothermal aging significantly degrades the mode-II fracture toughness of the composite bonded joints. The decrease is larger for the contaminated samples, which reveals that the combined effect is more severe than that of the two effects separately.

Furthermore, the results of the tensile mechanical testing performed by the University of Patras revealed the effect of each contamination scenario in the tensile performance of the scarfed samples. A single contamination of an adherend with TD-1 proved to be beneficial since the sample presented a higher failure load than the reference samples. This was attributed to the enhancement of the matrix properties due to its oxidation and the formation of carbonyl groups at the surface. However, the negative effect of the combined contamination was also demonstrated. The results showed that a combined contamination of thermal degradation and de-icing fluid has a negative effect on the mechanical performance of the scarfed repair joints, reducing the failure load by up to 38%.

In total, 378 test coupons were tested using mode-I and mode-II fracture toughness tests, while 136 samples were tested using a centrifuge and tensile tests, resulting in a total of 514 tested specimens.

Finally, regarding the numerical simulations, a composite panel stiffened with two T-stringers was simulated under compression using the LS-DYNA FE platform. The comparison concerning the maximum load for the contamination scenarios showed that all contamination scenarios had a negative influence on the load-bearing capacity of the stiffened panel. Also, as a result of the contamination, the debonding initiated earlier than in the reference case.

Based on the findings achieved here, in-process ENDT will be implemented to assess features characteristic either to the pre-bond adherend surfaces (see Chap. 3) or to the adhesive joints (see Chap. 4) that were manufactured following the described intentionally applied contamination scenarios.

References

1. Markatos DN, Tserpes KI, Rau E et al (2014) Degradation of mode-I fracture toughness of CFRP bonded joints due to release agent and moisture pre-bond contamination. J Adhes 90(2):156–173. https://doi.org/10.1080/00218464.2013.770720
2. da Silva LFM, Öchsner A, Adams RD (2011) Introduction to adhesive bonding technology. In: da Silva LFM, Öchsner A, Adams RD (eds) Handbook of adhesion technology, vol 2. Springer Berlin Heidelberg, Berlin, Heidelberg, pp 2–3

3. Charalambides MN, Hardouin R, Kinloch AJ et al (1998) Adhesively-bonded repairs to fibre-composite materials I. Experimental. Compos Part A: Appl Sci Manuf 29(11):1371–1381. https://doi.org/10.1016/s1359-835x(98)00060-8
4. Pantelakis S, Tserpes KI (2014) Adhesive bonding of composite aircraft structures: challenges and recent developments. Sci China Phys Mech Astron 57(1):2–11. https://doi.org/10.1007/s11433-013-5274-3
5. ComBoNDT "Quality assurance concepts for adhesive bonding of aircraft composite structures by advanced NDT" (2015–2018) Project funded from the European Union's Horizon 2020 research and innovation programme under grant agreement No 636494
6. Tornow C, Schlag M, Lima LCM et al (2015) Quality assurance concepts for adhesive bonding of composite aircraft structures—characterisation of adherent surfaces by extended NDT. J Adhes Sci Technol 29(21):2281–2294. https://doi.org/10.1080/01694243.2015.1055062
7. Markatos DN, Tserpes KI, Rau E et al (2013) The effects of manufacturing-induced and in-service related bonding quality reduction on the mode-I fracture toughness of composite bonded joints for aeronautical use. Compos B Eng 45(1):556–564. https://doi.org/10.1016/j.compositesb.2012.05.052
8. ENCOMB "Extended Non-Destructive Testing of Composite Bonds" (2010–2014) Project funded from the European Union's Seventh Framework Programme under grant agreement No 266226
9. AITM 1-0053—Airbus Industry Test Method (2006) Carbon fibre reinforced plastics, determination of fracture toughness energy of bonded joints, Mode I (AITM 1-0053)
10. Ebnesajjad S (2008) Adhesives technology handbook, 2nd edn. William Andrew Inc., Norwich, NY
11. Pearson RA, Blackman BRK, Campilho RDSG et al (2012) Quasi-static fracture tests. In: da Silva LFM, Dillard DA, Blackman B et al (eds) Testing adhesive joints: best practices, A221. Wiley-VCH Verlag GmbH & Co, KGaA, Weinheim, Germany, pp 163–191
12. ISO 9022-12 (2015) Optics and photonics—environmental test methods—Part 12: Contamination 37.020 (ISO 9022-12:2015)
13. Budhe S, Banea MD, Barros Sd et al (2017) An updated review of adhesively bonded joints in composite materials. Int J Adhes Adhes 72:30–42. https://doi.org/10.1016/j.ijadhadh.2016.10.010
14. Creemers F, Geurts KJ, Noeske M (2016) Influence of surface contaminations on the quality and bond strength of structural adhesive joints. In: European adhesion (EURADH) conference
15. Gause RL (1989) A noncontacting scanning photoelectron emission technique for bonding surface cleanliness inspection (NASA TM-100361 Technical Memorandum)
16. Moutsompegka E, Tserpes KI, Polydoropoulou P et al (2017) Experimental study of the effect of pre-bond contamination with de-icing fluid and ageing on the fracture toughness of composite bonded joints. Fatigue Fract Eng Mater Struct 40(10):1581–1591. https://doi.org/10.1111/ffe.12660
17. Tserpes KI, Markatos DN, Brune K et al (2014) A detailed experimental study of the effects of pre-bond contamination with a hydraulic fluid, thermal degradation, and poor curing on fracture toughness of composite-bonded joints. J Adhes Sci Technol 28(18):1865–1880. https://doi.org/10.1080/01694243.2014.925387
18. Hollander JM, Jolly WL (1970) X-ray photoelectron spectroscopy. Acc Chem Res 3(6):193–200. https://doi.org/10.1021/ar50030a003
19. ASTM International D5573-99 (2005) Standard practice for classifying failure modes in fiber-reinforced-plastic (FRP) Joints 83.120 (ASTM D5573-99). www.astm.org
20. AITM 1-0006—Airbus Industry Test Method (1994) Carbon fibre reinforced plastics, determination of interlaminar fracture toughness energy—Mode II (AITM 1-0006)
21. Hoffmann M, Tserpes K, Moutsompegka E et al (2018) Determination of adhesion strength of pre-bond contaminated composite-to-metal bonded joints by centrifuge tests. Compos B Eng 147:114–121. https://doi.org/10.1016/j.compositesb.2018.04.014
22. Rietz U, Lerche D, Hielscher S et al (2015) Centrifugal adhesion testing technology (CATT)—a valuable tool for strength determination. J Adhes Soc Jpn 51(s1):293–297. https://doi.org/10.11618/adhesion.51.293

23. LUM GmbH (2018) Technical specification od Adhesion Analyser LUMiFrac. https://www.lum-gmbh.com. Accessed 25 June 2018
24. Beck U, Reiners G, Lerche D et al (2011) Quantitative adhesion testing of optical coatings by means of centrifuge technology. Surf Coat Technol 205:S182–S186. https://doi.org/10.1016/j.surfcoat.2011.02.016
25. DIN EN 2823 European Association of Aerospace Industries (1998) Aerospace series—fibre reinforced plastics—determination of the effect of exposure to humid atmosphere on physical and mechanical characteristics (DIN EN 2823)
26. Choi DM, Park CK, Cho K et al (1997) Adhesion improvement of epoxy resin/polyethylene joints by plasma treatment of polyethylene. Polymer 38(25):6243–6249. https://doi.org/10.1016/S0032-3861(97)00175-4
27. Ochoa-Putman C, Vaidya UK (2011) Mechanisms of interfacial adhesion in metal–polymer composites—effect of chemical treatment. Compos A Appl Sci Manuf 42(8):906–915. https://doi.org/10.1016/j.compositesa.2011.03.019
28. Cysne Barbosa AP, Fulco PAP, Guerra ESS et al (2017) Accelerated aging effects on carbon fiber/epoxy composites. Compos B Eng 110:298–306. https://doi.org/10.1016/j.compositesb.2016.11.004
29. Collins DH, Freels JK, Huzurbazar AV et al (2013) Accelerated test methods for reliability prediction. J Qual Technol 45(3):244–259. https://doi.org/10.1080/00224065.2013.11917936
30. Ewert U, Jaenisch GJ, Osterloh K et al (2011) Performance control: nondestructive testing and reliability evaluation. In: Czichos H, Saito T, Smith L (eds) Springer handbook of metrology and testing. Springer, Berlin, Heidelberg
31. Johnson WS, Butkus LM (1998) Considering environmental conditions in the design of bonded structures: a fracture toughness mechanics approach. Fatigue Fract Eng Mater Struct 21(4):465–478. https://doi.org/10.1046/j.1460-2695.1998.00533.x
32. Liljedahl CDM, Crocombe AD, Wahab MA et al (2007) Modelling the environmental degradation of adhesively bonded aluminium and composite joints using a CZM approach. Int J Adhes Adhes 27(6):505–518. https://doi.org/10.1016/j.ijadhadh.2006.09.015
33. Pitt S, Jones R, Peng D (2012) Characterization of the durability of adhesive bonds. Fatigue Fract Eng Mater Struct 35(11):998–1006. https://doi.org/10.1111/j.1460-2695.2012.01688.x
34. Mubashar A, Ashcroft IA, Critchlow GW et al (2011) A Method of predicting the stresses in adhesive joints after cyclic moisture conditioning. J Adhes 87(9):1061–1089. https://doi.org/10.1080/00218464.2011.600675
35. Bowditch MR (1996) The durability of adhesive joints in the presence of water. Int J Adhes Adhes 16(2):73–79. https://doi.org/10.1016/0143-7496(96)00001-2
36. Moutsompegka E, Tserpes KI, Brune K et al (2017) The effect of pre-bond contamination with fingerprint and ageing on the fracture toughness of composite bonded joints. In: 7th EASN international conference on innovation in European aeronautics research, 26–29 September, Warsaw, Poland
37. Floros IS, Tserpes KI, Löbel T (2015) Mode-I, mode-II and mixed-mode I+II fracture behavior of composite bonded joints: experimental characterization and numerical simulation. Compos B Eng 78:459–468. https://doi.org/10.1016/j.compositesb.2015.04.006
38. Tserpes KI, Peikert G, Floros IS (2016) Crack stopping in composite adhesively bonded joints through corrugation. Theoret Appl Fract Mech 83:152–157. https://doi.org/10.1016/j.tafmec.2015.10.003

3

Surface Quality Assessment

Mareike Schlag, Kai Brune, Hauke Brüning, Michael Noeske, Célian Cherrier, Tobias Hanning, Julius Drosten, Saverio De Vito, Maria Lucia Miglietta, Fabrizio Formisano, Maria Salvato, Ettore Massera, Girolamo Di Francia, Elena Esposito, Andreas Helwig, Rainer Stössel, Mirosław Sawczak, Paweł H. Malinowski, Wiesław M. Ostachowicz, and Maciej Radzieński

Abstract This chapter introduces various extended non-destructive testing (ENDT) techniques for surface quality assessment, which are first characterized, then enhanced, and finally applied to assess the level of pre-bond contaminations intentionally applied to carbon fiber reinforced plastic (CFRP) adherends following the procedures described in the previous chapter. Based on two user cases comprising different scenarios that are characteristic of either aeronautical production or repair, the detailed tests conducted on two types of sample geometry, namely flat coupons and scarfed pilot samples with a more complex shape, form the basis for applying the advanced ENDT procedures for the monitoring of realistic and real aircraft parts, as will be described in Chap. 5. Specifically, the reported investigations were performed to assess the surface quality of first ground and then intentionally contaminated CFRP surfaces using the following ENDT tools: the aerosol wetting test (AWT), optically stimulated electron emission (OSEE), two differently implemented approaches based on electronic noses, laser-induced breakdown spectroscopy (LIBS), Fourier-transform infrared (FTIR) spectroscopy, laser-induced fluorescence (LIF), and laser vibrometry.

M. Schlag (✉) · K. Brune · H. Brüning · M. Noeske
Fraunhofer Institute for Manufacturing Technology and Advanced Materials IFAM, Wiener Str. 12, 28359 Bremen, Germany
e-mail: mareike.schlag@ifam.fraunhofer.de

C. Cherrier · T. Hanning · J. Drosten
Automation W+R GmbH, Messerschmittstraße 7, 80922 Munich, Germany

S. De Vito · M. L. Miglietta · F. Formisano · M. Salvato · E. Massera · G. Di Francia · E. Esposito
ENEA – National Agency for New Technologies, Energy and Sustainable Economic Development, TERIN-FSD, Portici Research Center, P.le E. Fermi 1, 80055 Portici, Italy

A. Helwig · R. Stössel
Airbus Defence and Space GMBH, Willy-Messerschmitt-Straße 1, 82024 Taufkirchen, Germany

M. Sawczak · P. H. Malinowski · W. M. Ostachowicz · M. Radzieński
Institute of Fluid–Flow Machinery, Polish Academy of Sciences, Fiszera 14, 80-231 Gdańsk, Poland

Keywords ENDT for surface quality assessment · Aerosol wetting test · Optically stimulated electron emission · Electronic noses · Laser-induced breakdown spectroscopy · Fourier-transform infrared · Laser-induced fluorescence · Laser vibrometry

3.1 Introduction

In the previous chapters, we detailed that one of the reasons inhibiting the certification of adhesive bonding for primary aircraft structures is the sensitivity of the bondline integrity to the presence of defects that can affect the strength of the joints. These defects are not accessible to visual monitoring during the bonding process. Furthermore, the most critical defects arising during the manufacture of adhesive joints are those that are not detectable by currently available NDT methods. This has led us to develop extended NDT (ENDT) methods capable of detecting such effects on CFRP adherends, whereby we evaluate their efficiency and assess their applicability benefits based on the analysis of specimens with increasingly complex sample geometries, starting from flat coupon samples that exhibit distinct levels of intentionally applied contaminations [1] and moving toward CFRP panels with more complex/curved geometries that might even exhibit multiple contaminations. The present chapter describes the respective contributions of the individual partners of the ComBoNDT consortium [2], thereby detailing the different specialist approaches within a jointly developed concept for quality assurance.

The subsequently detailed ENDT procedures for surface quality assessment constitute an essential input into the framework for the overall concept for the quality assessment of adhesively bonded joints described in this book. The results presented in this chapter were achieved in the research conducted into the applied ENDT methods. Looking ahead, we would like to highlight here that in certain contamination scenarios of the aircraft production and repair user cases, these pre-bond in-process methods were sensitive to impacts that were shown to affect a strength reduction of successively manufactured adhesive bonds and, therefore, could potentially be utilized to identify not-in-order (NIO) adherends during a surface quality assessment procedure.

3.2 Aerosol Wetting Test (AWT)

In this section, we introduce the aerosol wetting test (AWT) as a tool for surface quality assessment and detail how its performance was enhanced in the ComBoNDT project for the in-process monitoring of CFRP adherends.

3.2.1 Principle and Instrumentation

AWT allows the inline monitoring of the surface state, specifically through an inspection of its wettability.

Throughout the last decade, various aspects of this technology have been enhanced, whereby the most recent advances were achieved in the ComBoNDT project. Hereby, both the hardware and software were adapted in order to achieve more relevant and more reliable results in terms of measurement, data evaluation, and post-processing.

3.2.1.1 Measurement Principle

AWT allows the monitoring of a surface state by analyzing its wetting properties. In its most common implementation, an aerosol of ultra-clean water is sprayed onto the surface using an ultrasonic spray nozzle. Depending on the wetting behavior and wettability of the surface, the droplet pattern, diameter, and distribution vary, as exemplified in Fig. 3.1. The wetting behavior is then automatically analyzed based on the images recorded by a camera system and image processing algorithms.

This surface inspection method has various advantages, which are briefly:

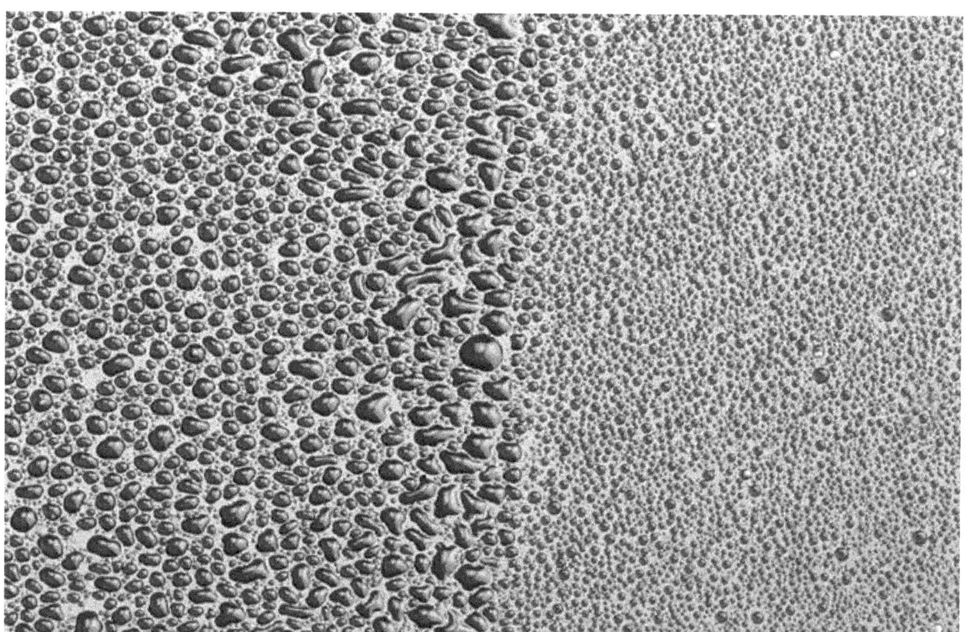

Fig. 3.1 Aerosol wetting test (AWT) droplet pattern influenced by the local wettability of the substrate surface: a more wettable surface region (left) with less and larger droplets compared to a less wettable surface region (right) exposed to the same amount of droplets from a water aerosol generated by an ultrasonic nozzle

- The inspection speed and the size of the inspected surface.

AWT is particularly adapted to the measurement of the properties of the droplet pattern at the edge of a part. The inspection is conducted on a 30 mm wide area at a speed of 6 m per minute; therefore, it is especially suitable for inspecting specific, narrow bonding areas. In practice, each image is separated from the previous one and then the evaluation is performed. The values in this book are given based on calibrated image dimensions, such as the base area (30 × 30 mm).

- Its low impact on the surface.

This method is non-destructive and has a minimal influence on the investigated part. Approximately 0.2 μL of ultra-clean water is deposited by the spray per square centimeter. After drying (which takes less than a minute for most substrate materials), the surface typically does not show any residue since the water used is ultra-clean.

- Its simplicity of use and implementation.

The system enables the inspection of various parts with only a few limitations. The measurement can take place in various environments (for example, on a production line), on various materials, and it only needs a standard energy supply.

- The simplicity of achieving the results.

Once the images are captured, various image processing algorithms and decision-making processes run simultaneously. If the calibration of the system has been flawlessly achieved, the result can be integrated into a simple IO/NIO signal for the inspector, saving the more complex data for later analysis.

3.2.1.2 Software Enhancement

During the ComBoNDT project, various software modifications were made to the existing system, most significantly to the image processing algorithms. Once the images of the droplet patterns have been generated by the camera, the main task, which is also the hardest, is detecting the droplets (and their lateral boundaries) and separating them on the image from the background formed by the material texture.

At the beginning of the project, this was accomplished by a very straightforward image processing step (thresholding followed by morphological operations); however, it emerged that this approach is extremely unstable if there are variations of the surface properties (texture, structure…).

In a first step, this rather simple image processing method was replaced with a more complex one that can nevertheless be considered standard image processing. This enhancement primarily facilitates the detection of droplets on surfaces with slight variations in color or also light intensity. This first step was fully integrated into the research system and already is implemented on the system for inline detection.

However, for some complex surfaces or distinct contamination scenarios resulting in more varied and unpredictable droplet appearances and patterns, even this enhanced image processing was not sufficiently effective. Hence, further improvements were made, with the detection and evaluation of the droplet pattern being done by a convolutional neural network (CNN). We trained the network on various datasets generated by the AWT system (various materials, various contaminations, and/or activation of the surface). With a wide set of samples, the network was trained to separate the image pixels belonging to a droplet from the pixels belonging to the background.

We visualize the various stages of the up-to-date image processing in the table displayed in Fig. 3.2.

Following this use of a CNN to detect the droplets, further classical image processing approaches were integrated for an easier decision-making process. Consequently, a sequence of various filters was applied to the image, and local values for the standard parameters (e.g., wetted area percentage, deposited water droplet diameter, drop count to evaluate the droplet number density) were calculated and graphically displayed, as shown in Fig. 3.3.

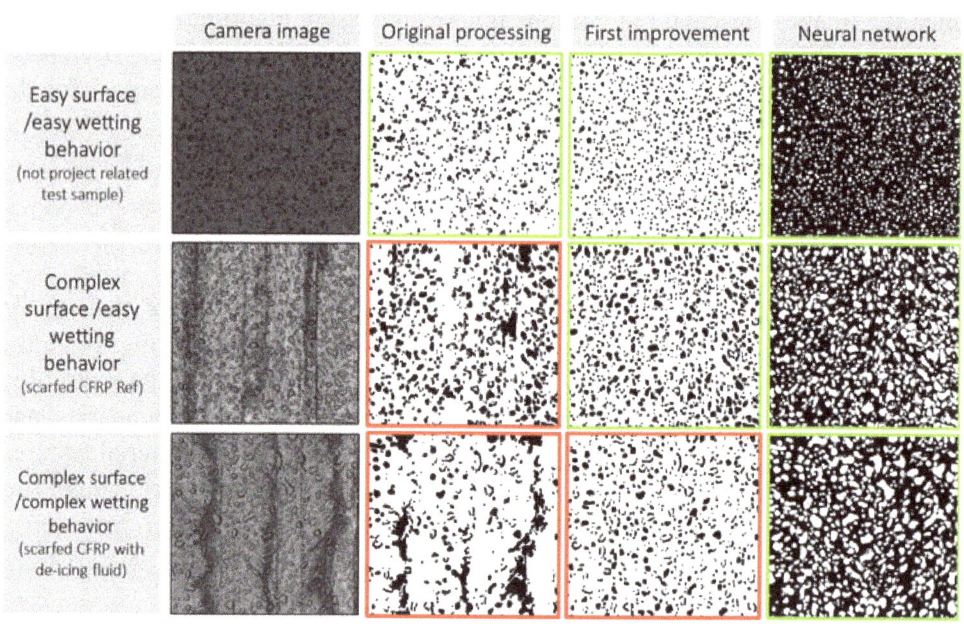

Fig. 3.2 Various AWT image processing generations on various surfaces with distinct complexities. Droplet images with green frames highlight a successful droplet detection; some droplet detections (marked by a red frame) were not successful enough for a meaningful analysis

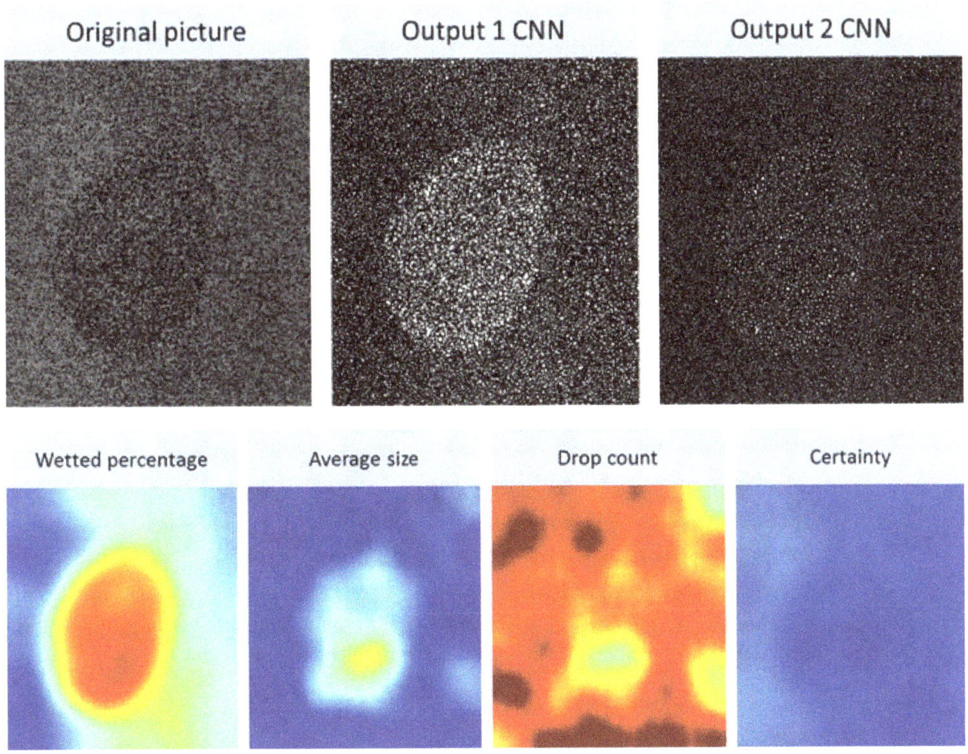

Fig. 3.3 Example output of the so-called heatmap after convolutional neural network (CNN) processing. The displayed example was obtained within a fingerprint contamination scenario

3.2.1.3 Hardware Enhancement

Some improvements were also made to the hardware, mainly concerning the AWT measurement head, which was fully redesigned. This was motivated by various considerations:

- Improving the optical system properties.
- Facilitating the use of a robot to perform the measurements on parts with complex geometries, which are presented in Chap. 5.

Specific hardware with high computing power was used in order to integrate the CNN image processing.

3.2.1.4 Up-to-Date Measurement Apparatus

By the end of the project, the updated AWT system consisted of three main components, namely the measurement head, the electrical cabinet, and the processing computer system, as presented in Fig. 3.4.

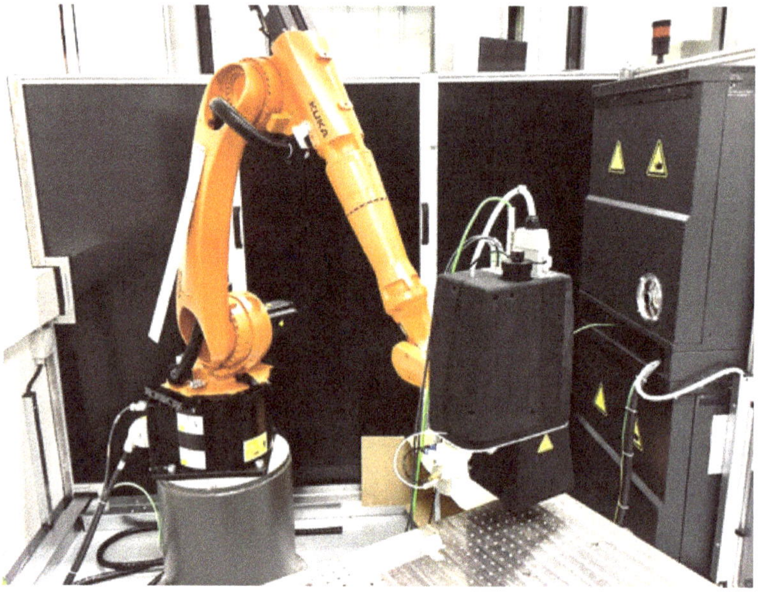

Fig. 3.4 Updated robot-aided bonNDTinspect AWT measurement head imaged during the full-scale demonstration (detailed in Chap. 5) conducted at Fraunhofer IFAM in Bremen

3.2.2 AWT Results

Here, we present our findings achieved by applying AWT to assess distinct contamination scenarios for CFRP adherends with different shapes.

3.2.2.1 AWT Results for the Coupon Level Samples

In the following investigations, the various test results for each of the distinct contamination scenarios and three different contamination levels are compared to the results for the clean ground reference samples, which differ depending on the respective production or repair user case.

The results shown here are always presented for those two droplet pattern features out of the four subsequently listed ones that showed the best differentiating correlation with the respectively applied contamination levels. The four evaluated features are as follows:

- Average droplet diameter.
- Wetted percentage, i.e., the percentage of the surface covered by water.
- Number of droplets per surface area, i.e., the droplet density.
- Average droplet compactness, whereby the compactness of each droplet is calculated by determining the area to perimeter ratio.

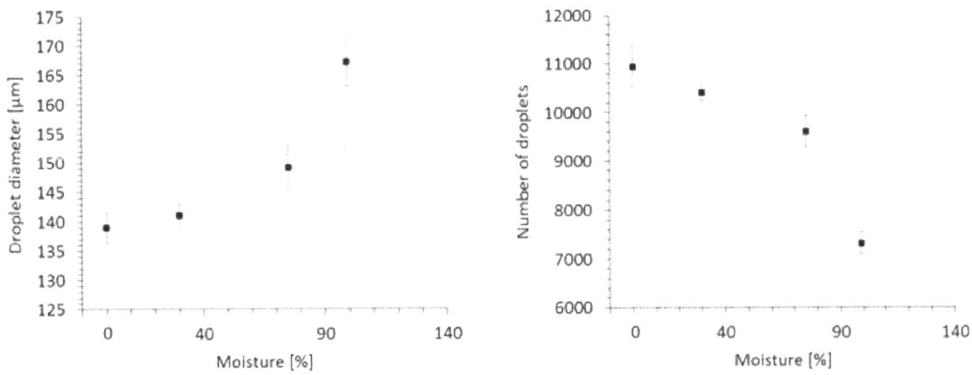

Fig. 3.5 Evaluation results for the AWT features of droplet diameter and number of droplets as obtained for the CFRP samples from the MO scenario

3.2.2.2 Detection of the Moisture of CFRP Substrates (P-MO)

The moisture contamination of CFRP parts from the production user case was successfully revealed and distinguished for the three different contamination levels achieved by material exposure to environments with distinct relative humidities, as shown in Fig. 3.5.

3.2.2.3 Detection of Release Agent (P-RA)

Release agent contamination was successfully revealed and distinguished for the three different contamination levels achieved by depositing distinct amounts of a silicon-containing release agent onto the surface, as shown in Fig. 3.6.

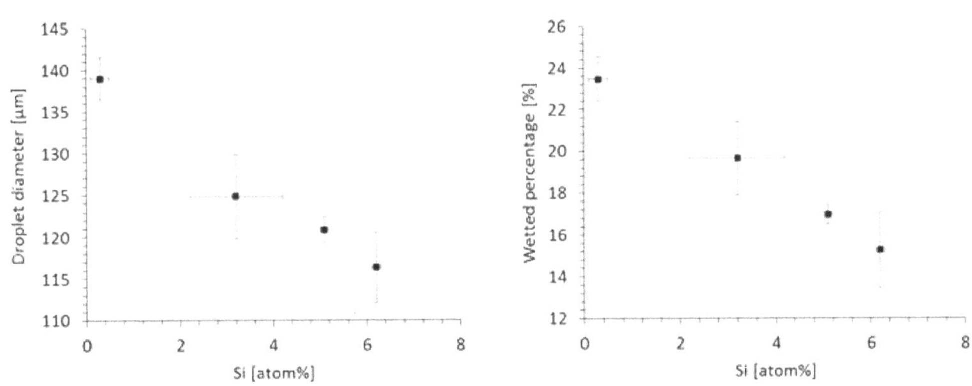

Fig. 3.6 Evaluation results for the AWT features of droplet diameter and wetted percentage as obtained for the CFRP samples from the scenario involving a silicon-containing release agent deposited at atomic concentrations, characterized by XPS investigations

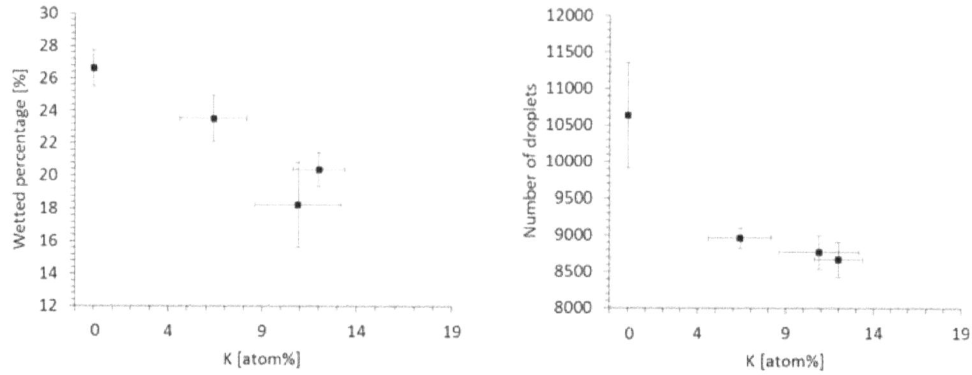

Fig. 3.7 Evaluation results for the AWT features of wetted percentage and number of droplets as obtained for the CFRP samples from the de-icing fluid scenario involving a potassium-containing de-icing fluid deposited at atomic concentrations, characterized by XPS investigations

3.2.2.4 Detection of Fingerprints on Production Samples (P-FP)

The fingerprint contamination within the production user case was contrasted and detected using AWT, although discrimination between the different contamination levels could not be achieved. For locally deposited contaminations, such as the fingerprints used here, the detection was also successfully achieved with local image processing algorithms, which allowed the characterization of the size and position of the contamination.

3.2.2.5 Detection of De-icing Fluid (DI)

The de-icing fluid contaminations were clearly revealed by AWT when comparing the droplet patterns obtained on intentionally contaminated CFRP specimens with those of clean reference samples. However, the different contaminations levels could barely be differentiated from each other, see Fig. 3.7.

3.2.2.6 Detection of Fingerprints on Repair Samples (R-FP)

Contaminations applied to CFRP surfaces for the fingerprint scenario within the repair user case were barely detected with AWT, and the distinct contamination levels were not differentiated.

3.2.2.7 Detection of Thermal Degradation (R-FP)

The effects of the thermal impact on CFRP coupon specimens were clearly detected by AWT. However, the different contamination levels were only barely differentiated.

		No detection	Partial detection	Clear detection	Partial discrimination of contamination level	Discrimination of contamination level
Production	Moisture (MO)					X
	Release agents (RA)					X
	Fingerprint (FP)			X		
	RA + FP				X	
Repair	De-icer (DI)				X	
	Fingerprint (FP)		X			
	Thermal degradation (TD)				X	
	TD + DI				X	

Fig. 3.8 Summary of the detection capacity (marked X) observed for the flat CFRP coupon samples for the production or repair user cases (first column) within the respective contamination scenarios (second column)

3.2.2.8 Summary of the Performance of AWT for Contaminated CFRP Coupon Samples

The summary presented in Fig. 3.8 shows the detection capacity of AWT for various contamination scenarios. The samples considered here are the flat CFRP coupon samples, which may not be directly relevant for the evaluation of the performance of the AWT method in terms of real parts.

3.2.3 AWT Results for the Pilot Level Samples

The AWT results obtained for the scarfed CFRP pilot samples are presented in the following. They cannot be directly compared with the results for the coupon samples or those of the realistic parts. Indeed, the variations between part type, geometry, and surface have a significant influence on the AWT measurement and evaluation results.

Therefore, we perform comparisons only among similar parts of the same type. The analytical process was set up following the objective of detecting surface contamination on the respective part.

3.2.3.1 Pilot Samples for the Production User Case with Combined Release Agent and Fingerprint Contamination

Regarding the surface constitution/texture, the pilot samples of the production user case showed a strongly structured surface and were slightly curved. The AWT droplet detection was partially successful with the first image processing improvement, and it was successful in all cases using neural network processing, see Fig. 3.9.

In this case, the geometrically strongly structured background as well as the combination of two contamination processes involving two types of deposited contaminants led to a more difficult differentiation of the various contamination levels, see Fig. 3.10.

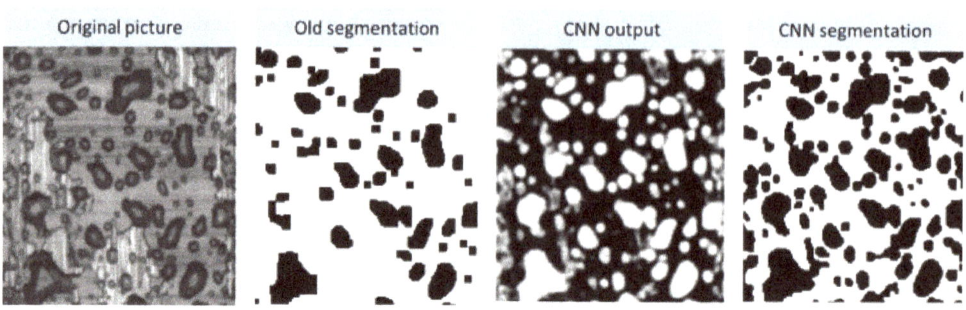

Fig. 3.9 Overview of the findings from different AWT droplet image evaluation and processing procedures for the production pilot samples, starting from an original image (left), after performing the droplet segmentation with standard image processing (labeled "old segmentation"), and after training a CNN

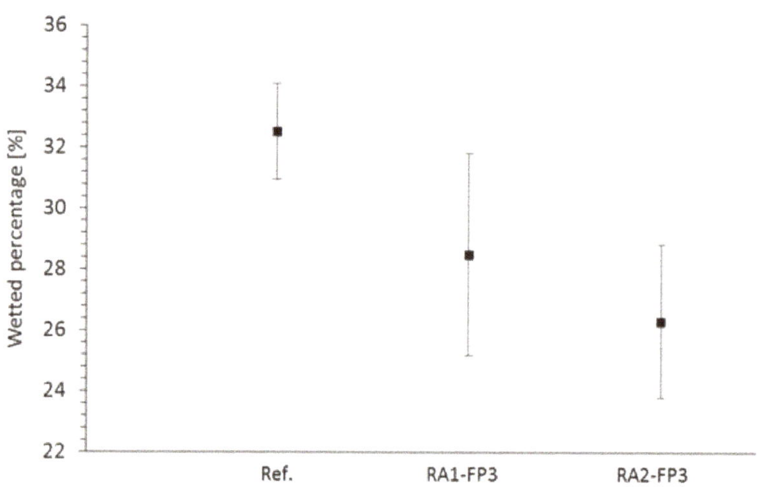

Fig. 3.10 AWT findings when evaluating the feature of wetted percentage for distinct levels of the intentionally deposited combined "release agent (RA)+fingerprint (FP)" contamination within the production user case with pilot CFRP samples

3.2.3.2 Repair Scenario Pilot Samples with Combined "Thermal Impact+De-icing Fluid" Contamination

We would like to reiterate here that in contrast to the smooth CFRP specimens of the respective production user case, the pilot samples within the repair user case were shafted CFRP samples. In the following, we focus on the contamination scenario that is based on distinct levels of combined thermal degradation and subsequent de-icing fluid contamination.

Similar to the production pilot samples, the detection of the applied water droplets following the AWT procedure was challenging, even though the surface did not show a very strong texture/structure. However, the deposited waterborne de-icing fluid contamination (as similarly observed on coupon samples) leads to a very good spreading of the droplets. Therefore, identifying individual droplets during the detection step of the AWT data evaluation process was harder since the droplets appeared to be "open" on the captured images instead of being assigned a roundish and "closed" contour. This type of droplet is not detectable using standard image processing. Once again, for such a scenario, the processing of the images with the CNN proved to be quite efficient, see Fig. 3.11.

Both the detection of contamination and the differentiation from the reference CFRP surface state were easily achieved through the observation of the AWT "droplet diameter" feature. The differentiation of various contamination levels was also possible here, see Fig. 3.12.

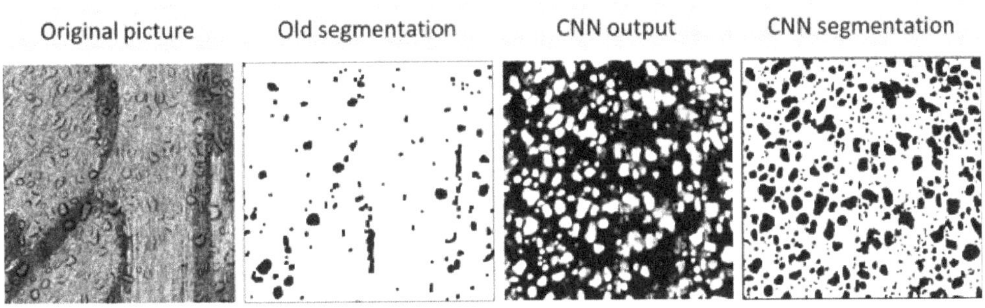

Fig. 3.11 Starting from the captured original AWT image, this sequence was obtained from distinct droplet segmentation approaches using standard image processing ("old segmentation", left) and convolutional neural network ("CNN segmentation", right) when applying AWT to intentionally contaminated CFRP pilot samples within the repair user case

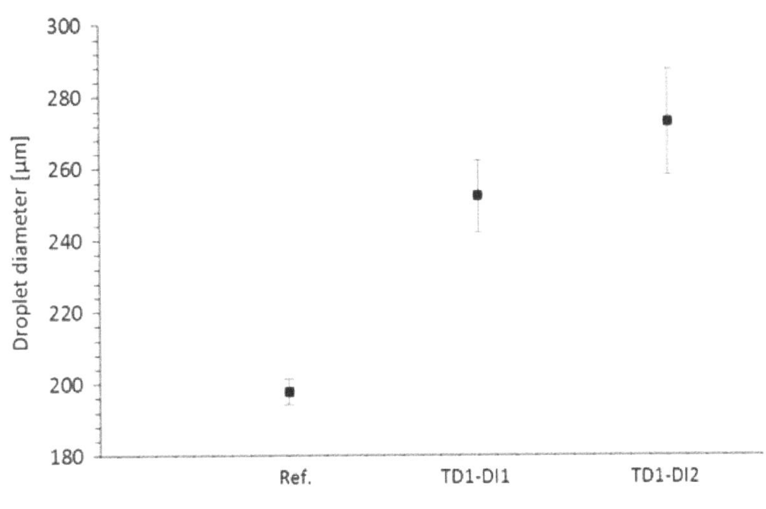

Fig. 3.12 AWT findings when evaluating the "droplet diameter" feature obtained for the CFRP pilot samples prepared within the combined "thermal impact+de-icer" contamination scenario

3.2.4 AWT Performance in Inline Surface Quality Assessment

In order to establish and enhance the performance of AWT when inspecting the surface states of CFRP parts relevant for specific aeronautical user cases, we iteratively determined and advanced the abilities of the system on flat coupon samples, pilot samples, and realistic parts and then conducted an assessment of the potential inline application of the technology. As explained, this technology relies on a comparative assessment of the surface state. Hereby, the very powerful systematic AWT inspection procedure, based on a convolutional neural network (CNN), must be taught to differentiate contaminations. Once the system had been trained to correctly detect the droplets and to differentiate the contaminations, its use in inline applications was straightforward.

The inline application allows fast and non-destructive monitoring and classification of the surface states and clearly exceeds the performance of the NDT approaches used so far (e.g., the water break test). Most contaminations investigated during the ComBoNDT research project in distinct scenarios with various degrees of contamination were successfully detected, and even the relevant contamination levels investigated here could be differentiated.

In conclusion, we developed sensitive and productive AWT procedures that not only facilitated the differentiation between the surface states of clean and intentionally contaminated parts but also permitted discrimination between distinct levels of contamination for several contamination scenarios. As the significance of AWT investigations and the thus obtained findings are based on contrasts in the wetting behavior of the inspected surfaces, our investigations plausibly indicate that AWT is a very surface-sensitive technique that enables a significant differentiation between clean surfaces and surfaces with sub-monolayer and monolayer contamination. However, AWT does not allow discrimination between surface states composed of a few or several molecular layers of contaminants. Finally, the integration of the technology in inline applications without major constraints was achieved.

3.3 Optically Stimulated Electron Emission (OSEE)

In this section, we introduce optically stimulated electron emission (OSEE) as a tool for surface quality assessment and detail how its performance was enhanced in the ComBoNDT research project for the in-process monitoring of CFRP adherends. Using OSEE for the surface inspection of carbon/epoxy composite and CFRP substrates has been described in several works over the last decades [3, 4], and the method is presently gaining visibility in surface quality assessment prior to bonding [1, 5, 6].

3.3.1 Principle and Instrumentation

OSEE is a surface analytical technique that relies on recording the photocurrent emitted from a sample surface region illuminated with UV light, typically under environmental conditions that prevail in cleaning or adhesive bonding processes. There are two modes of operation. During "microscopic" OSEE mapping, the sample is scanned using an electron collector, typically by applying lateral movements with a speed ranging from 1 mm to 1 cm per second. Meanwhile, during "spectroscopic" local measurements, the photocurrent is measured at fixed positions upon varying and recording the hold-up time. We performed our OSEE experiments under ambient conditions using an SQM300 surface quality monitor (purchased from Photo Emission Tech., Inc. (PET), USA).

Regarding the principle of an OSEE measurement in more detail, during the local inspection, regions of the sample surface are exposed to UV light from a mercury vapor lamp with prominent emission maxima at 4.9 and 6.7 eV. As the work function of the respective substrate surface amounts to approximately 5 eV, the emission maximum at the higher energy of 6.7 eV essentially contributes to the photoelectrons emitted by the sample surface, and the emitted electrons exhibit kinetic energies of less than approximately 2 eV. A sub-micrometer information depth of this method is observed when investigating the surface of a solid sample covered with a film exhibiting a thickness of some nanometers. This allows for the sensitive detection of thin films on substrates that do not significantly trap electric charges upon photoelectron emission. The interaction of the emitted photoelectrons with the ambient atmosphere is dominated by an electric field effective at the collector of the sensor to an extent that permits sensor-surface distances in the millimeter range during OSEE measurements. Carefully controlling the distance between the sensor and the surface is a prerequisite for effectively applying the setup presented in Fig. 3.26, see Fig. 3.13.

To mount the substrates for analysis, we equipped a commercial OSEE device with an electrically conductive and earthed moving table upon which we positioned the analyte sample using movements in two perpendicular horizontal directions (x and y) under the sensor. The vertical distance between the sample surface and the sensor (z) was set using a micrometer screw attached to the holder of the sensor. The enhanced OSEE setup developed during the ComBoNDT research project permitted vertical sensor movements with the aid of an electric motor, and the most advanced OSEE procedure was achieved with robot-aided three-dimensional sensor positioning. In OSEE implementations depending on a moving table for positioning the CFRP specimens, a surface scan was performed, with the table programmed to move according to a certain step size and number of steps in both horizontal directions, defined by the user through the machine-associated software. As the scan advanced, a photocurrent was obtained for each part of the scanned surface, and an apparently dimensionless value (which actually indicates a current converted into a voltage [7], given in centivolts), henceforth denoted as the OSEE signal, was indicated on a display and digitally recorded. Finally, a digital worksheet with the emission values for the entire

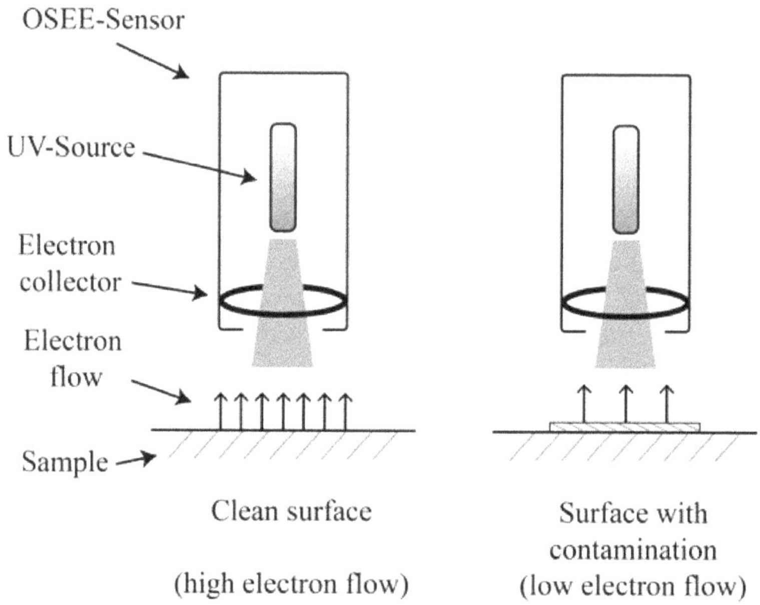

Fig. 3.13 Diagram showing the components and the principle of optically stimulated electron emission (OSEE) for measuring a photocurrent to reveal the surface state of a substrate

analyzed sample, i.e., an OSEE map, was obtained as a result of the test. For further evaluation of these maps especially for non-localized contaminations, the mean value of all the data points in the map together with its standard deviation was calculated.

3.3.2 OSEE Results

In the following, we report the OSEE enhancements and findings obtained in the ComBoNDT research project, in which the consortium partners at Fraunhofer IFAM performed the in-process monitoring of CFRP adherends with different shapes that are relevant for distinct technologically relevant user cases and which had undergone an intentional application of various contamination scenarios.

3.3.2.1 OSEE Results Obtained on CFRP Coupon Level Samples

First, we detail the advancements of the OSEE technique. Then, we report the respective OSEE results for the production user cases, characterized by a grayish abrasive dust obtained during the grinding of the CFRP surface, and the repair user cases, characterized by black abrasive dust obtained during the grinding of the CFRP surface. When assessing the surface quality of the coupon level CFRP specimens, we used OSEE to investigate a set of three 10 cm wide square samples to examine the surface state. For each sample, a surface scan was performed using a 6 mm wide aperture at a constant sensor-surface distance, with the table being programmed to move according to 15 steps with a width of 5 or 6 mm (in both horizontal directions), as defined by the user through the machine-associated software.

In going beyond the commercially available state of the art, and thereby increasing the technology readiness level of the OSEE technique, three principal advances were performed within the ComBoNDT project. First, we aimed at improving the reliability of the technique by considering the influence of topography, especially the sample-sensor distance, on the sensor signal and by controlling or also avoiding electrostatic charging effects. Second, we sought an adaption of the device setup and control systems to the automated scanning of CFRP surfaces used in real manufacturing processes within the production or repair user cases. Third, we manufactured the electronic parts for the OSEE adaptation (e.g., serial interface, power supply, and relay board) as defined by Fraunhofer IFAM within the project. The achieved advancements are presented in Fig. 3.14.

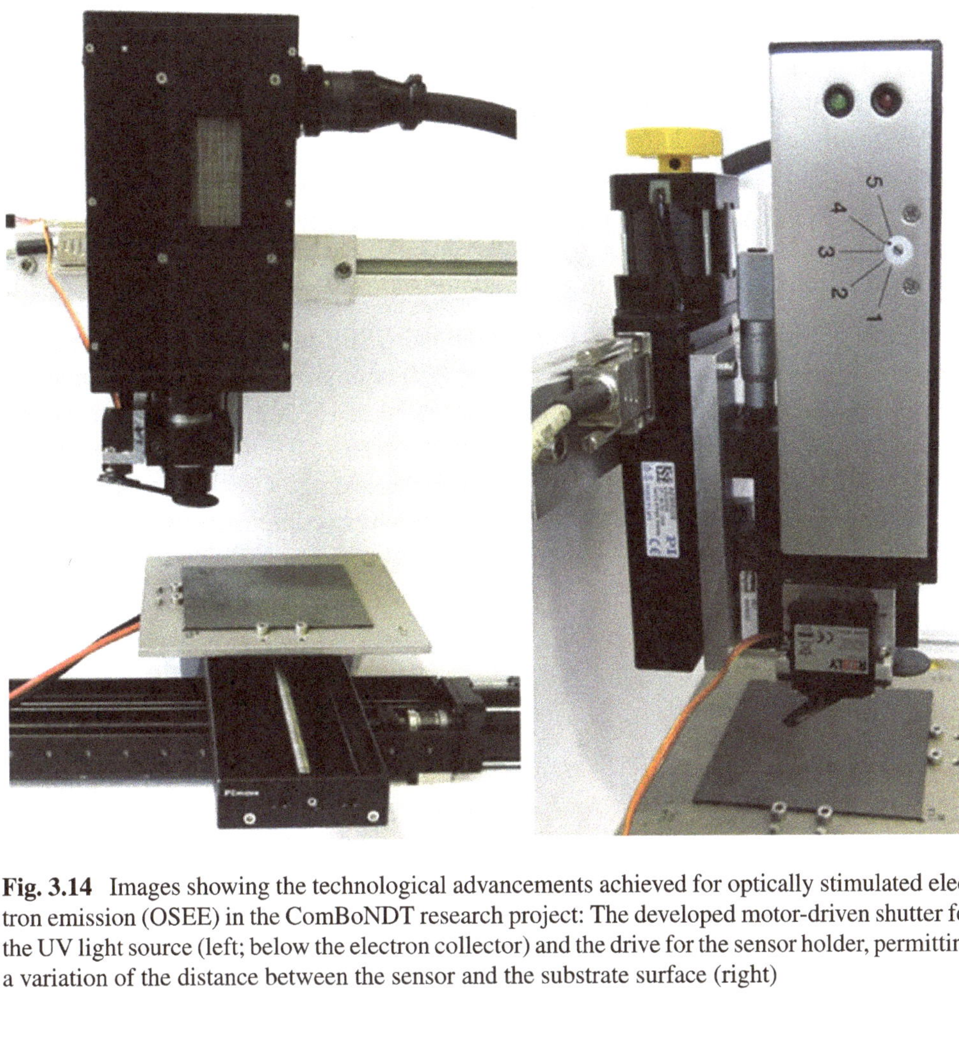

Fig. 3.14 Images showing the technological advancements achieved for optically stimulated electron emission (OSEE) in the ComBoNDT research project: The developed motor-driven shutter for the UV light source (left; below the electron collector) and the drive for the sensor holder, permitting a variation of the distance between the sensor and the substrate surface (right)

In terms of the abovementioned "microscopic" and "spectroscopic" operation modes of OSEE, a hybrid operation mode was developed and implemented. Based on rapidly opening a shutter in the light path of the UV source after having reached a measurement position (by laterally moving the newly developed x, y scanning table), an instant time-dependent ("in-time") sample mapping was facilitated that allowed combining the x, y mapping option while recording the local charging behavior or also recording the local height (i.e., sensor-surface distance) dependence of the OSEE signal. Specifically, by introducing an automatized variation of the sample-sensor distance based on a third precision drive (for the vertical z-direction), the variation of the OSEE signal upon changing the sensor-surface distance became possible. This also enables an assessment of surface topographies that are more complex than flat surfaces.

Production user case based on CFRP coupons

For the production user case based on CFRP coupon specimens, we investigated three different contamination scenarios, and the respectively obtained OSEE results were compared to the reference surface state of the production samples (P-RE):

- First, different levels of surface contaminations with a silicone-based release agent (P-RA) were prepared, as described in Chap. 2, and the degree of contamination was quantified by XPS analysis.
- Second, for the production fingerprint scenario (P-FP), samples were contaminated by different amounts of a synthetic sweat formulation according to DIN ISO 9022-12, as also detailed by Moutsompegka et al. [8].
- Third, a moisture scenario during production (P-MO) was considered.

Figure 3.15 presents the respectively obtained OSEE maps for the three samples prepared following the clean reference (P-RE) scenario. With the instrumental settings applied, an average OSEE intensity of 754 ± 152 a.u. (arbitrary units) was obtained.

In the following, we present plots showing the average OSEE intensity for the three contamination scenarios P-RA, P-FP, and P-MO in comparison to that obtained for P-RE.

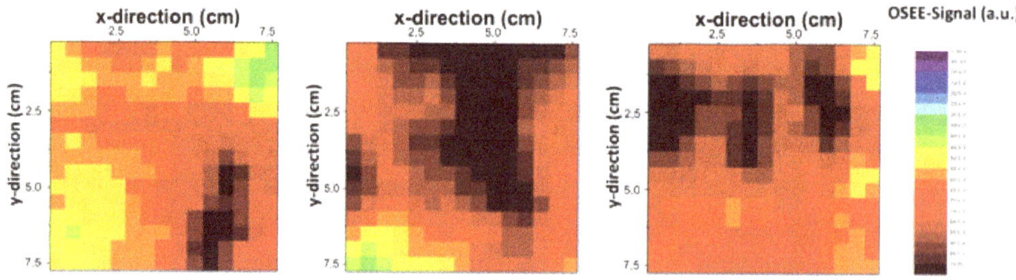

Fig. 3.15 Optically stimulated electron emission (OSEE) maps (15 × 15 pixels) for the three samples prepared following the clean reference scenario of the production user case (P-RE), based on CFRP coupon specimens resulting from a grinding process that was characterized by producing a grayish abrasive dust

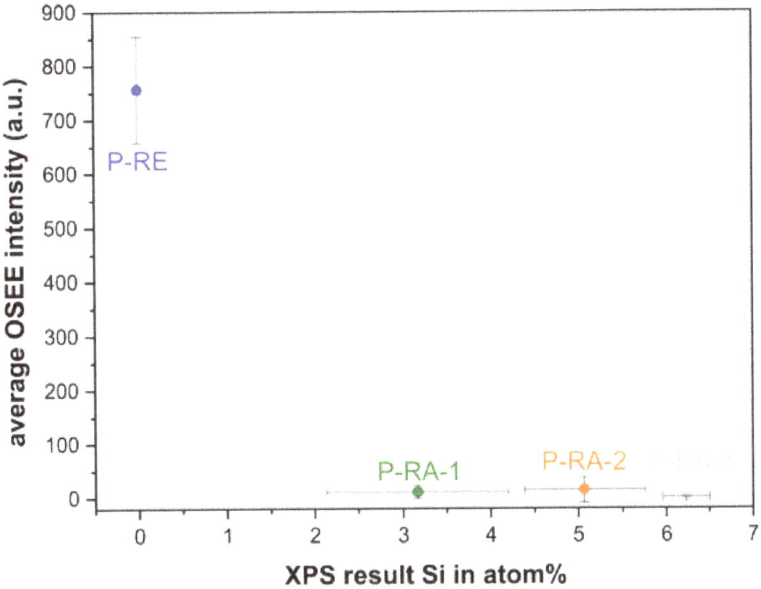

Fig. 3.16 Average OSEE intensities for the three samples within each sample set, P-RA-1, P-RA-2, and P-RA-3, as part of the P-RA scenario, based on different contamination degrees achieved with an intentionally deposited silicone-based release agent, compared to the OSEE results for the P-RE scenario

As may be inferred from Fig. 3.16, all the surfaces corresponding to contamination levels P-RA-1, P-RA-2, and P-RA-3 within the P-RA scenario can be clearly detected and differentiated from the surface state corresponding to P-RE.

As may be perceived from Fig. 3.17, the average OSEE intensities obtained for surfaces corresponding to moisture levels P-MO-1, P-MO-2, and P-MO-3 within the P-MO scenario can be clearly detected and differentiated from the clean CFRP surface state corresponding to P-RE.

As may be seen in Fig. 3.18, all the surfaces corresponding to contamination levels P-FP-1, P-FP-2, and P-FP-3 within the P-FP scenario can be clearly detected and differentiated from the surface state corresponding to P-RE, even when evaluating 7.5 cm × 7.5 cm wide areas, which exceeds the area covered by the respectively applied fingerprint (the size of a human thumbprint).

Repair user case based on CFRP coupons

For the repair user case based on the coupon level CFRP specimens, three different contamination scenarios were investigated, and the respectively obtained OSEE results were compared to the reference surface state of the repair samples (R-RE):

- For the first repair scenario (R-DI), the surfaces were intentionally contaminated with different amounts of de-icing fluid (also called de-icer, an aqueous potassium format solution).
- The second scenario (R-FP) involved contaminations with hydraulic Skydrol fluid [8], the main ingredients of which are phosphate esters.

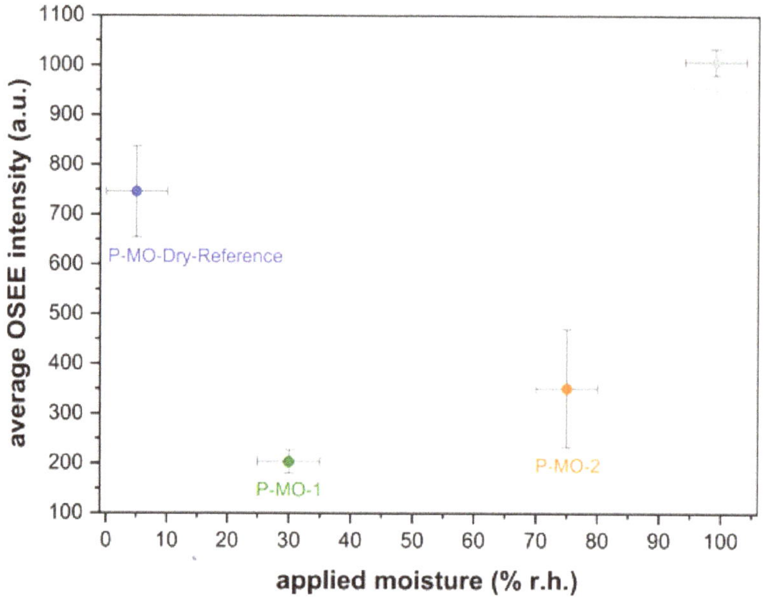

Fig. 3.17 Average OSEE intensities for the three samples within each sample set, P-MO-1, P-MO-2, and P-MO-3, as part of the P-MO scenario, based on different moisture degrees achieved through intentional exposure to distinctly humid environments, compared to the OSEE results for the P-RE scenario

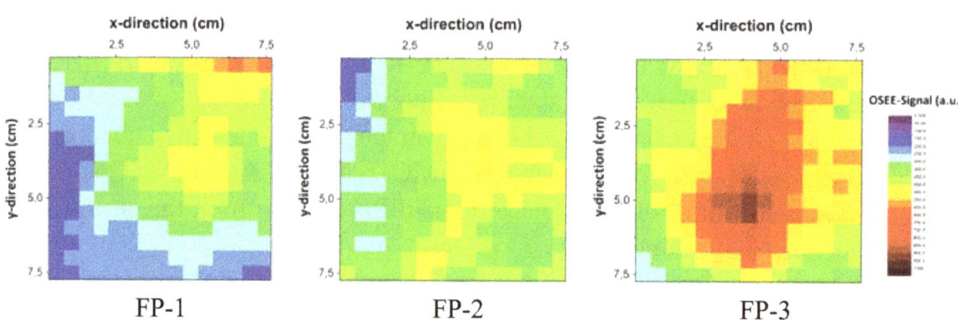

Fig. 3.18 Average OSEE intensities for the three samples within each sample set, P-FP-1, P-FP-2, and P-FP-3, as part of the P-FP scenario, based on locally applied contaminations of different degrees, achieved by intentionally depositing a sodium-containing synthetic sweat formulation, compared to the OSEE results of the P-RE scenario

- The third repair scenario considered a thermal impact affecting CFRP surface degradation during repair (R-TD).

We highlight here that the OSEE intensity values for the samples prepared according to the R-RE scenario showed significantly higher OSEE intensities than those prepared according to the P-RE scenario. We attribute this finding to a more profound grinding in the case of the R-RE samples, which led to a higher area ratio of the exposed carbon fibers; we expect carbon fibers to contribute a higher OSEE intensity than the polymer matrix of the composite material. The OSEE settings

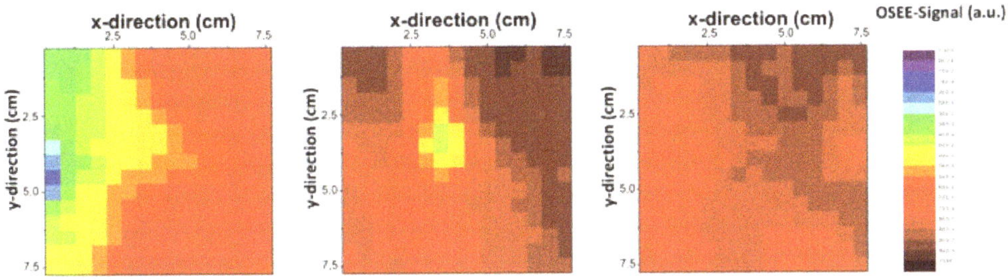

Fig. 3.19 Optically stimulated electron emission (OSEE) maps (15 × 15 pixels) for the three samples prepared following the clean reference R-RE scenario of the repair user case, based on CFRP coupon specimens resulting from a grinding process that is characterized by producing black abrasive dust, measured with adjusted OSEE settings

adjusted for the CFRP specimens of the R-RE scenario were applied for all the samples investigated within the repair user case.

Figure 3.19 presents the OSEE maps obtained for the three samples prepared following the R-RE scenario. With the instrumental settings applied, an average OSEE intensity of 739 ± 106 a.u. was obtained. We highlight here that, based on our OSEE investigations, the I-R-OSEE-RE-1 sample had presumably been polished more strongly on the right side than on the left. We note that any OSEE map obtained on ground CFRP substrates reveals both the lateral homogeneity of the sample surface (based on the standard deviation of the OSEE signal) and the depth of polishing (based on the intensity of the OSEE signal).

In the following, plots showing the average OSEE intensity of the three scenarios R-DI, R-FP, and R-TD (provided that they were investigated) are presented in comparison to that obtained for R-RE.

As may be inferred from Fig. 3.20, the surface states corresponding to contamination levels R-DI-2 and R-DI-3 within the R-DI scenario can be clearly detected and differentiated from the surface state corresponding to R-RE. However, the surface state of the sample with level R-DI-1 cannot be differentiated from that of the R-RE sample with the OSEE settings applied.

As shown in Fig. 3.21, when applying the used OSEE settings, the average OSEE intensities obtained for surfaces corresponding to the thermal degradation level R-TD-1 within the R-TD scenario cannot be clearly differentiated from the surface state corresponding to R-RE. Moreover, the six samples corresponding to the surface states R-TD-2 and R-TD-3 were not investigated by OSEE due to their visually perceivable strong deformation.

Figure 3.22 presents the respectively obtained OSEE maps for the three samples prepared following the R-FP scenario. For all three samples investigated, the OSEE intensities measured around the centers of the samples were diminished compared to the R-RE state. The lateral inhomogeneity within the OSEE maps of samples I-R-OSEE-FP-1-3 and I-R-OSEE-FP-2-3 (corresponding to two different levels of exposure within the R-FP scenario) is interpreted to result from the different spreading of the fluid applied with the fingerprint since that application was performed centrally

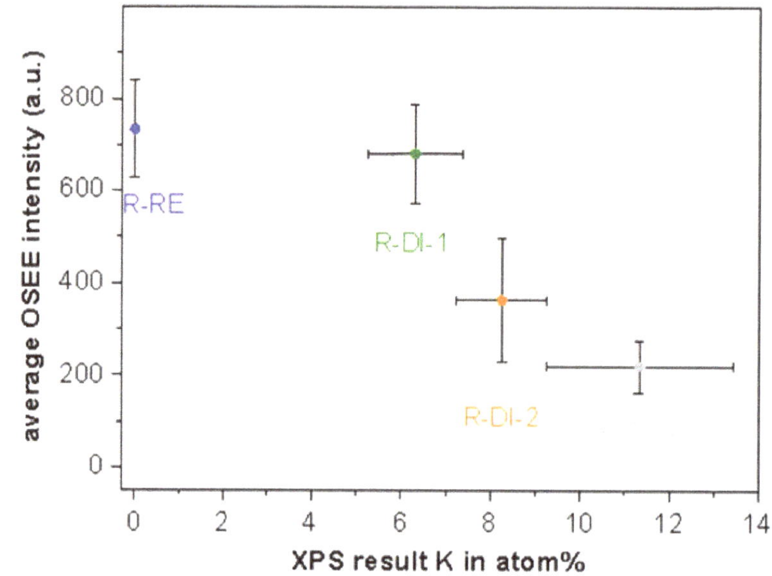

Fig. 3.20 Average OSEE intensities for the three samples within each sample set, R-DI-1, R-DI-2, and R-DI-3, as part of the R-DI scenario, based on locally applied contaminations of different degrees, achieved with an intentionally deposited potassium-containing de-icing fluid, compared to the OSEE results of the R-RE scenario

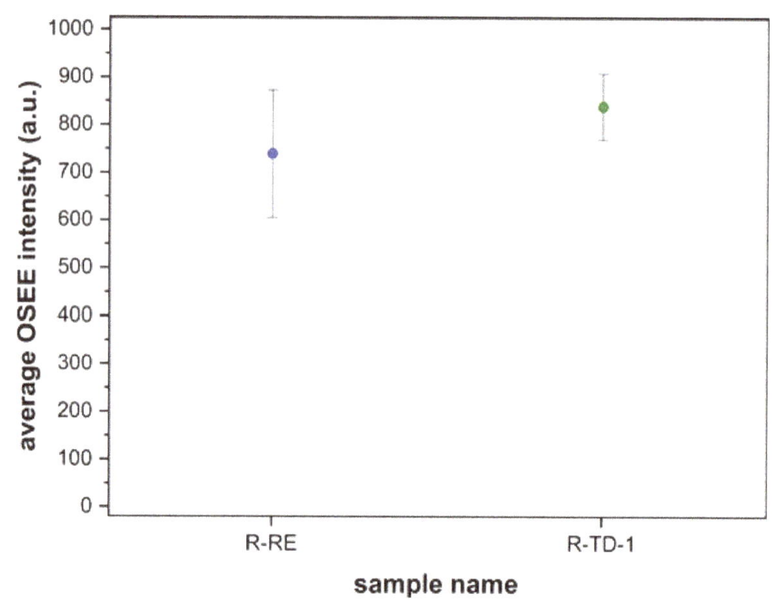

Fig. 3.21 Average OSEE intensities for the three samples within the sample set R-TD-1 compared to the OSEE results for the clean and freshly ground CFRP reference R-RE scenario. The three samples within each of the sets R-TD-2 and R-TD-3, as part of the R-TD scenario, were not investigated due to their visually perceivable strong deformation

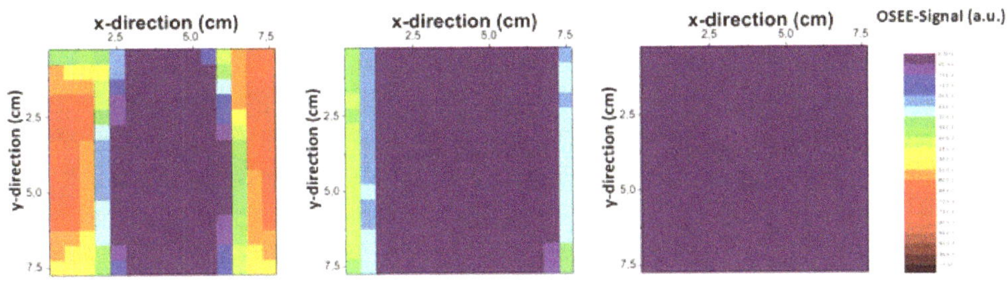

Fig. 3.22 Optically stimulated electron emission (OSEE) maps (15 × 15 pixels) for three samples prepared following three different levels of contamination within the R-FP scenario, based on locally applied contaminations of different degrees, achieved by intentionally depositing Skydrol hydraulic oil

within the OSEE mapping area using a fingerprint with the width of a human thumbprint.

As shown in Fig. 3.23, all the surfaces corresponding to contamination levels R-FP-1, R-FP-2, and R-FP-3 within the R-FP scenario can be clearly detected and differentiated from the surface state corresponding to R-RE, even when evaluating 7.5 cm × 7.5 cm wide areas, which exceed the area covered by the respective originally applied fingerprint, which was originally the size of a human thumbprint.

Combined contaminations

Combined contaminations include the RA+FP scenario within the production user case and the TD+DI scenario within the repair user case. In both cases, we investigated

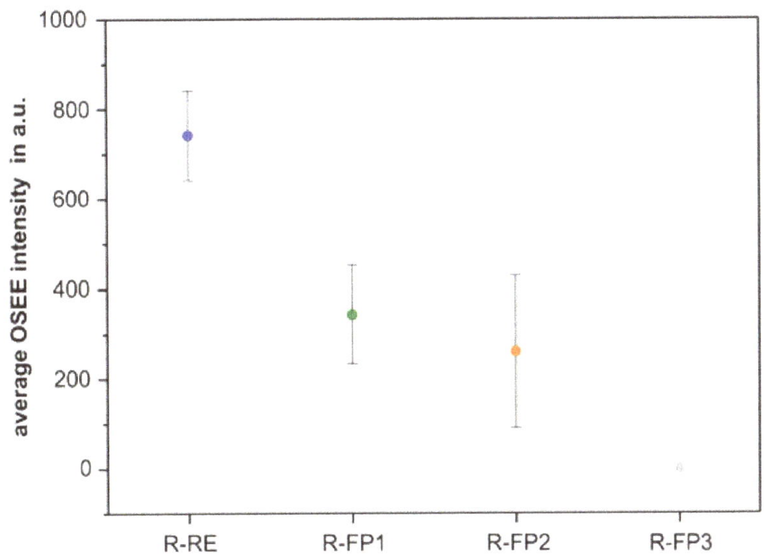

Fig. 3.23 Average OSEE intensities for the three samples within each sample set, R-FP-1, R-FP-2, and R-FP-3, as part of the R-FP scenario based on locally applied different contamination degrees, achieved by intentionally depositing Skydrol hydraulic oil, compared to the OSEE results of the clean CFRP reference R-RE scenario

two levels of contamination in addition to the respective references. On the surfaces of the P-RA+FP samples, we applied two distinct levels of release agent contamination (RA) before the application of the salt-based fingerprint in the center of the sample area. For the R-TD+DI samples, there were two levels of de-icing fluid contamination (DI) applied after the thermal degradation treatment (2 h in an oven at 220 °C).

As was shown in Sect. 2.1.2.2, samples contaminated following the P-RA scenario showed much lower OSEE signals than the clean reference CFRP samples prepared according to the P-RE scenario. The sample surface states comprising combined contaminations show similar OSEE intensities to the sample prepared within the P-RA-2 scenario; thus, these combined contamination levels can be clearly differentiated from the surface state corresponding to P-RE.

Specifically, the OSEE maps obtained for the samples prepared following the P-RA-1+FP3 (cf. Fig. 3.24) and P-RA-2+FP3 scenarios (cf. Fig. 3.25) reveal the positions where the fingerprints were applied since the respective regions show higher OSEE intensities than the surrounding surface regions.

Finally, within the considered repair user case, the OSEE results for surface states corresponding to R-RE were compared with the OSEE results obtained for sample

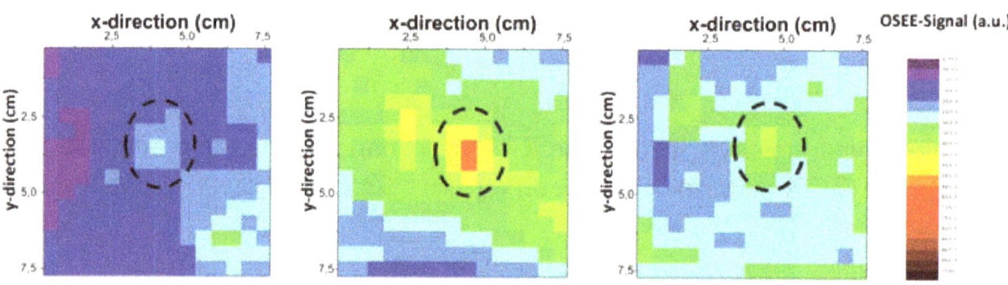

Fig. 3.24 Optically stimulated electron emission (OSEE) maps (15 × 15 pixels) for the three samples prepared following the P-RA-1+FP3 contamination scenario, measured with adjusted OSEE settings. Presumably fingerprinted (FP) regions showing higher OSEE signals are marked with dashed ellipses

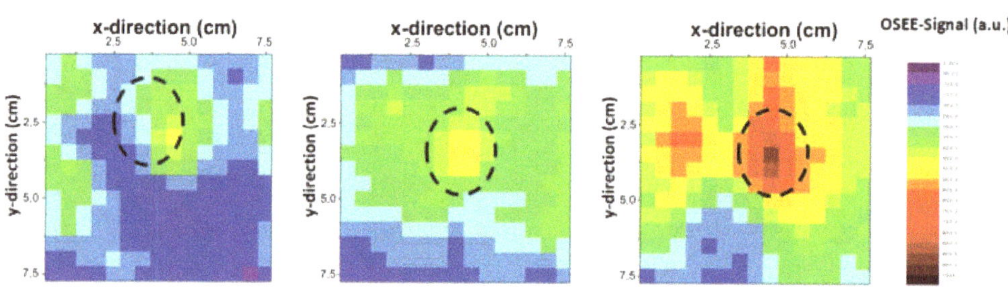

Fig. 3.25 Optically stimulated electron emission (OSEE) maps (15 × 15 pixels) for the three samples prepared following the P-RA-2+FP3 contamination scenario, measured with adjusted OSEE settings. Presumably fingerprinted regions showing higher OSEE signals are marked with dashed ellipses

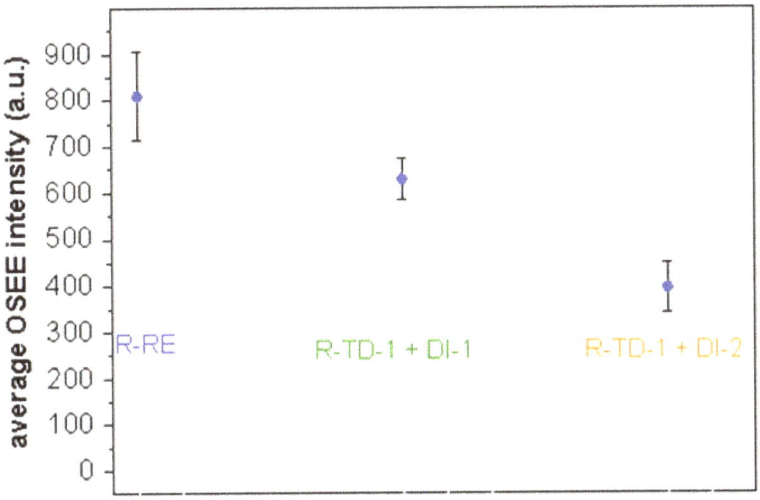

Fig. 3.26 Average OSEE intensities for the three CFRP coupon samples within each sample set of the R-TD-1+DI-1 and R-TD-1+DI-2 combined contamination scenarios within the repair user case; here compared to the OSEE results of the R-RE scenario

surfaces exhibiting combined contaminations following the R-TD-1+D-1 and R-TD-1+DI-2 contamination scenarios. The respective average OSEE intensities are compared in Fig. 3.26. The sample surface states with the combined contaminations can be differentiated from the surface state corresponding to R-RE. Somewhat remarkably, the surface states of samples within the R-TD-1+DI-1 scenario can be differentiated from those of the R-RE samples, in contrast to samples prepared following the R-DI-1 and R-TD-1 scenarios.

3.3.2.2 OSEE Results on Pilot Level Specimens

In the following, we present selected findings obtained when applying the available OSEE technology for assessing the surface quality of CFRP pilot level specimens from the production user case.

One key feature of the investigated pilot level specimens from the production user case was the peel ply surface of a CFRP part. This peel ply surface resembles a closed epoxy film that forms the outermost layer of the sample. For the OSEE technique to work, it is necessary that the studied samples show at least some amount of electron emission. However, probably due to the closed and electrically insulating terminal epoxy film, we did not observe any significant electron emission. The closed polymer film on top of subjacent carbon fiber layers appears to prohibit a measurable OSEE signal. We tried several approaches with the aim of achieving measurable signals, e.g., using the highest possible detector sensitivity setting and decreasing the distance to the sample in order to increase the flow of the UV light and decrease the distance for the electrons to travel. These changes have proved to be useful on other kinds of material surfaces in the past; nevertheless, with the closed epoxy film on this type of CFRP part, no signal was observed. The samples from the production scenario were therefore categorized as not measurable with the OSEE technique as long as charging compensation was not facilitated.

3.3.2.3 Technological Advancements of the OSEE Method Toward Industrial Automation

During the ComBoNDT research project, several enhancements were made to the setup to implement the existing OSEE sensor in an industrial environment. A key requirement for applying the OSEE technique when investigating specimens with curved surfaces is maintaining a constant height of the sensor (especially the essential component, i.e., the electron collector in the sensor head) above the CFRP surface. To achieve this, the OSEE sensor was adapted to be mounted on and used with an industrial robot.

Specifically, the main work here was the construction of an adapter plate for the mechanical connection of the OSEE sensor itself to the flange of a suitable industrial robot. At the Fraunhofer IFAM site in Bremen, a KUKA KR 20 device with a KUKA KR C4 control system was used. The flange with which such a robot is equipped follows an international norm, which was taken into account when making the layout for the holes on the robot side of the adapter plate. On the OSEE side of the adapter plate, a simple clamping mechanism was chosen to hold the OSEE sensor head in place. This clamping mechanism allowed easy access to the sensor head for the maintenance or direct manipulation of the sensor head itself. It also facilitated adjusting the height of the sensor compared to the sample surface in case that this crucial parameter needed to be changed without making adjustments to the programming and motor-driven positioning system running the robot itself.

The plate was made from aluminum, and all the necessary holes and cut-outs were manufactured using a CNC-controlled milling machine. The clamping mechanism was constructed in the same way as it was provided with the commercially available OSEE sample positioning system in the lab to enable an easy exchangeability of the used components. The resulting design of the OSEE sensor head connected to the industrial robot is shown in Fig. 3.27.

To ensure that the adapter plate worked correctly in terms of mechanical stability, several tests were performed. These examinations included test runs with different speed settings for the robot motion as well as different acceleration and brake settings. Not only was the mechanical stability of the mounted head itself checked every time, but also the repeatability of the motion performed by the robot was assessed. For these tests, extensive adjustments were made to the used programming within the KUKA KR C4 control system. These tests also ensured that the solutions established for the cable connections to the sensor head in terms of electrical power and sensor data worked as expected.

As described above, one key element of the OSEE technique is the need to ensure a constant distance between the measuring head and the sample. To achieve this goal, it was considered important to have a verification mechanism to constantly

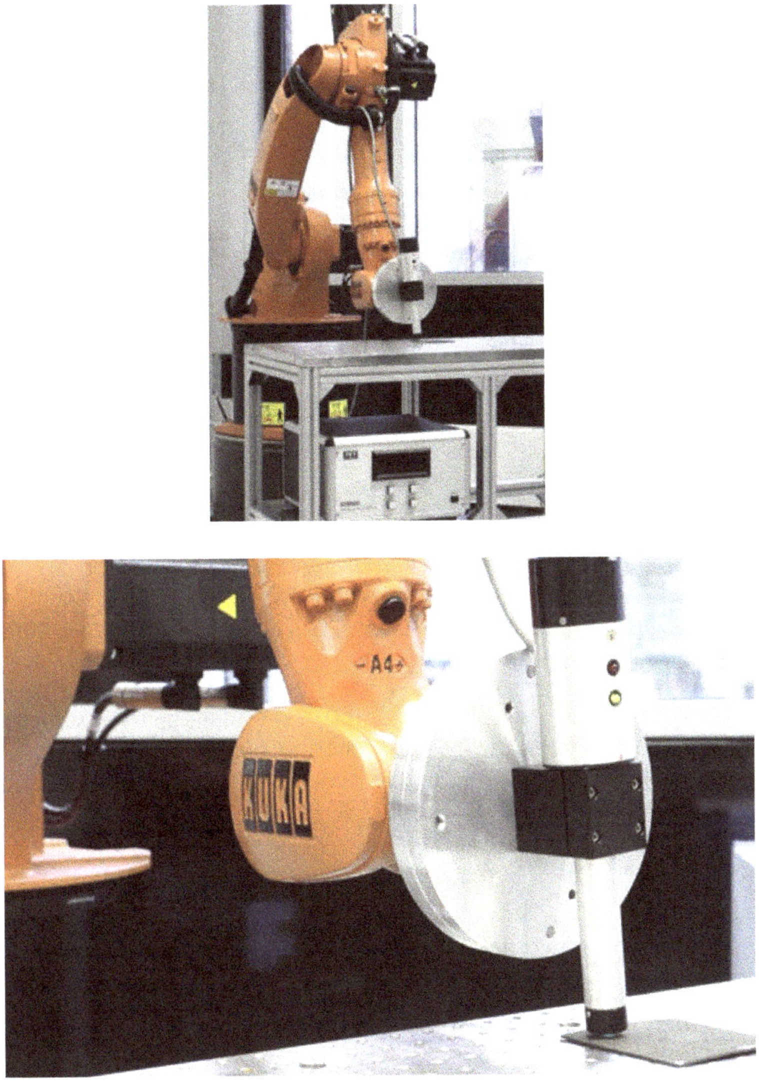

Fig. 3.27 OSEE sensor head mounted to a KUKA KR 20 standard industrial robot for test measurements on the CFRP pilot samples to be investigated in the ComBoNDT research project at the technical center at Fraunhofer IFAM in Bremen; bottom: A more detailed image of the clamping mechanism holding the sensor head in place

check this parameter and provide feedback to a control unit. Therefore, some further enhancements in terms of manipulating the OSEE head were made by incorporating a distance sensor into the measuring setup. For this purpose, a laser line scanner was used. This sensor device was mounted close to the OSEE measuring head with the laser line being projected directly onto the spot that was illuminated by the UV light of the OSEE sensor and from which the OSEE captured the emitted electrons for the photocurrent measurements. This enabled safeguarding of the constant evaluation of the distance between the OSEE sensor and the sample. The laser line scanner can

Fig. 3.28 Setup comprising a laser-based distance sensor for use in conjunction with the OSEE sensor head and an industrial robot; a laser line scanner (on the left) is attached to the manipulator that holds the OSEE measuring head in such a way that the setup enables constant checking of the distance between the sensor and the sample and the measuring of the optically stimulated electron emission at a constant distance between the sensor and the CFRP surface

also be connected to the programmable logic controller (PLC, henceforth referred to as SPS), which pilots the whole measuring process, as displayed in Fig. 3.28.

With these modifications, the precision of the robot motion was all that limited the areas accessible for the scheduled use of the OSEE device as an ENDT technique. As long as it is possible to follow the surface geometry of a part with a constant distance, the OSEE facilitates the inspection of large surfaces at a high speed.

3.3.3 Performance in Inline Surface Quality Assurance

In this section, we summarize the findings highlighting the performance of the OSEE-based ENDT technique for the surface quality assessment of CFRP specimens in distinct user cases.

In a first and trendsetting approach within the ComBoNDT research project, we technologically advanced OSEE and successfully applied this ENDT technique to map CFRP test coupons prepared by intentionally depositing distinct contaminants within several contamination scenarios, among them scenarios comprising combinations of contaminations that might have adverse (and, thus, possibly compensatory) effects on the OSEE signal. Specifically, the reference (RE) states of the production (P-RE) and repair (R-RE) user cases were clearly differentiated from each other, and the suitability of using OSEE to assess the results and reproducibility of a CFRP surface grinding process was indicated. When investigating samples from production scenarios, we applied adjusted OSEE settings that differed from the ones applied for samples from the contamination scenarios of the considered repair use case. In a nutshell, when comparing the sample surface states obtained within different contamination scenarios of the production user case with the ones prepared following the P-RE scenario, OSEE enabled all sample sets from the P-RA, P-MO, and P-FP scenarios as well as the sets with combined contaminations (P-RA-1+FP3 and P-RA-2+FP3) to be differentiated from P-RE. Moreover, within the contamination scenarios of the repair user case, OSEE enabled all sample sets from the R-FP scenarios, most of the samples from the R-DI scenarios (namely those with higher levels of contamination than R-DI-1) as well as sets with combined contaminations (R-TD-1+DI-1 and the scenario R-TD-1+DI-2) to be differentiated from R-RE. However, the surface states of the samples from the R-TD scenarios and samples of the R-DI-1 scenario were not differentiated from those of the R-RE samples. Clearly, if the adhesive joints manufactured based on these adherend surfaces showed inferior mechanical properties compared to joints prepared following the R-RE scenario, then these findings would stand in opposition to recommending OSEE as the ENDT method of choice in this user case. Finally, when referring to the lateral homogeneity of the CFRP test coupons, OSEE indicated several peculiarities: The inhomogeneous grinding of one coupon sample from the repair scenario, the spreading of fingerprinted fluid within the R-FP scenarios, and the position of fingerprints within samples exposed to combined contaminations in the P-RA-1+FP3 and P-RA-2+FP3 scenarios.

Aiming at assessing the surface quality of the pilot level CFRP specimens, we further improved the OSEE technique to permit its use in an industrial environment by incorporating a laser line scanner in the measuring setup to precisely control the distance between the sensor and the studied sample, which might have a more complex shape than a flat sheet. Concerning the investigations conducted on the pilot samples from the production user cases, the OSEE method encountered some severe limiting challenges concerning its use as a measuring system for these kinds of samples. In the case of the pilot samples of the production user case, the closed epoxy film of the peel ply surface prevented a measurable OSEE signal from being obtained. ENDT procedures using OSEE as a measurement technique evaluate relative changes to a given reference signal, e.g., the one characteristic of a clean CFRP surface. Therefore, the procedure was not suitable for assessing the surface quality of the pilot samples of the production user case as the surfaces of the reference substrate did not provide a sufficient signal strength. In our opinion, the encountered challenges, in the form they are presented here, do not appear to be unsolvable if further development is invested in a more customized embedding of the powerful OSEE technique in a setup that enables the elevated positive charging of electrically non-conductive analyte surfaces to be counteracted.

Moreover, we achieved an integration of the OSEE technology in inline applications, e.g., a robot-aided approach, without fundamental constraints. We present an approach to assembling the OSEE sensing technique in a setup that provides further information on the geometrical constellation of the measurement situation. We suggest that a real-time height adjustment and an improved method of signal detection facilitating an automated adjustment of the signal range could provide new ways to use this technique, even on materials that present the challenges observed during the ComBoNDT project. Within the project, it was not possible to implement these improvements. However, as techniques to change the sensitivity of analog–digital converters become more and more available, these challenges could be tackled in a future project.

We would like to underline that OSEE is a highly sensitive inspection method that allows the detection of even very low levels of contamination [1]. Often, a monolayer of a contaminant on an otherwise well-emitting material can be safely detected, as was demonstrated here, e.g., in the case of release agent deposits that we diagnosed using x-ray photoelectron spectroscopy. Nevertheless, establishing OSEE-based ENDT procedures may encounter difficulties with some of the technical surfaces found in industrial applications that, for example, rely on manually performed abrasive surface pretreatment. Such technical surfaces, which are often laterally inhomogeneous in their surface composition or surface conditions, show a considerable variance in their OSEE signal. This may make it difficult to locally differentiate between contaminations and substrate effects. In other words, assessing merely the feature of optically stimulated electron emissivity does not enable the identification of the substantial source of signal variations and provides a signal contrast that is not sufficiently specific to assess analytical challenges, such as the material-related identification of (mixed) contaminants or the essential elements thereof. We estimate that these challenges can be compensated for to a certain level using modern data analysis methods for such applications that benefit from the high sensitivity of the OSEE technique.

3.4 Electronic Nose

In this section, we introduce the electronic nose technique, highlighting two distinct instruments serving as ENDT tools for surface quality assessment and detailing how their performance was enhanced in the ComBoNDT research project for the in-process monitoring of CFRP adherends representative of close to potential aeronautical user cases.

3.4.1 Principle and Instrumentation

Electronic noses (e-noses) are chemical multi-sensor devices, based primarily on microsensors, which are capable of conducting low-cost analyses of complex gas mixtures through chemical fingerprinting. Coupling an array of chemical sensors to pattern recognition algorithms is an idea dating back to the 1980s and began significant development during the 1990s. Due to low overall costs in terms of purchase, operation, and maintenance as well as interesting research results, they looked—and still look—very promising and appealing for the development of industrial-grade ENDTs [9]. A significant number of commercial platforms have been available since 2000 for the comprehensive detection of components in mixtures as well as identification and/or quantification in medical diagnostics, environmental monitoring, and the food industry [10]. Despite some success stories, increasing the application of this technology to a wider range of industrial scenarios and user cases is currently obstructed by technological limitations [11]. Apart from the well-known specificity and sensibility issues that affect single sensors, the variability of fabrication outcomes severely hampers the use of a shared calibration function for e-noses. Thus, ad hoc calibration procedures need to be implemented for each individual e-nose device. The development of calibration transfer strategies could help overcome these limitations but, currently, there are no definitive results [12, 13]. Drift effects (mostly due to aging and poisoning) in sensor devices as well as environmental parameters have affected the mass adoption of e-noses in the high-value production industry because they jeopardize operative requirements. Presently, most research contributions in advancing e-noses still target the headspace analysis of liquid/solid samples in controlled environments, which may amount to underestimating the challenges of on-field applicative scenarios. From the point of view of today's user, we would like to highlight that on the one hand, specific requirements such as high reliability, fast response, and the possibility to be operated by a chemically non-expert workforce can be, in principle, met by up-to-date e-nose platforms. On the other hand, some instrumental operative requirements, including the need for operation in uncontrolled or even harsh environments [14], may still prove very challenging due to the abovementioned lags and issues.

Conversely, and from a prospective point of view, we observe that along with the requests for novel integrated health monitoring systems, the need for new NDT technologies that should be capable of coping with the new challenges brought by the lightweight aircraft industry is steadily growing [15, 16]. The adoption of lightweight composite materials like CFRP for primary structural components is a major trendsetting milestone and may contribute to a significant reduction of per-mile-passenger transportation costs. It is estimated that this could allow for a dramatic increase in cost efficiency for ground operations (up to 50%), a reduction in fuel usage (up to 20%), and consequently, a CO_2 emission rate reduction at the fleet level of up to 15% [2]. CFRP parts are assembled and joined by adhesive bonding, a critical and special process that requires the adherend panel surfaces to have a high grade of cleanliness. If not, the mechanical properties of the assembly itself could be severely compromised, leading to a high risk of potential structural failure [17]. Surface contamination can be caused by various processes that occur during the assembly or operative life of CFRP panels. Adherends with improperly removed deposits of contaminants, such as hydraulic oil or de-icing fluid, may come in contact with a CFRP surface during aircraft operation, while the release agents used in CFRP molding processes during composite production can severely affect the mechanical properties of a resulting adhesively bonded joint. Such deposits may produce chemical damage to the surface or create a physical screen to the adhesion, while the thermal degradation of the CFRP adherend surface can affect the adhesion properties of the panel, thus compromising the mechanical strength of the bond. Several authors have measured the mechanical parameters of adhesive bonds based on contaminated CFRP panels (e.g., [18]). In particular, Tserpes et al. showed a reduction in the fracture toughness (G_{IC}) in excess of 25% for Skydrol®500-B contamination and of more than 60% for significant release agent contamination [19]. Specifically, a 7% silicon-containing substance (calculated as at.% based on XPS investigations) remaining on the surface after demolding can lead to a total lack of adhesion [20].

Bearing in mind these production–technological challenges and the instrumental current state of the art, we anticipate that customized e-noses may represent a suitable technology that can contribute to addressing the lack of a verified procedure for assessing bond quality, which has ultimately been slowing down the adoption of CFRP for primary structures. ENEA and Airbus, both partners in the H2020 ComBoNDT research project [2], have been and continue to be focused on developing e-nose ENDT solutions for surface cleanliness checks to be applied in composite pre-bond quality assurance. In the ComBoNDT project, this endeavor required the design of ad hoc sampling and measurement subsystems, the careful selection of an ad hoc sensor array as well as the design of proper machine learning algorithms to cope with the requirements of this safety-critical application field. The subsequently reported work accounts for the in-project development process, which aimed to reach and demonstrate a high TRL for the use of the e-nose as a detector of surface contaminations before bonding.

3.4.1.1 Generic E-nose Architecture Considering CFRP Contamination Scenarios

The basic idea underlying any e-nose design is a close coupling between an array of chemical sensors and a pattern recognition system. The latter can be focused on detection, classification, or quantification [21, 22]. The sensor array is usually designed to focus on broad sensitivity and diversity in order to augment the chemical fingerprinting capabilities. Metal oxide (MOX) sensors, polymeric sensors, and electrochemical sensors, often combined with photoionization detectors (PID) and/or ion mobility spectrometers (IMS), have been widely employed in e-nose devices. Hybrid arrays are also found in the literature, always with the aim of enhancing the diversity of potential applications and user cases.

In more detail, MOX sensors are chemiresistors, i.e., their electric resistance changes as a consequence of their interacting with the environment in which they are deployed. These sensors and their signals are non-specific since they are responsive to a wide variety of volatile organic compounds (VOCs) and environmental conditions (e.g., humidity). Their signal dynamics and sensing window are perhaps the largest in the gas sensor realm, and they are cheap and easy to integrate on electronic boards. On the other hand, the signal stability and response repeatability are the principal drawbacks for this family of gas sensors. An array of MOX sensors can be successfully implemented in a closed chamber of an e-nose, working in a "differential mode" to overcome the limited repeatability and poor signal stability. In this so-called differential mode, sensors are exposed to filtered air before and after being exposed to the air analyte sampling. Features linked to the signal variation, which occurs when the sensor resistance is disturbed by an odor sample, can represent a repeatable odor pattern. A sensor equipped with PID can be considered as a VOC exposure meter. This detector is based on the photoionization of a gas by means of an UV lamp, and it can detect VOC particles ranging from sub-ppm to thousands of ppm. PID-based sensors cannot discriminate VOC species; they only account instantaneously for the species photo-ionized in the excitation process overall. In this way, they can be useful in understanding the integral level of the odor exposure on the e-nose sensor array.

Similar to a nose in the biological sense, the structure and flow conditions in the pneumatic section of an e-nose are of paramount importance in co-determining its final performance. Forced flow is usually adopted in benchtop scale solutions, while open sensing is generally adopted in battery-operated, long-term deployments of smart multi-sensor systems. In the first case, especially in benchtop devices, the sensor array is first exposed to pure synthetic or filtered air in order to assess the baseline results. Sensor baseline generation under clean air conditions is a vital aspect of e-nose performance, and a new technique of sensor baseline estimation without the need for an external gas supply is also available [23].

Following e-nose data acquisition, pattern recognition subsystems are employed, which are primarily designed to provide classification capabilities. Both supervised and unsupervised pattern recognition designs have been widely explored. Among the plethora of different supervised designs, k-nearest neighbors algorithms (k-NN), support vector machine (SVM), neural networks, and partial least squares discriminant analysis (PLS-DA) systems are the most commonly adopted. While most of these approaches are strongly non-linear and aim at modeling significant non-linearities found in the multivariate sensor models, PLS-DA combines a linear transformation to reduce the number of evaluated dimensions with the discrimination capabilities of Fisher discriminant analysis. Dimensionality reduction is also usually tackled by principal component analysis (PCA), resorting to the first principal components. Recently, this approach has been reported less often, essentially because discriminant characteristics may be embedded in relatively low variance components.

Indeed, for the design of the e-nose pattern recognition subsystem, feature extraction and selection comprise one of the most important steps. Designing appropriate features and selecting the most informative combination is in fact one of the main performance drivers. Steady-state responses, when appropriately normalized for the reduction of baseline drift issues, can be sufficient for obtaining a discriminant fingerprint for several analytes and mixtures. However, most of the time-dynamic features based on the sensor response during exposure, presentation, and flushing transients are essential for obtaining an adequate classification performance [24, 25]. With respect to ongoing developments, we may state that quantification issues and problems are primarily tackled using non-linear approaches exploiting the regression capabilities of neural networks, support vector machines, and Gaussian processes as well as other data-driven approaches. Recently, the analysis of dynamic behavior has attracted interest also for quantification problems, exploiting a model that can

Table 3.1 Graded concentration ranges of contaminants intentionally applied to CFRP reference (RE, for production or repair) substrates as identified and defined by partners of the ComBoNDT research project for aeronautical user cases in a way that yielded a loss of bond strength of up to 30% of the initial strength

Contamination scenario	Contaminant	Concentration ranges
Production	Release agent (RA)	3–8% (Si at.%)[a]
	Moisture (MO)	0.4–1.4% (mass uptake)
	Fingerprint (FP)	0.2–0.7% (Na at.%)[a]
Repair	De-icer (DI)	6–12% (K at.%)[a]
	Hydraulic fluid (FP)	<0.5 g/m^2

[a]Atomic surface concentrations (in at.%) for the listed species were measured by XPS analysis

take into account the intrinsic dynamic behavior of the sensor response toward target gases and non-target interfering substances.

In the specific framework of the quality assessment of CFRP structures through ENDT methods focusing on surface cleanliness, the essential steps were determining and listing the chemical targets to be considered. Within the contemplated aeronautical user cases, this basic requirement was identified and defined by partners of the European Union (EU) ComBoNDT research project, a consortium that includes Airbus, the main EU aerospace industry stakeholder. This list is based on production or repair user cases and comprises hydraulic fluid, water (humidity), fingerprints applied unconsciously and locally by workers, release agents, and de-icing fluid, and is further complemented by thermal impacts and even damages. More details about this topic can be found in Chap. 2. The contamination scenarios are then divided according to the specific workplace at which the contamination can occur, namely within production or repair areas.

The interaction of these chemically nameable substances with CFRP structures can undermine the composite adherends at different levels, eventually affecting the mechanical strength of the adhesive bond between CFRP panels. Specifically, release agents are silicon(and siloxane)-based formulations used during the molding and demolding process of composite panels, and they can penetrate up to hundreds of nm into the CFRP panel matrix. Another source of the potential weakening of adhesive bonds stems from the presence of sodium chloride residue left by occasional fingerprints. Therefore, a fingerprint simulant prepared according to DIN ISO 9022-12, containing sodium chloride, urea, ammonium chloride, lactic acid, acetic acid, pyruvic acid, and butyric acid in demineralized water was added to the contaminant list. Additionally, the hydraulic fluid considered in this study was a fire-resistant phosphate ester-based liquid that, under certain conditions, can release phosphoric acid and alcohols. Finally, the considered de-icing fluid was a potassium formate-based formulation involved in runway or aircraft de-icing. Moisture exposure of CFRP parts as well as thermal impact and damages (TD) were also investigated.

Different contamination levels (as detailed in Table 3.1) were determined in a way that yielded a loss of bond strength of up to 30% of the initial strength (with respect to using two reference specimens, called RE, as adherends).

In addition to the flat coupon level samples (see Chap. 2) that had undergone these intentional contamination procedures comprising individual contaminants, according to the high TRL required for the measurement techniques, we also tested two e-nose setups against CFRP samples that had mixtures of two different contaminants applied to them. In particular, we investigated CFRP panels contaminated with a combination of release agent and (artificial sweat) fingerprints for the production user case (P-RA-FP) and thermally impacted panels that had additionally been contaminated by an application of de-icing fluid within the repair user case (R-TD-DI).

Additionally, user cases based on curved pilot level CFRP samples in which contamination was applied on both convex and concave surfaces were assessed by e-nose inspection. Finally, the e-nose systems were tested for the monitoring of realistic parts as part of joint tests oriented toward TRL assessment (defined within ComBoNDT).

3.4.1.2 Details of the ENEA E-nose

During the ComBoNDT project, the consortium partner ENEA continuously developed its e-nose approach featuring two main successive versions aimed at handling the increasingly high demands of the project's contamination scenarios, starting from planar coupons and reaching the realistic parts level. A description of these two main versions is given here.

The first version, the ENEA e-nose ver.1 (SNIFFI, [26]), is a closed-chamber gas sensing system. It is equipped with a chamber containing the sensor array, into which the air analyte sample enters by means of an air pump. Figure 3.29 shows the main components and functional connections of the ENEA e-nose ver.1.

The embedded architecture is organized in different layers, each one requiring different expertise, as shown in Fig. 3.30.

The sensor array is based on six four-pin slots that can host up to six heated chemo-resistive commercial sensors. During the ComBoNDT project, three couples of MOX Figaro Sensor family (Tgs2600, Tgs2602, and Tgs2620) were used. This commercial sensor array can be completed with up to six prototype sensing films using plug-and-play USB-like transducers that have been specifically designed to be mounted on a custom sensor board, as shown in Fig. 3.31. In particular, ENEA developed the graphene chemo-resistive devices chosen to complete the array. The features characterizing the environmental status and total VOC contamination of the chamber volume are measured by means of temperature and humidity sensors (Sensirion SHT75) and a PID, i.e., a photoionization detector (PID-AH, AlphaSense Inc.). The sensor chamber is built using a Teflon polymer to ensure a base sensor board with good electrical insulation as well as an aluminum cover to ensure good heat dissipation, see Fig. 3.31.

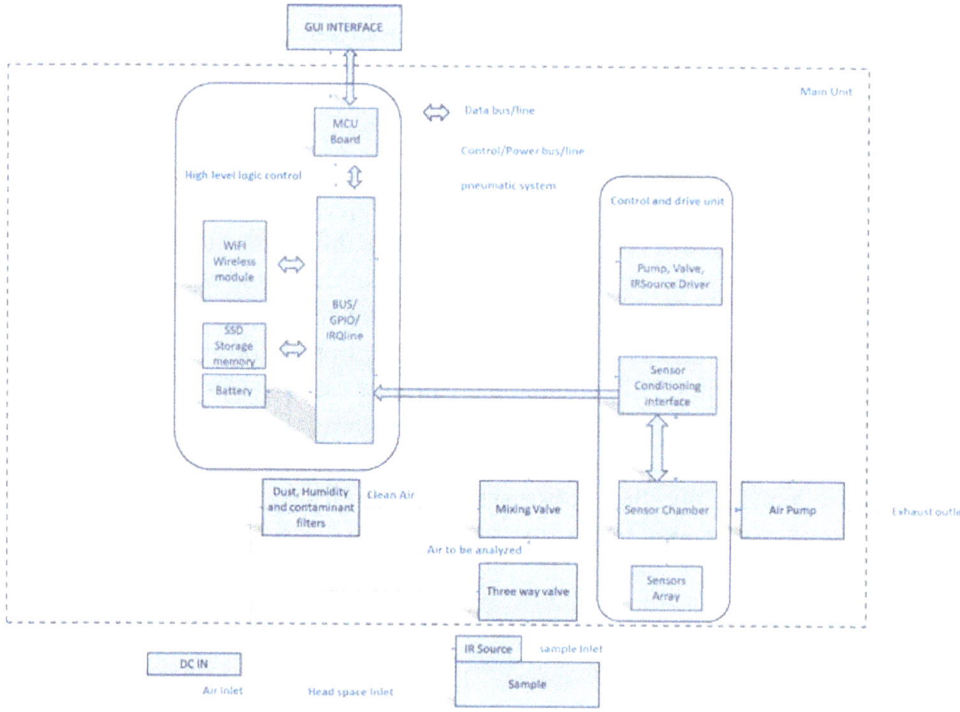

Fig. 3.29 Functional block diagram of the system representing the ENEA e-nose version ver.1 developed in the ComBoNDT research project for CFRP surface quality assessment

Fig. 3.30 Embedded layer-wise architecture of the ENEA e-nose ver.1 system

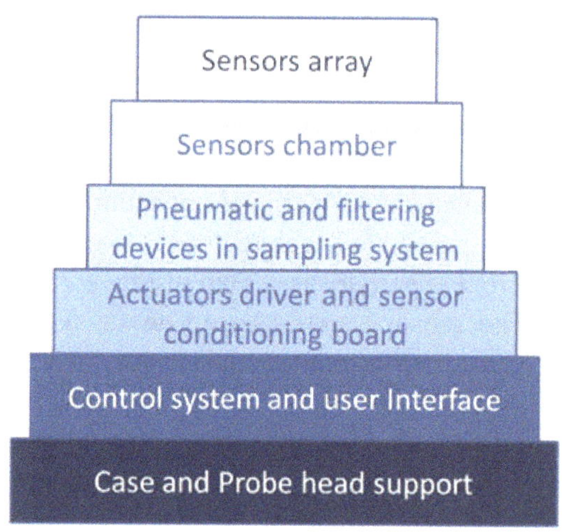

The ENEA e-nose ver.1 has two operating modes, allowing for switching between two different inlets to perform different application-oriented sampling tasks (as detailed in the block diagrams in Figs. 3.29 and 3.32):

- A front inlet for the air sampling of odors coming from a headspace (e.g., a bottle, an air sampler bag) (see the photograph in Fig. 3.33).

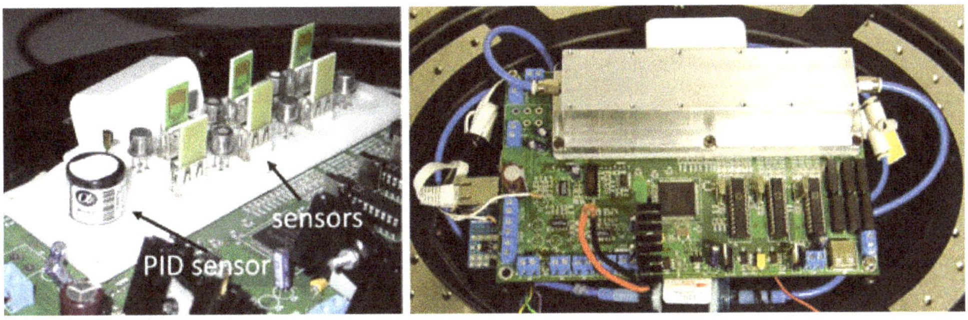

Fig. 3.31 Sensor chamber on the sensor board of the ENEA e-nose ver.1 system; left image: open state with the 12 chemiresistors, photoionization detector (PID) and temperature (T) and humidity (RH) sensors; right image: closed state

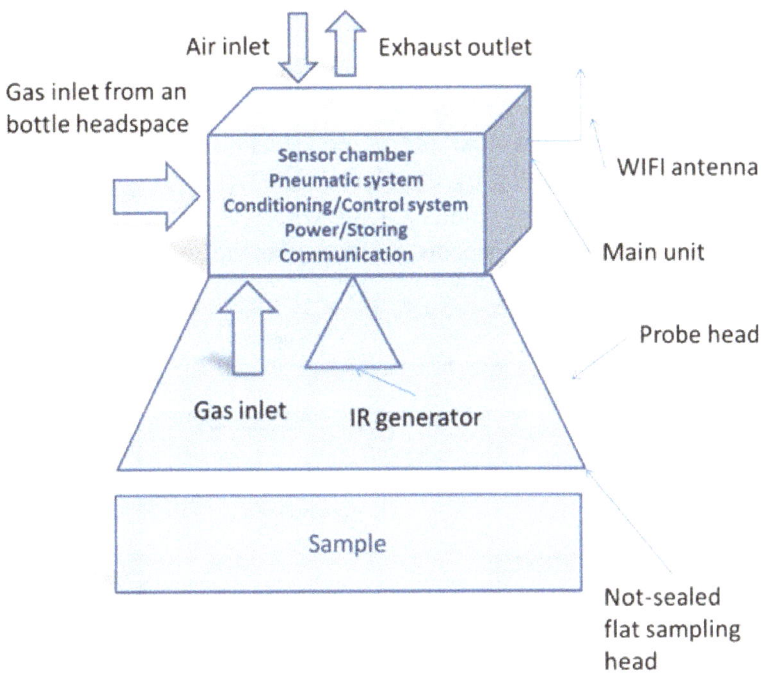

Fig. 3.32 Block design of the ENEA e-nose ver.1 system

- A bottom inlet, heated by a 20 W halogen lamp, to perform analysis focusing on surface contamination (see the image in Fig. 3.33).

Finally, this e-nose has a line of purified air that is used to conduct cleaning phases, maintain the gas sensors in a stable steady state, and properly dilute sampled air.

A high-level logic control layer based on low-cost open-source hardware drives the actuators and the sensor conditioning board. The graphic user interface (GUI) (based on a web server) allows the following operations to be performed: creation, saving, and recalling of measurement recipes; monitoring of sensors' real-time status' wireless Wi-Fi access; and transmission of the sensor log. The repeatability of the

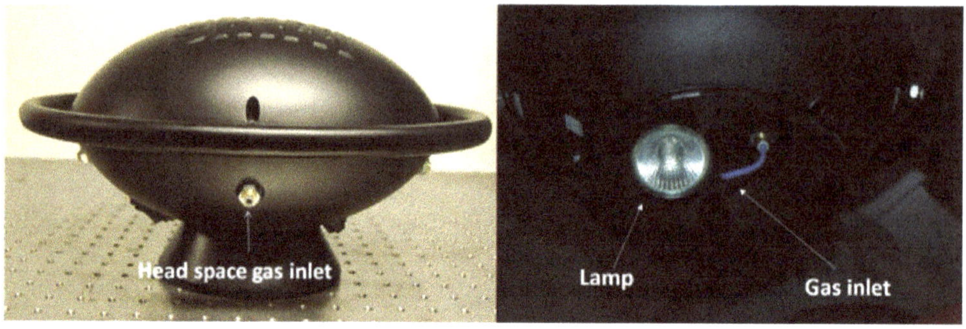

Fig. 3.33 Air inlets of the ENEA e-nose ver.1; on the left: a front inlet for headspace analysis (bottle or sampler bag); on the right: a bottom inlet and a spotlight halogen lamp to analyze surface contamination

sampling analysis is guaranteed by automated measurement cycles. A standard cycle can be repeated many times and is divided into four phases, the duration of which can be set by the user, as presented in Fig. 3.34.

We further developed the e-nose prototype (ver.1) described thus far to match the higher TRL targets foreseen by the ComBoNDT project objectives. Based on the analysis of the preliminary results for the CFRP coupon samples, we deemed it necessary to improve several aspects, in particular the sampling operating mode.

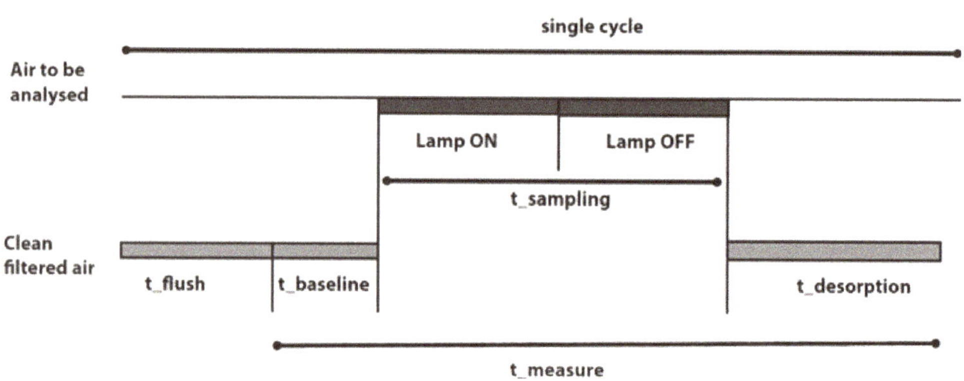

Fig. 3.34 Outline of a single generalized measurement cycle of the ENEA e-nose ver.1

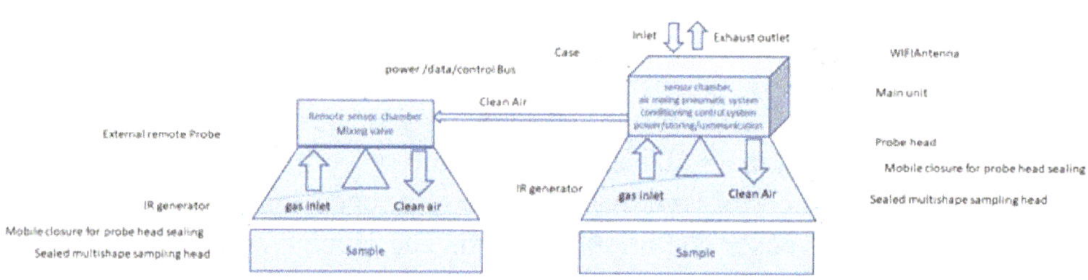

Fig. 3.35 Block design of the advanced e-nose prototype, ENEA e-nose ver.2 [27]

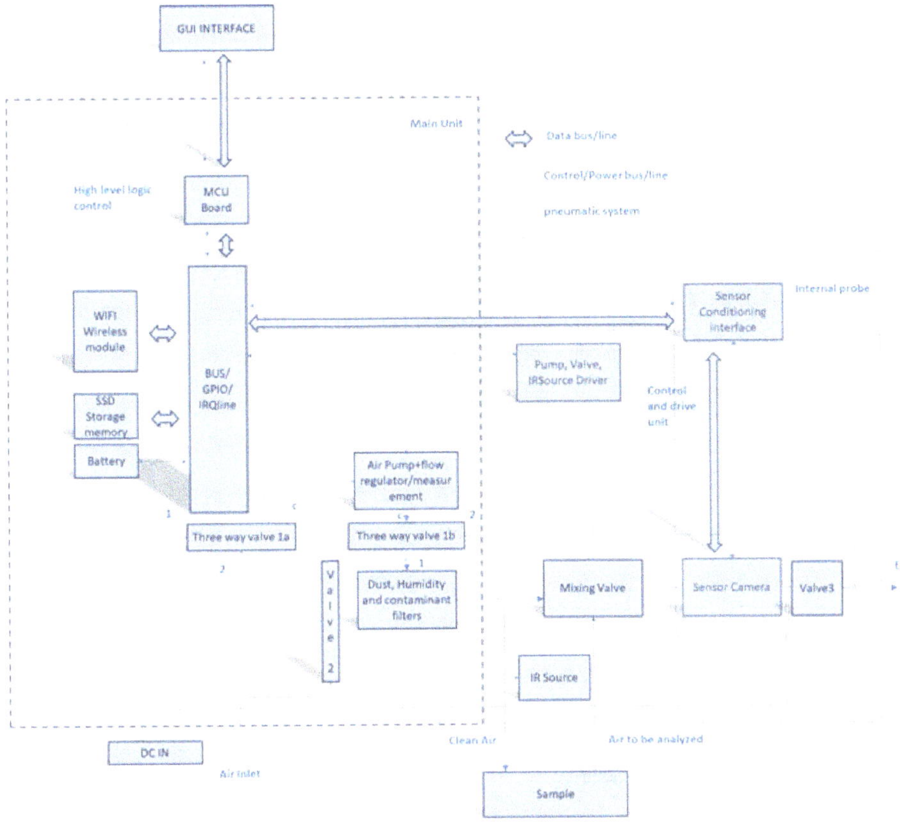

Fig. 3.36 Functional block diagram of the advanced e-nose prototype, ENEA e-nose ver.2

Figures 3.35 [27] and 3.36 present the design of the resulting second version, the e-nose prototype ver.2.

Aiming to analyze CFRP surface contamination without external interference, we first designed a new sampling head that allows the sampling area to be isolated from the outside, as detailed in Fig. 3.37. In this "sealed" sampling area, a flow of clean reference air is directed toward the sample surface while irradiation from an infrared emitter stimulates the desorption of the analytes from the investigated surface. The air sample is then pumped toward the sensor chamber. We paid special attention to the cleaning phase of the sampling chain. To improve the desorption capabilities, a 50 W halogen lamp was used in this new device. In addition, a sampling head suitable for curved samples was designed and realized. The sensor chamber and sensor array design were essentially unchanged, whereas the control actuator drive and sensor conditioning board were modified according to the new setup of the sampling head. The control actuators also offer the possibility of driving a remote sampling head.

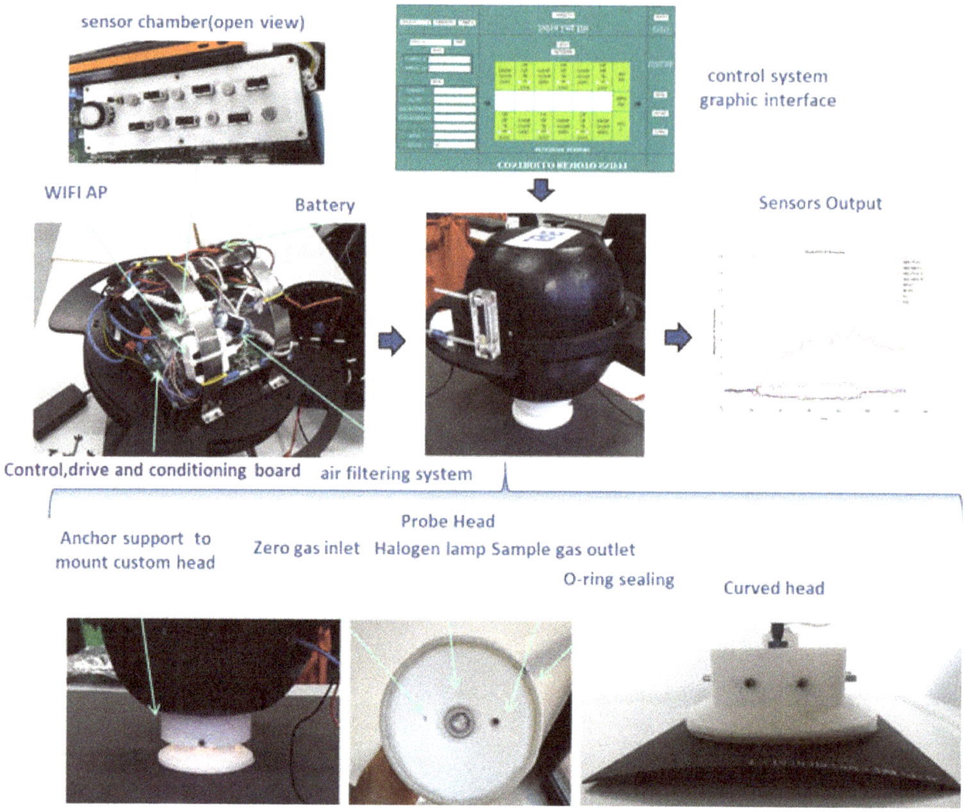

Fig. 3.37 Details of the advanced ENEA e-nose ver.2 prototype showing an open view of the sensor chamber, the GUI of the control system, details of the sampling head and the probe head, and a plot with an exemplary sensor signal outputs

3.4.1.3 Details of the Airbus E-nose and Desorber Device

Unlike the ENEA e-nose system, the Airbus approach relies on a custom-made e-nose in combination with an optimized sample-taking apparatus. The core part of this system is shown in Fig. 3.38 and consists of a ruggedized version of a multi-purpose e-nose marketed by the AIRSENSE Analytics company under the name

Fig. 3.38 Portable e-nose from AIRSENSE Analytics

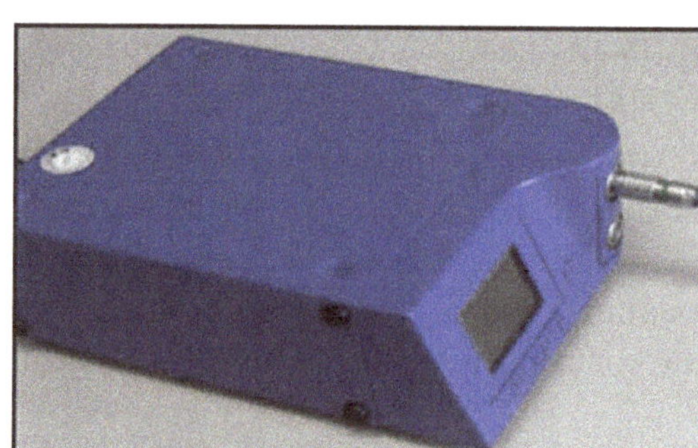

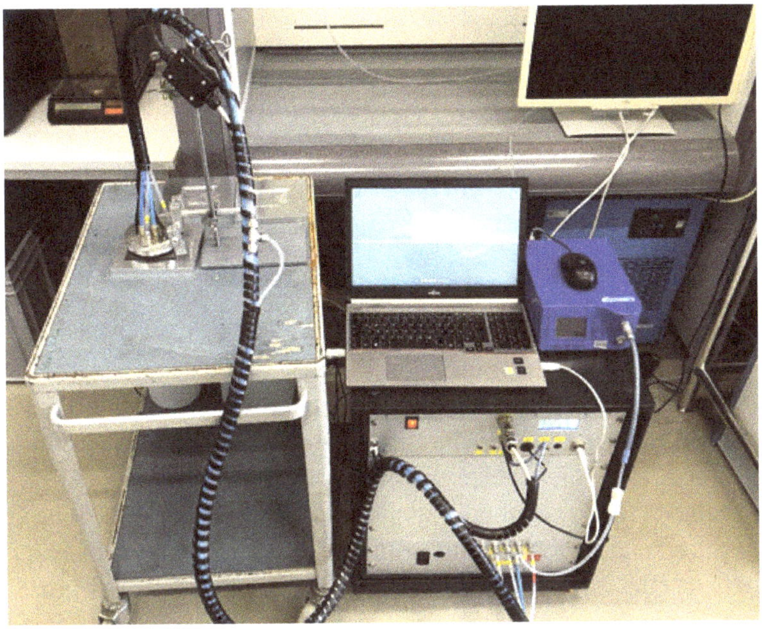

Fig. 3.39 E-nose system and desorber device—image of the assembly (20″ rack) in the laboratory

PEN (Portable Electronic Nose) [28]. This system contains an array of ten different MOX gas sensors. All sensors are operated according to their specific supplier recommendations.

The system was retrofitted with temperature sensors and a humidity sensor and was integrated into a mobile rack. Some further ruggedization was necessary to enable shop floor compatibility. To equip this multi-purpose core system for its intended use of contamination detection, it was fitted with an external sampling head called a "desorber device" (see Fig. 3.39) invented by Airbus and advanced to TRL six during the ComBoNDT project.

The integrated desorber system features a fully automated measurement process. In principle, the same procedure as that used for the ENEA system and depicted in Fig. 3.34 may be used. The system controls all relevant measurement parameters automatically, and the status of the system is displayed on a liquid-crystal display (LCD). The desorber system controls the sample heating and operates the humidity sensors and the pyrometer, which were integrated into the desorber device sampling head presented in Fig. 3.39.

To take the step from the laboratory to shop floor conditions, we conducted investigations to replace the synthetic air cylinders as a zero-gas supply. The system had to be adapted to work with oil-free compressed air, which is easily available on the production floor. Hereby, a particle filter, charcoal filter, and hydrocarbon oxidizer were deemed necessary to ensure sufficient clean air quality.

3.4.2 E-nose Methodology

In this section, we first detail the experimental methodology for assessing e-nose data as well as the pattern recognition technique. Finally, we present the obtained results for the CFRP specimens from the distinct user cases, according to their shape and the respectively considered contamination scenarios.

3.4.2.1 Experimental Methodology

The ENEA e-nose ver.1 was used to conduct measurements on CFRP surfaces by means of two different methods. The first was the standard measurement method, named the "0 method", which makes use of only the IR irradiation of the sample. The second one, named the "PC method", enhances the extraction of volatiles by treating the surface with a suitable low-boiling solvent. The process involves first spraying a few milliliters of ethanol over the surface of the sample with an airbrush and soon thereafter performing the e-nose measurement. The adoption of the PC method was motivated by the need to improve the desorption of the volatiles. In principle, this treatment also enables differentiation between different surfaces and surface states, also on the basis of their capability to retain and desorb the solvent (e.g., ethanol) as a function of the surface contamination. The measurements carried out with the advanced ENEA e-nose ver.2 were accomplished using only the standard method described above. For all measurements, the time exposure to the IR emitter was tuned to heat the CFRP to a temperature above 120 °C (but below 150 °C to avoid permanent CFRP damage). To ensure the repeatability of the results, a measurement strategy was adopted, which was mainly intended to avoid the presence of residual contamination inside the sample line after the sensing phase. To this end, a blank sample was analyzed periodically (at least once per hour), which also allowed the repeatability of the sensing responses to be verified. A cleaning procedure was implemented to prevent residual contamination in the sampling line.

To control and perform the right measurements, an assessment of the data is necessary to visualize the output sensors' features describing the odor pattern of the sample immediately after a measurement cycle. This offers the possibility to rapidly compare measurements and thus to understand if the results are reliable or if a cleaning phase of the system or a change in the measurement cycle parameters is necessary. Solid-state sensors can show a certain tendency to signal drift, and this could jeopardize the comparison among different samples. To overcome this issue, we processed the raw sensing responses to extract parameters that could be used to compare different samples. Essentially, each sensing response R was normalized to its own baseline R_0, and the percentage of the relative response R_{sp} was calculated as follows:

$$R_{sp} = \frac{\max|R - R_0|}{R_0}$$

Subsequently, a signal differential was performed (always relative to the baseline value) and the maximum was calculated. This is the same as having calculated the rate of the sensing responses and the maximum rate of the response itself. In our experience, these parameters are less affected by sensor drift and saturation, although they contain the same discrimination features of the sensor response [29]. Finally, a simple comparison was performed by calculating the ratios of the maximum response rate of each sensor with respect to that of the fastest sensor.

As a result of this signal elaboration, after the measurement cycle, the operator can see the entire sensor performance, as shown in Fig. 3.40, see Fig. 3.41.

The presented histogram gives an overview of how the individual sensors react to the analyzed odor. The red bars represent the adsorbing phase, while the blue ones

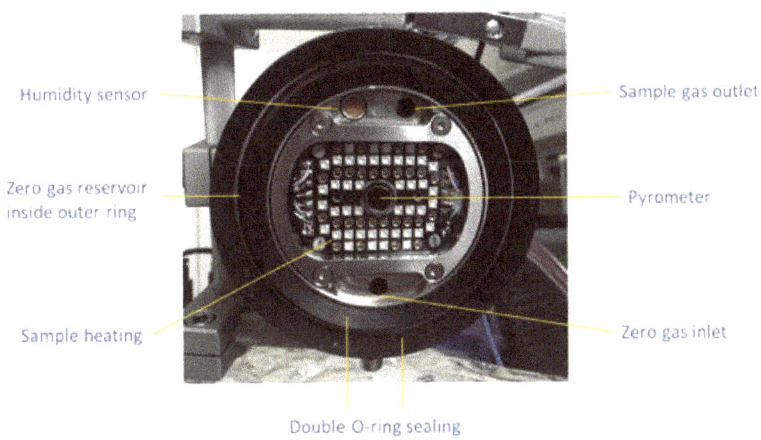

Fig. 3.40 Underside view of the desorber device sampling head. Excitation and heating are possible via infrared (IR) irradiation or ultraviolet light-emitting diodes (UV LEDs)

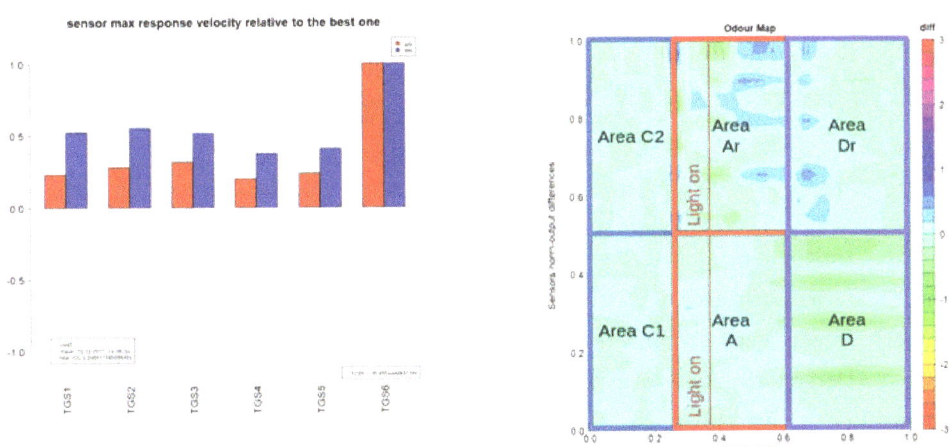

Fig. 3.41 The e-nose sensor response rate and an odor map obtained during a measurement cycle; on the left: a histogram of the sensor reaction to the odor during the adsorbing phase (red) and desorbing phase (blue); on the right: a contour map of the sensor signal distortion during a measurement cycle

refer to the desorbing phase. Moreover, a contour map is shown, which is useful for the e-nose technical specialist to understand if there are anomalies in the measurement cycle and if the recorded sensor signal distortion is well suited to the odor pattern. Signal distortions can be related to changes in odor intensity or abnormalities due to imperfect air filtering, or they can be ascribed to an odor fingerprint. The map then represents the differences between the signal shape of each individual sensor and that of the others. The map is divided into areas. The C (labeled C1 and C2) areas depict the sensor behavior earlier on, during the baseline phase in filtered air; an optimal behavior is the lowest distortion (light blue color). Areas labeled with A (or Ar) and D (or Dr) display the signal distortion intensity in the adsorbing and desorbing phase, respectively. Performing such feature visualization allows an experienced specialist to validate measurements and helps to prevent an applied pattern recognition technique from being confused by bad measurements. We would like to stress that this approach is completely modular and can be easily adapted to any sensor array.

Pattern Recognition Technique

Before actually tackling the pattern recognition task, the feature extraction process took place based on the measured e-nose data. From each sensor signal, five features were computed. The first three are specifically computed by averaging sensor response value through a specified time range. The obtained value then underwent a baseline normalization procedure being divided by the average sensor response recorded during the baseline acquisition time.

In Table 3.2, we present the computed feature array, including the steady-state and dynamic features. After the feature extraction, the data were normalized on a per column (sensor) basis to zero mean and unitary variance. We would like to highlight that several algorithms were implemented in order to select relevant features, including linear discriminant analysis, an entropy-based procedure (tree classifier), and a feature ranking algorithm based on K-nearest neighbors (***relief-f***).

Table 3.2 Description of the extracted features; the first five are computed for each of the seven e-nose sensors and the last two are from included temperature and humidity sensors

# Feature	Description	Computation time range (s)
Feature 1	Steady-state response (wrt averaged baseline)	[150, 180]
Feature 2	Steady-state response—IR Off (wrt averaged baseline)	[181, 200]
Feature 3	Desorption status (wrt averaged baseline)	[250, 300]
Feature 4	Uptake derivative	[60, 120]
Feature 5	Desorption derivative	[210, 250]
Feature 6	Temperature	[0, 330]
Feature 7	Relative humidity (RH)	[0, 330]

With the subsequently detailed e-nose data obtained for CFRP specimens, preliminary data analysis was performed using principal component analysis (PCA) to further reduce dimensionality and extract most of the signal's variance. The centers of the clusters, computed as the mean of the scores related to each class, were highlighted as well as their standard deviation (1-sigma) ellipses. This representation aimed to show a preliminary qualitative evaluation of the (inner and outer) variance distribution for contamination classes through cluster localization and spatialization. Furthermore, PCA was used as an anomaly detection method by relying on the first PC projections extracted from a distribution obtained for reference samples. In this way, a threshold level was set, delimiting a 2-sigma coverage on the supposed multivariate Gaussian distribution of the reference samples. Each sample whose representation fell outside this limiting border was detected as a contaminated sample. In some cases, more powerful pattern recognition schemes were implemented to cope with the subtleties of the different tasks. Details are given in the following sections. Finally, shallow neural networks were used, instead, for regression tasks aimed at estimating the level of contamination.

3.4.2.2 E-nose Results on Coupon Level Samples

In the following, we detail the findings achieved using the Airbus and ENEA e-nose setups and related procedures for flat CFRP coupon level samples in the ComBoNDT research project.

Airbus E-nose Setup

The received CFRP samples were measured in a random order using the device setup described above. Before starting a detailed multivariate data evaluation, the gathered raw data were normalized using a mean center function around zero followed by a scaling of the variables to their unit standard deviations. With this normalized data matrix, a PCA was done, and a PLS-DA was performed for classification purposes.

Figure 3.42 provides an insight into the PLS-DA results obtained when investigating intentionally contaminated CFRP coupons using the Airbus e-nose setup. On the 2D score plot of the latent variables (LV) LV1 and LV2, three different clusters of findings can be identified from the e-nose measurements. Dark blue rectangles form the cluster of the reference sample measurements (RE). In the second cluster, pink stars belong to samples that have been contaminated with de-icing fluid and the light blue triangles represent the combined repair samples (R-Combi). All other performed measurements form a third cluster on the 2D view. The respective abbreviations for the contamination scenarios were already introduced in Table 3.1.

By turning to a 3D analysis by adding the third latent variable (LV3), discrimination concerning the third cluster of measurements becomes feasible. When we investigated for even higher numbers of latent variables (LV4 and LV5), improved separation of the production scenario samples occurred (results not shown here).

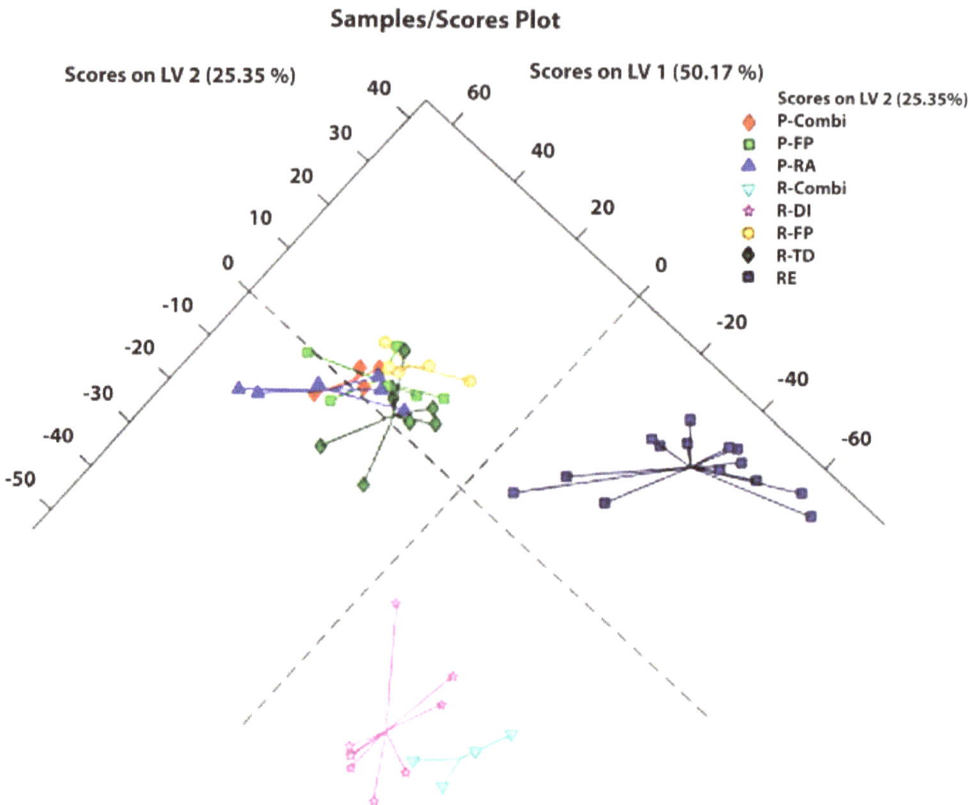

Fig. 3.42 LV1/2 score plot of the e-nose measurements for flat CFRP coupon level samples. Scores based on features from distinct contamination scenarios are color-coded; details are given in the text

After the multivariate data analysis of the obtained measurement data, we offer the following summing up statements for the ENDT investigations of CFRP coupons using the Airbus e-nose setup:

- Clean reference samples (RE) and samples contaminated with de-icing fluid (DI) cluster in their own quadrant in the LV1 versus LV2 score plot and can be easily and clearly identified.
- By evaluating LV3, thermally damaged samples (TD) and repair scenario fingerprint samples (R-FP) can be separated and identified.
- Spatial separation of release agent (RA) samples and production fingerprint (P-FP) samples is possible, but the degree of separation is rather low (LV4, LV5).
- The "clean sample or dirty sample" decision (that is, related to the "sample ready to bond or not ready bond" decision) can be made for all the investigated contaminations.

3.4.2.3 ENEA E-nose Setup

Before detailing the findings obtained with the ENEA e-nose for the ENDT inspection of CFRP coupon level samples, we report in Table 3.3 the number of panels and corresponding features (37) sampled in the frame of the production user case by means of two different sampling methods.

Concerning the fingerprint (FP), moisture (MO), and release agent (RA) contamination scenarios within the production use case, the results presented in Fig. 3.43 which were obtained with the 0 method for the sampling of analytes did not reveal the capability to distinguish between reference and contaminated samples, at least in the considered PCA subspace.

Based on these preliminary results, we decided not to go further in the investigations of CFRP specimens from the production use case but instead focused on coupon panels from the defined repair user case to test the performance of e-nose ver.1. In

Table 3.3 Number of sample measurements with the CFRP coupon sample-based production use case for both the applied e-nose sampling methods; more details are given in the text

	Production use case
0 method	25
PC method	44

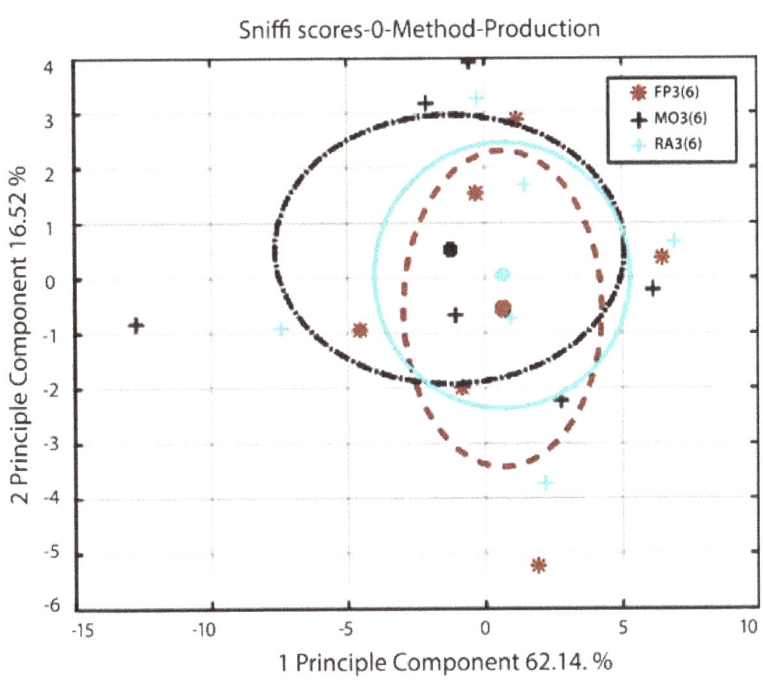

Fig. 3.43 E-nose findings when performing PCA on the 0 method sampled coupon level CFRP panels

Table 3.4 Number of sample measurements with the CFRP coupon sample based on the repair use case; more details are given in the text

	Repair use case
0 method	25
PC method	41

Table 3.4, we present the number of samples taken for the analysis of specimens from the repair user case.

Assuming a two-class classification problem (FP/Skydrol versus ALL), we performed a preliminary data analysis using PCA so as to highlight the capabilities of the e-nose ver.1 to discriminate fingerprint/Skydrol hydraulic oil contamination at each contamination level from any other contaminated and reference CFRP samples. In this way, both sampling methods showed a limited capability to discriminate FP/Skydrol contaminated samples from the others in the PCA subspace. Meanwhile, the PC sampling method enhanced the separation capabilities, obtaining a clear separation for FP level 3 contaminated samples, as shown in Fig. 3.44.

Encouraged by these results, we performed a feature selection step for the samples investigated and recorded with both methods. Based on the ranking feature algorithm ***relief-f***, ten relevant features were extracted for the PC method. Specifically, a subset

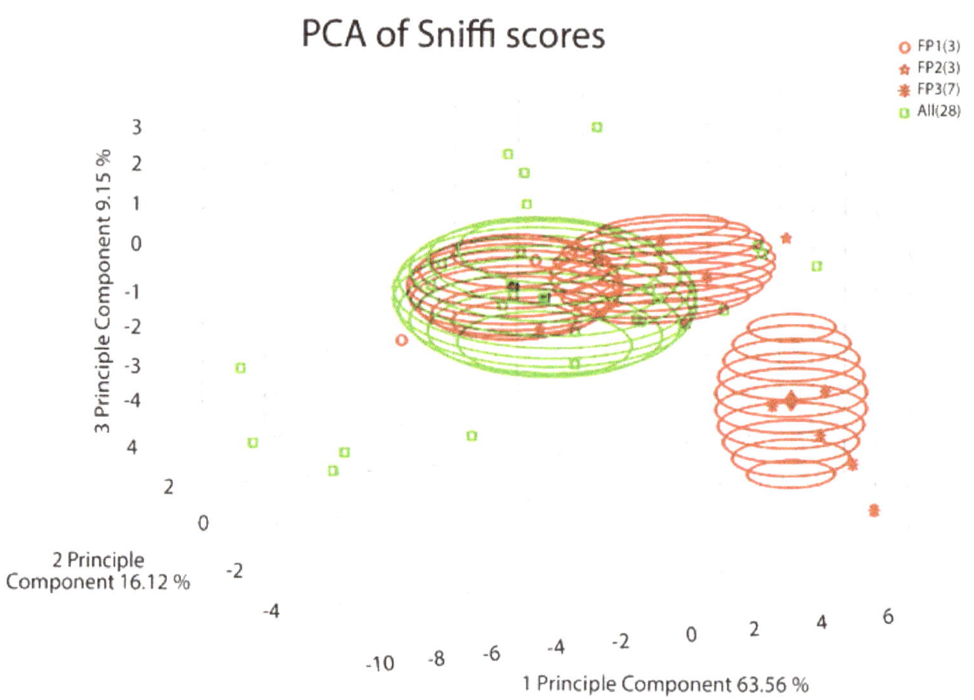

Fig. 3.44 E-nose findings when performing PCA on the PC method sampled flat CFRP panels

of steady-state and desorption rate features for MOX sensors was selected by the algorithm. By using the selected feature vector along with a logistic regression classifier, the ***receiver operator characteristic*** (ROC) curve was drawn (see Fig. 3.45). Its area under the curve (AUC) performance indicator was then computed, obtaining a value of 0.84. In this way, we have shown that a total correct classification (CC) rate of 78% can be achieved, at a false negative rate of 31% (Table 3.5).

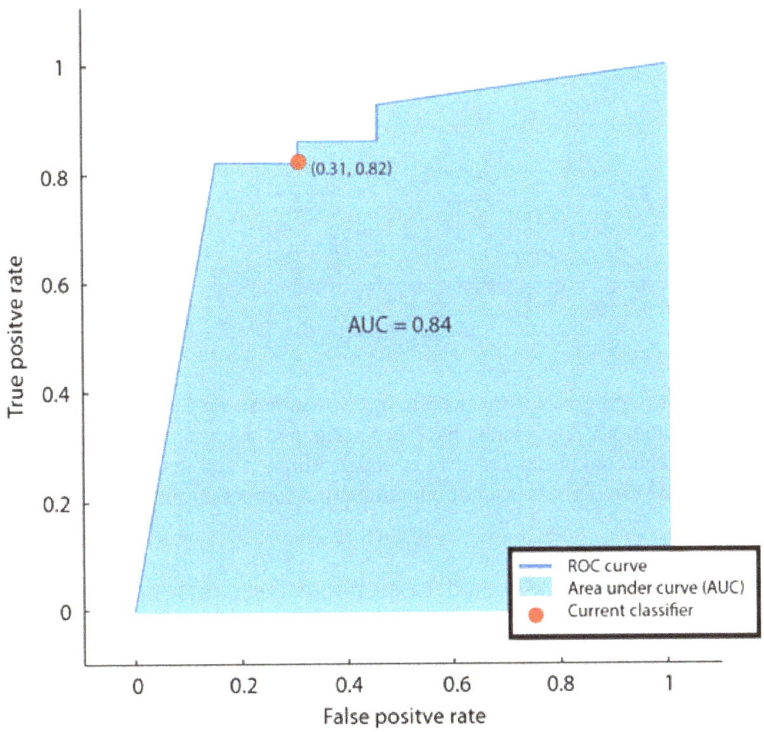

Fig. 3.45 AUC of the logistic regression classifier when evaluating the e-nose findings for the CFRP coupons

Table 3.5 Parameters of the classifier performance assessment for evaluating e-nose data obtained for CFRP coupons. One class representing FP-contaminated samples (at all contaminated levels)

Classification Parameter	Rate
Correct Classification	78%
False Negative	31%
False Positive	35%
AUC	0.84

Confusion Matrix			
True class	0	23	5
	1	4	9
		0	1
		Predicted class	

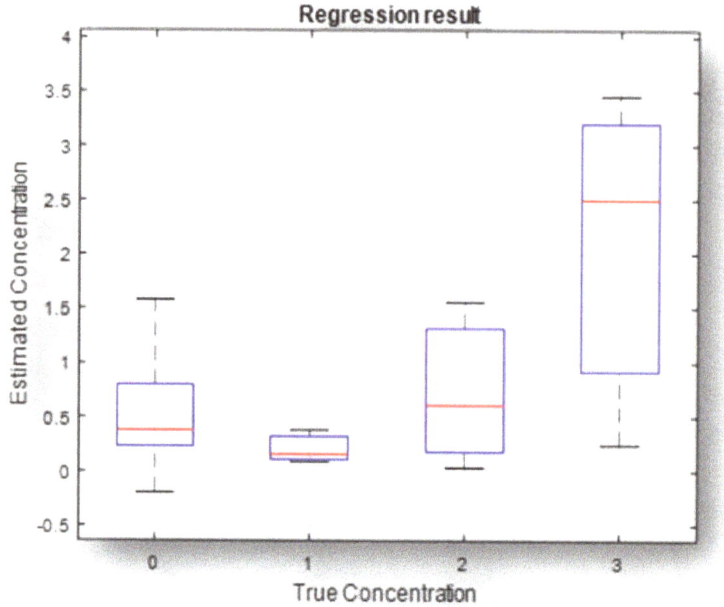

Fig. 3.46 Neural network regression results obtained with e-nose ver.1 data, considering all samples (note 0 level contamination). The results have been obtained using a leave-1-out cross-validation procedure. Note that the contamination level is underestimated and that the regressor is basically capable to discriminate the highest level of contamination samples from uncontaminated samples

Further analyses were conducted to compute the contamination quantification capabilities of the e-nose ver.1. In particular, a simple multivariate regression algorithm (FFNN) was used to estimate the level of contamination by a linear continuous encoding of the level score ranging from 0 to 3. Results shown in Fig. 3.46 show that the selected regression algorithm when applied to all e-nose ver.1 captured data is not capable to directly estimate the contamination level of a sample. Actually, it is only capable to discriminate the third (highest) FP contaminant level, and its mean absolute error (MAE) evaluation resulted in a 0.69 value.

In Fig. 3.47, we show the results obtained for the correctly identified contaminated samples. In this case, the regression method offers much better performance, highlighting a progressive behavior when evaluating contamination levels of increasingly contaminated samples. The MAE score, in this case, reaches a 0.5 value. This suggests that experimental conditions variability severely hampers the inherent discrimination and quantification capability for some of the samples. However if the conditions favor the correct measuring of the samples, then its identification as contaminated and the consequential contamination level quantification may be successful.

Samples recorded with the 0 method were screened with a discriminant analysis (DA) approach in order to select the most discriminative features. In particular, we restricted our analysis to the three features obtaining the best DA scores. Specifically, all the selected features were uptake derivatives of MOX sensor responses. Two of these were selected for the final classification task, namely the uptake derivatives

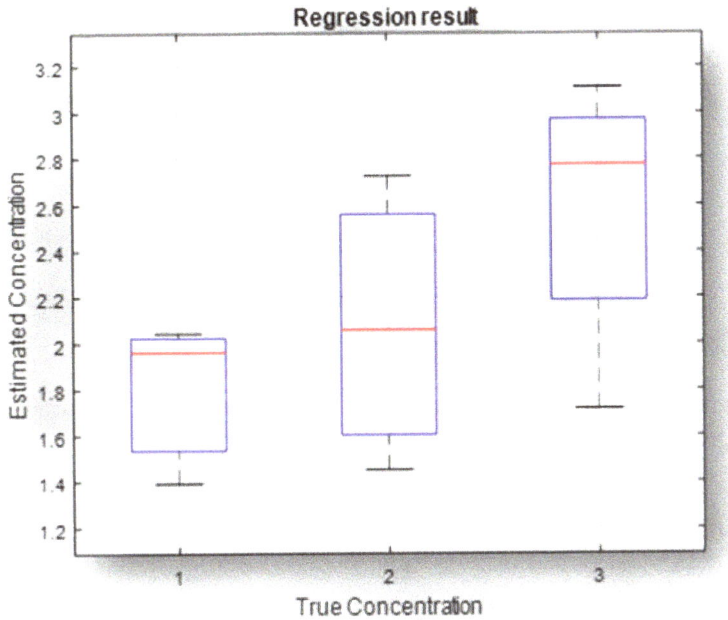

Fig. 3.47 Regression results obtained with the ENEA e-nose ver.1 when considering correctly identified contaminated samples

with respect to MOX1 and MOX5, which were used in the classification task to differentiate the FP/Skydrol contaminated coupons from the CFRP reference (RE) samples. This selection allowed a simple tree classifier to achieve a correct classification rate of 86.44% (false negatives FN = 0%, false positives FP = 27%, as listed in Table 3.6).

Table 3.6 Confusion matrix as obtained by a classification tree (CT) when discriminating uncontaminated reference samples from FP/Skydrol contaminated samples

Confusion Matrix			
True class	0	8	3
	1	0	11
		0	1
	Predicted class		

Fig. 3.48 E-nose with the desorber device imaged with the sampling head approaching CFRP parts with different sample geometries

3.4.2.4 E-nose Results for the Pilot Level Samples

Subsequently, we detail the findings achieved, respectively, with the Airbus and ENEA e-nose setups and related procedures for curved CFRP pilot level samples in the ComBoNDT research project.

Airbus E-nose Setup

The received CFRP samples were measured in a random order with the above-described device setup. The images in Fig. 3.48 show the desorber device sampling head taking analyte samples from CFRP parts with different sample geometries.

CFRP Pilot Level Samples from the Production User Case

The considered CFRP pilot level samples of the related production user case were curved specimens with convex and concave surfaces. All surfaces could be measured successfully with the e-nose desorber device. The contemplated contamination scenario was implemented by depositing on the surface a combination of release agent (RA), at two different concentration levels (RA1 and RA2), and fingerprints (FP) containing artificial sweat, whereby the contaminants were applied to both sides of the sample. We measured all the samples only once to avoid errors due to differential outgassing of contamination volatiles. With the normalized data matrix, a PLS-DA was performed for classification purposes.

In Fig. 3.49, we present the results of the respective PLS-DA analyses. The graph on the left side of this figure shows that the applied algorithm enables the contaminated samples to be distinguished from the reference samples with a clean surface. The right-hand graph shows the results, highlighting the additional distinguishing

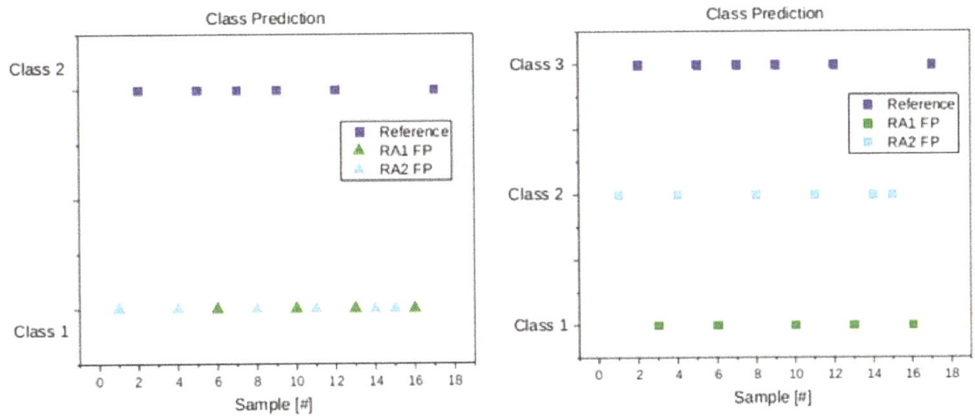

Fig. 3.49 Results of the class prediction of the PLS-DA of data obtained by e-nose measurements on the CFRP pilot samples from the production user case; left: prediction of two classes (clean reference class 2; contaminated class 1); right: prediction of three classes, including a discrimination between the two different contaminant concentrations

among the contaminated sample sets into two classes, providing the information on different concentrations of contaminants and thus facilitating a quantification.

CFRP Pilot Level Samples from the Repair User Case

The pilot level CFRP samples of the repair user case were prepared by introducing a scarfed surface. We successfully measured all surfaces with the desorber device. The applied scenario affecting the sample surfaces was a combination of a thermal impact (TD) and the deposition of contaminants based on de-icing fluid (DI).

We observed that all measured samples showed a very high sensor response when compared to the CFRP coupons. Specifically, even the scarfed reference samples reacted as if they were coupon samples with certain contaminations. This behavior could originate, on the one hand, from the thermal damage of the sample caused by the scarfing process or, on the other hand, from a solvent or chemical substance being used during or after the scarfing process, e.g., to clean the sample. Based on these observations, to experimentally verify these presumptions, some samples (reference, thermal impact (TD), and the "TD-DI" combination) were placed in a vacuum oven at 120 °C for one day. After this desorption process, the samples were measured again.

Figure 3.50 displays the results of the PLS-DA analysis. We infer that after applying the PLS-DA algorithm, we are able to distinguish among thermally damaged samples, samples with additional de-icing fluid contamination, and samples with a clean surface.

Performing a cross-validation step with the obtained data, the confusion table of cross-validation presented in Table 8 is returned.

Table 3.7 shows that the CFRP surface obtained after applying DI contamination can be clearly distinguished from the surfaces of the other samples and that thermally impacted samples are identified correctly in eight out of eleven cases. A small amount of incorrect classification exists between reference samples and thermally damaged samples. Considering this error, it should be noted that the reference samples are also scarfed samples and might bear an odor characteristic of thermal damage due to the mechanical scarfing treatment.

ENEA E-nose Setup

The pilot sample set was characterized by the advanced ENEA e-nose ver.2.0. The measurements were performed according to the standard method, i.e., the 0 method sampling procedure. Our classification task within the production user case aimed to differentiate the contaminated CFRP panels from the reference ones to achieve a GO/NOGO classification. Reference samples were labeled as belonging to the RE class, while the contaminated class, including all panels (RA1FP3 and RA2FP3) having undergone combined contamination, was tagged as the RAFP3 class. For this task, a PLS-DA approach was selected. Training was performed on a normalized

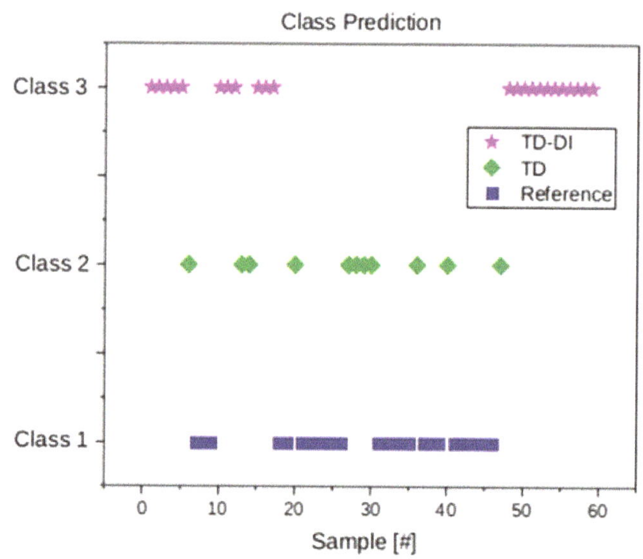

Fig. 3.50 Results of the class prediction of the PLS-DA of data obtained by e-nose measurements on the repair user case pilot level CFRP samples

Table 3.7 Confusion matrix obtained when cross-validating data from the e-nose measurements on CFRP pilot level specimens from the repair user case comprising specimens prepared following three different contamination scenarios (RE, TD, and TD-DI); more details are given in the text

Actual class	RE	TD	TD-DI
Predicted as RE	22	3	0
Predicted as TD	3	8	0
Predicted as TD-DI	0	0	23

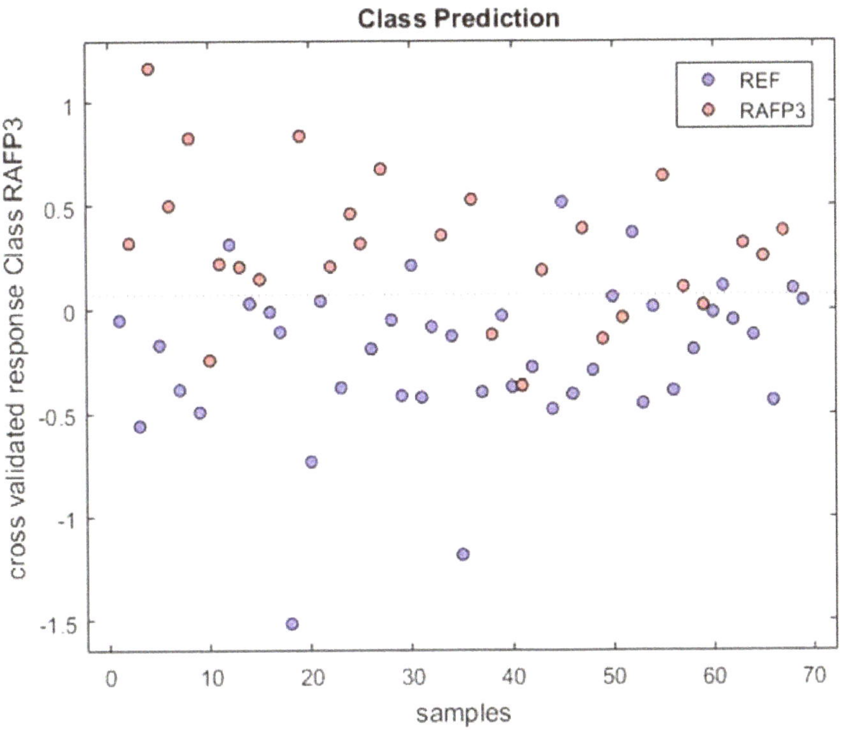

Fig. 3.51 Computed cross-validated response for the class comprising contaminated (RAFP3) CFRP pilot level specimens versus the reference samples (REF). The red dotted line indicates the contaminated class threshold crossing which will determine a sample classification as a contaminated one

data matrix of 37 features and 69 total panels. An internal leave-one-out cross-validation procedure[1] was used to select the optimal number of components, while external cross-validation with 20% of samples used as a test set at each validation run was used for the performance estimation. The results demonstrated a 0.83 correct classification rate level.

Table 3.7 shows the results of the classification of the 69 samples. The classification threshold was set by the PLS-DA method, see Fig. 3.51.

The trained classifier provides the wrong estimations for just 12 out of the 69 samples, as confirmed by the confusion matrix shown in Table 3.8.

Furthermore, the value of the area under the ROC curve (AUC) (presented in Fig. 3.52) approaches unity, confirming the classifier performance.

These results show how the advanced second version of the e-nose was finally implemented to achieve a good performance; also in the production user case, it was capable to identify the contaminated samples among the pilot samples.

---

[1]This procedure selects all samples for training purposes except one, set apart for evaluating the performance of the trained algorithm. The process is repeated for a number of times equal to the dimension of the originating sample set. In this case, the classification rate obtained by the test samples is used to forecast the performances of several models, each one using a different set of principal components.

Table 3.8 Confusion matrix obtained for the e-nose measurements on CFRP pilot level specimens of the production user case comprising specimens from two different contamination scenarios (RE and RAFP3); more details are given in the text

Actual class	RE	RAFP3
Predicted as reference	36	6
Predicted as contaminated	6	21

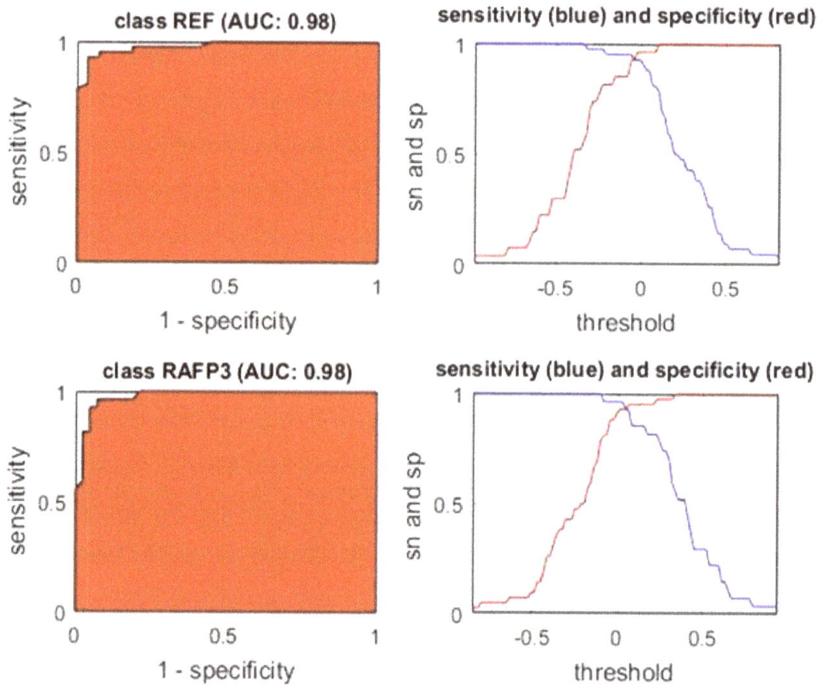

Fig. 3.52 ROC curves (left) and plots of sensitivity (blue) and specificity (red) values for the e-nose data as the class threshold is changed (right) for the classes comprising non-contaminated (upper) and contaminated (lower) CFRP pilot level specimens of the repair user case

3.4.3 *Final Remarks*

Regarding the Airbus e-nose setup, during the different test campaigns, the device proved its capability in terms of contamination detection on the provided samples and chosen application cases. During the coupon sample investigations, the system showed its ability to distinguish the different contaminations in both the production and repair user cases. At the pilot sample level, the desorber measuring head managed to take air-tight odor samples from concave and convex real-world geometries while dealing with new and previously unknown scarfed surfaces. During the three-day test event on technologically realistic parts at IFAM, the detection system substantiated its ability to also function in surroundings other than a gas testing facility and once again showed great sensitivity in detecting volatile chemical compounds.

Quantitative determination of moisture content using the e-nose system in combination with the desorber device is feasible for the desired material. The achieved error of prediction of less than 0.0763 wt% is very encouraging. Enlarging the dataset with additional measurements will further reduce the error of prediction to values of RMSECV (about 0.035 wt%). One big advantage of the e-nose in the combined method of operation is that information on the chemical surface condition (clean or contaminated) can be obtained at the same time with a very good value in terms of the moisture content.

If more precise measurements of the moisture content are required, the measuring setup of the e-nose system can be changed to a pure moisture measurement setup. Under this setup, the resolution of moisture detection will be increased, and the error of prediction can further be reduced.

At the coupon level, the first e-nose version by ENEA was not capable of dealing with the challenges of the production user case, although it was able to reach a 78% correct classification rate for FP detection in the repair user case. After the coupon level tests campaign, the ENEA e-nose was significantly improved by developing a new sampling head and filtering subsystems. The second version proved its capability to detect contaminated samples in both the production and repair user cases at the pilot and realistic parts challenge levels. In particular, the second version was able to achieve more than 80% accuracy in detecting RA-FP mixed contamination samples at the pilot level and obtained a perfect score during the realistic parts testing event in Bremen.

Based on the analysis of these results, e-nose technology appears very close to the maturity stage for the detection of surface contamination prior to the bonding phase. Of course, the e-nose methodology can only be effective whenever residual volatile compounds are present on the surface under analysis. The results of the ComBoNDT campaign are significant for the detection task and are encouraging with respect to the possibility to distinguish different contaminants, even when the concentration level was close to what we would expect to find in real-world scenarios. Quantification capabilities at this level, however, still appear to be difficult to achieve. Currently, the primary limitations of the techniques are due to measurement times, which can slow down the screening of large surfaces.

In summary, during the ComBoNDT project, two different approaches to contamination detection using e-nose were followed. On the one hand, a custom-made-of-the-shelf system (Airsense, Airbus E-Nose) was employed to reach a high TRL in a very short time. The system had to be retrofitted with additional sensors, and in order to achieve stable results on a shop floor environment without the influence of interfering odors from solvents and volatile compounds, a desorber device also had to be developed. On the other hand, the e-nose system from ENEA was developed almost from scratch and went through several optimization steps. Both approaches proved their eligibility and showed outstanding detection capabilities, as described above.

Both approaches offer advantages. The ENEA approach can be considered as an open-source system. The measurement setups (e.g., duration, flow, temperature) as well as sensor signal feature extraction and all other parameters are easy to set and change, whereas the custom-made device can only be controlled or influenced in a way that the manufacturer allows by providing an insight into the device's firmware. The Airbus approach, employing a working gas sensor system, saved a lot of time, which was used to work on a smart sample-taking device. With the integration of the e-nose and desorber device in one system, interfering influences from a shop floor environment can be ignored. The implementation of signal processing and multivariate data analysis could easily be achieved, tested, and adapted using the "open-source" ENEA system. Measurement control, data recording, feature extraction, and multivariate data analysis were successfully performed, displaying the results on a GUI designed and programmed by ENEA at the end of the project.

As already mentioned above, both approaches have their advantages, and the greatest advancements in the technology were achieved in the areas where research and experience could improve the system via further hardware and software developments. The only conclusion must be to combine these enhancements—smart sample taking and total system control—in an integrated common device to further advance sensor system performance.

3.5 Laser-Induced Breakdown Spectroscopy (LIBS)

In this section, we introduce laser-induced breakdown spectroscopy (LIBS) as a tool for surface quality assessment and detail how its performance was enhanced in the ComBoNDT project for the in-process monitoring of CFRP adherends.

3.5.1 Principle and Instrumentation

LIBS is a spectroscopic method for elemental analysis that is routinely used to determine the elemental compositions of solids, liquids, and gases. For surface technology applications, a high-power laser pulse is focused onto the sample, whereby a small amount of material (typically hundreds of ng to a few μg) evaporates and forms a micro-plasma above the surface [30]. In this way, surface species are excited, and due to the following relaxation process, element-specific radiation is emitted. The emitted light is subsequently separated by its wavelength and is detected using a (often high-resolution) spectrometer; then, the measured intensities of the optical emission are evaluated using dedicated software for qualitative and quantitative analysis. Quantification to assess the surface composition is possible through, e.g., peak ratios in combination with a suitable calibration of the method [6]. For the detection of surface contaminants with LIBS, the contaminant needs to contain (atoms of) the elements that are detectable by means of LIBS and that are not (or only in a very defined amount) part of the clean surface. Detection limits can be in the range of ppm but depend on the material, contaminant, and experimental setup, and they need to be

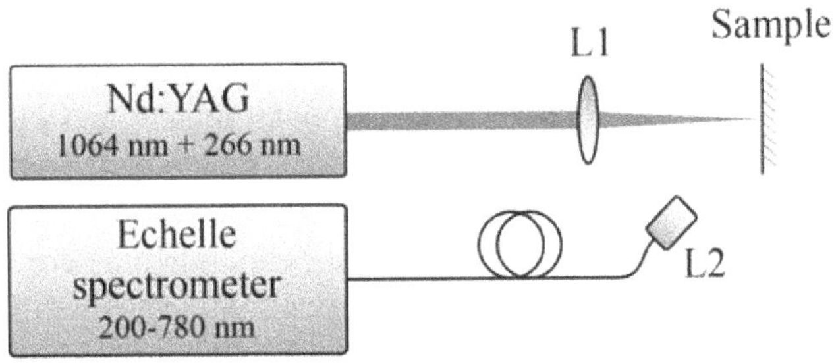

Fig. 3.53 Diagram showing the measurement principle in LIBS

determined separately for each combination of materials and species to be detected [1].

In comparison to conventional surface analysis methods, LIBS requires relatively short measurement times in the order of a few seconds. Measurements can be performed under atmospheric conditions without the need for sample preparation. In addition, LIBS can be adapted to inline applications [31].

Regarding the findings reported here, the LIBS measurements were performed with the LIPAN 4000 system from LLA Instruments GmbH, Berlin, Germany. Laser pulses from a Q-switched Nd:YAG laser with a 1064 nm laser wavelength, a pulse width of 6 ns, and a repetition rate of 20 Hz were used for excitation. In addition, tests were also performed with another Nd:YAG laser system that emits laser light with a wavelength of 266 nm at a rate of 10 Hz with a 7 ns pulse width. Figure 3.53 gives an overview of the LIBS setup.

Spectra were obtained and analyzed using an Echelle spectrometer (LLA Instruments GmbH, Berlin, Germany), which allows simultaneous detection of wavelengths from 200 to 780 nm with a spectral resolution of a few pm. The spectrometer was combined with an ICCD camera (1024 * 1024 pixels). Measurements were typically done at laser energies ranging from 95 to 180 mJ, depending on the laser light wavelength. The control of the spectrum recording and the evaluation of the spectra were performed using the ESAWIN software developed by LLA Instruments GmbH. The software has a large database that allows the automated identification of many relevant atomic emission lines in the optical spectral region.

3.5.2 LIBS Results

In the following, we report the LIBS advancements and findings obtained in the ComBoNDT research project, in which the consortium partners at Fraunhofer IFAM performed the in-process monitoring of CFRP adherends of different shapes that are relevant for distinct technologically relevant user cases and had undergone the intentional application of various contamination scenarios.

3.5.2.1 LIBS Results on Coupon Level Samples

We investigated three different contaminants and clean reference samples on flat 10 cm × 10 cm coupon level samples for a production user case comprising distinct contamination scenarios. In the following, the results for CFRP surface states obtained by applying the different contaminants are presented in comparison with the clean reference samples. Coupon samples with different amounts of moisture are not discussed due to the inability of LIBS to detect this contaminant (i.e., water in CFRP).

Release agent (RA) contamination scenario:

The silicon-containing release agent Frekote® 700NC was used in the RA scenario. For the contaminated samples, silicon emission lines were detected with LIBS. In Fig. 3.54, the LIBS spectra of a clean CFRP sample and an RA-contaminated sample are shown together. The relevant atomic emission lines (carbon and silicon) used for sample evaluation are marked.

Three correspondingly prepared RA-contaminated CFRP coupon samples for each degree of contamination (named 1, 2, or 3) and three clean reference samples were investigated with a 1064 nm single laser pulse energy of 180 (±10) mJ and 60 LIBS measurements on each sample specimen. The mean values and 95% confidence intervals were calculated from the three samples, respectively, and the resulting relative LIBS intensities (given as Si/C intensity ratios) are correlated to the respective XPS results in Fig. 3.55. The lowest level of contamination, namely I-P-RA-1, is

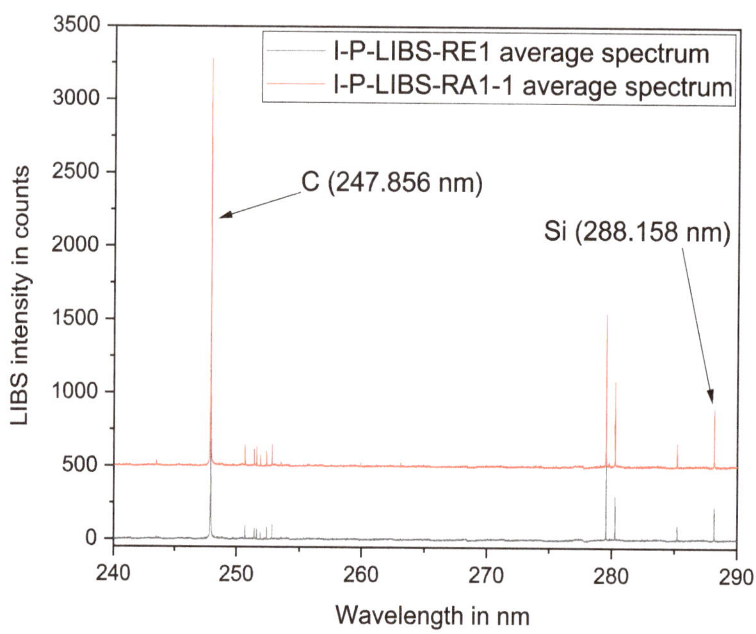

Fig. 3.54 LIBS spectrum obtained from a clean (black line) and an RA-contaminated CFRP coupon sample (red line) with the indication of relevant emission lines for carbon and silicon species

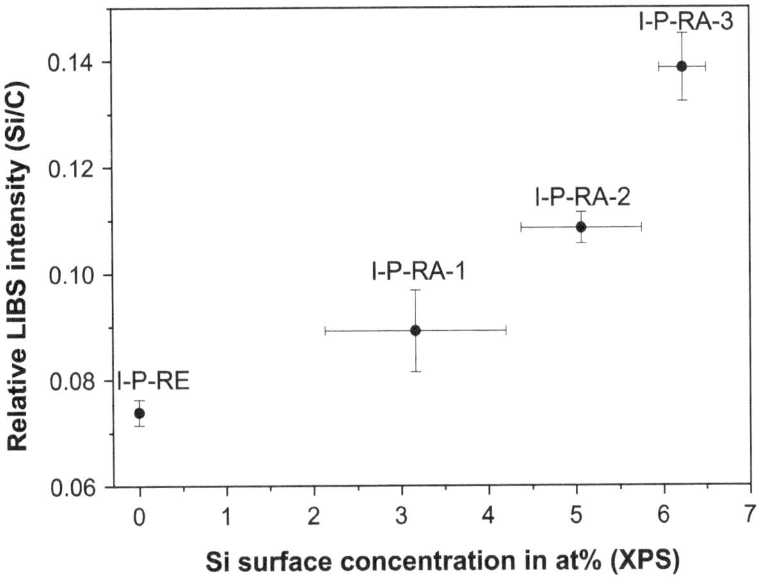

Fig. 3.55 Correlation between the 1064 nm LIBS relative intensities (Si/C) and silicon surface concentrations (in at.%) measured with XPS

clearly detectable compared to the clean reference sample and can additionally be differentiated from the two subsequent contamination levels (I-P-RA-2 and I-P-RA-3).

The same set of CFRP samples was investigated using the same LIBS setup coupled with a different laser using an excitation wavelength of 266 nm. The single laser pulse energy was reduced to 95 (±10) mJ. The results are shown in Fig. 3.56. An increased Si/C ratio for the contaminated samples is observed and both the reference sample and the three contamination levels can be clearly detected and differentiated.

The detection limit using the 266 nm laser for excitation is expected to be even lower than the contaminant surface concentration on the tested composite sample, with approximately 3 at.% (XPS). To increase the silicon concentrations on the CFRP surface, a differentiation of the level of contamination is better using 1064 nm for the plasma excitation. We explain this phenomenon by achieving different information depths depending on the laser excitation: Using the 1064 nm light, the information depth is comparably high. The CFRP adherend surface contributes to a great extent to the plasma emissions (we observe a large carbon signal and a comparably low silicon signal intensity). When using the 266 nm laser for plasma generation, the information depth is lower and the surface-near regions (e.g., deposited contaminants) contribute significantly more strongly to the measured signal. This results in an increased Si/C ratio and enables very surface-sensitive measurements.

Fingerprint (P-FP) contamination scenario:
LIBS investigations were performed on the CFRP adherends with an analogous procedure, as applied in the case of the samples from the RA scenario that had been locally contaminated with an artificial fingerprint (FP) solution (see Chap. 2).

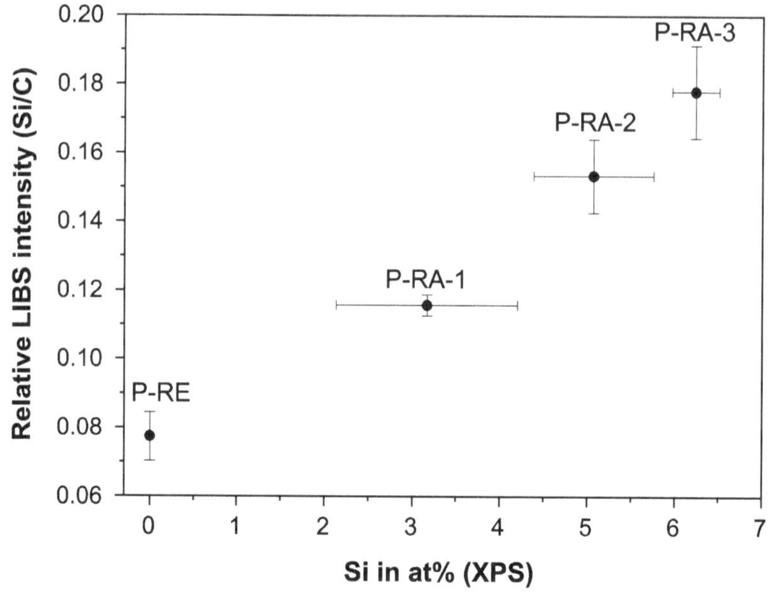

Fig. 3.56 Correlation between the 266 nm LIBS relative intensities (Si/C) and silicon surface concentrations measured with XPS

Using mean values from an area of 3.6 cm × 2 cm (1600 LIBS shots), the lowest level of contamination (named I-P-FP-1) is clearly detectable compared to the clean reference sample with both laser excitation wavelengths (1064 nm and 266 nm), see Fig. 3.57 for 1064 nm and Fig. 3.58 for 266 nm. Differentiation of the different contamination levels is, to some extent, possible with a laser wavelength of 1064 nm.

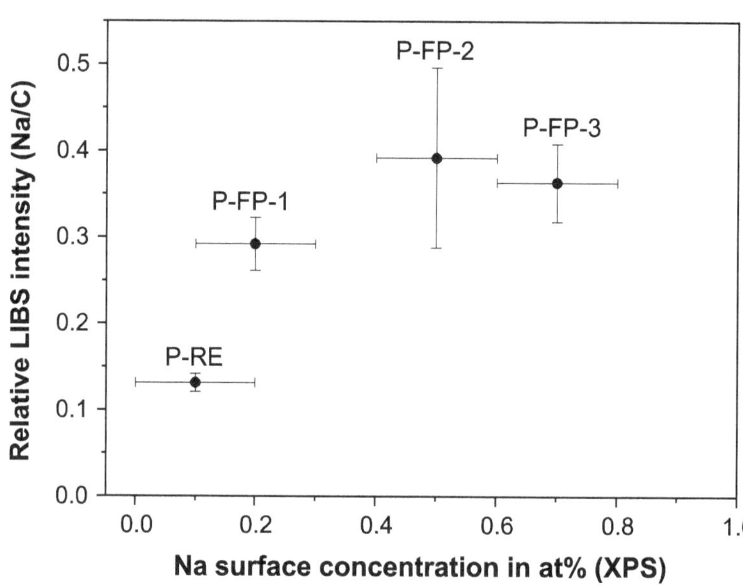

Fig. 3.57 Correlation of the 1064 nm LIBS relative intensities (Na/C) with sodium surface concentrations measured with XPS

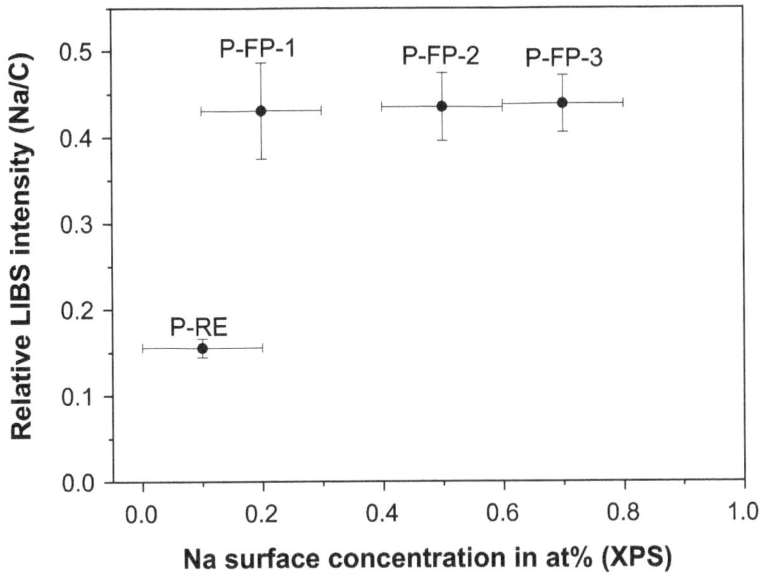

Fig. 3.58 Correlation of the 266 nm LIBS relative intensities (Na/C) with sodium surface concentrations measured with XPS

Just as for the RA scenario, for measurements using a 266 nm laser for plasma excitation, the information depth is comparably lower and the surface-near regions (e.g., contaminants) contribute significantly more strongly to the measured signal. This results in a clearer detection of contaminants on the I-P-FP-1 samples and a more significant discrimination from the reference CFRP surface state, but it does not allow for a differentiation of the three contamination levels (FP-1 to FP-3). The chosen evaluation method calculates the mean values from the areas with (fingerprinted region) and without (surrounding areas) contamination.

In this case, as in any case of punctual contamination, we suggest improving the detection result for contaminated regions by evaluating every single LIBS measuring spot and plotting the result in a space-resolved 2D diagram (hereafter named a map). Clean and contaminated areas on the sample can, thus, be identified, and the risk of missing small spots of contaminants (due to averaging comprising spots from surrounding and not contaminated regions) is reduced. Half of an artificial fingerprint and part of the clean surrounding areas were measured and evaluated, and the results are shown in Fig. 3.59 (for an excitation wavelength of 1064 nm) and Fig. 3.60 (266 nm), respectively. Green areas indicate regions with Na/C signal intensity ratios as found on a clean surface, while orange and red color-coded (with darker colors referring to a higher Na/C ratio) areas indicate an increased Na/C ratio and, thus, FP-contaminated areas.

Regarding the production user case, similar to the contamination scenarios within the repair user case, we investigated three different contaminants and clean reference samples on flat 10 cm × 10 cm CFRP coupon level samples. In the following, the results for the surface states based on the different contaminants are presented in a comparison with the findings for the clean reference samples. Coupon samples with

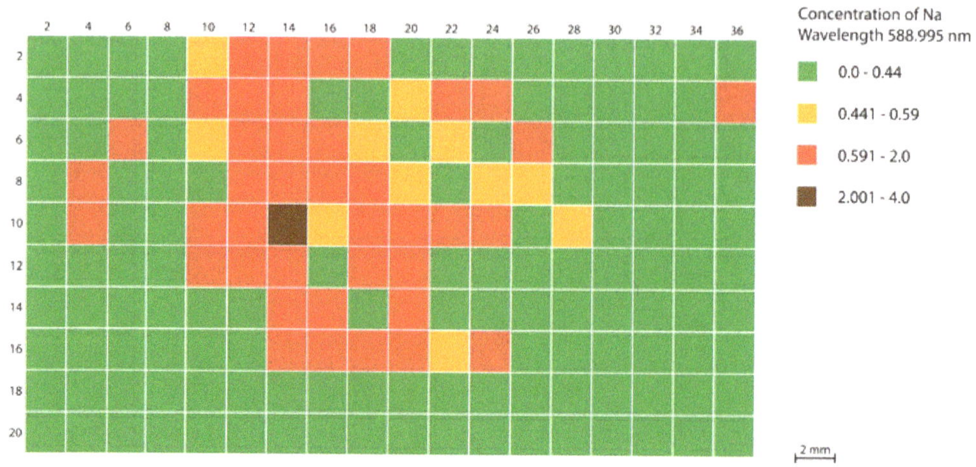

Fig. 3.59 LIBS map showing a space-resolved 2D diagram of the 1064 nm relative signal intensities (Na/C)

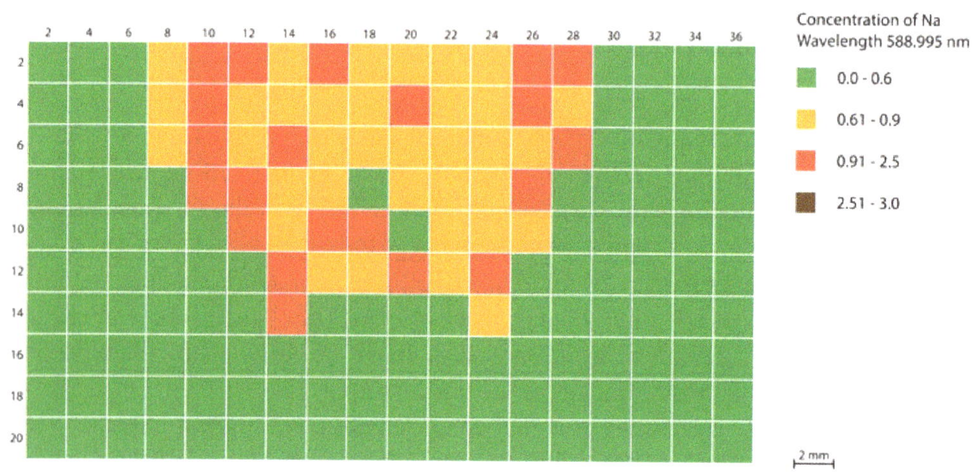

Fig. 3.60 LIBS map showing a space-resolved 2D diagram of the 266 nm relative signal intensities (Na/C)

different degrees of thermal impact and resulting degradation are not discussed due to the lack of a contamination-specific tracer element, which is essential for LIBS to detect different surface states.

De-icing fluid (DI) contamination scenario:

The de-icing fluid (DI) applied in this contamination scenario contained potassium as a tracer element, which enabled the contaminant detection and quantification with LIBS. Measurements with a 1064 nm laser wavelength were performed with an approximately 180 mJ single pulse energy and 60 measurements in an area of 4.5 × 9 cm. The LIBS intensities (K/C) correlate well with potassium concentrations measured with XPS, as shown in Fig. 3.61. We achieved a detection and differentiation of different levels of contamination with this set of settings. Using 266 nm as the excitation wavelength (approximately 95 mJ single pulse energy and 64 LIBS measurements in an area of 4 cm × 4 cm), DI contamination was also successfully detected; see Fig. 3.62. Differentiation of the higher contamination levels DI-2 and

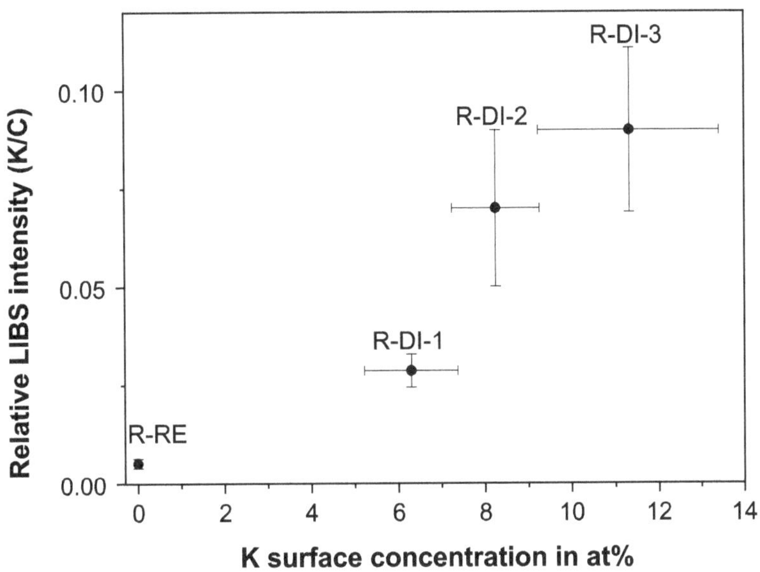

Fig. 3.61 Correlation between the 1064 nm LIBS relative intensities (K/C) and potassium surface concentrations measured with XPS

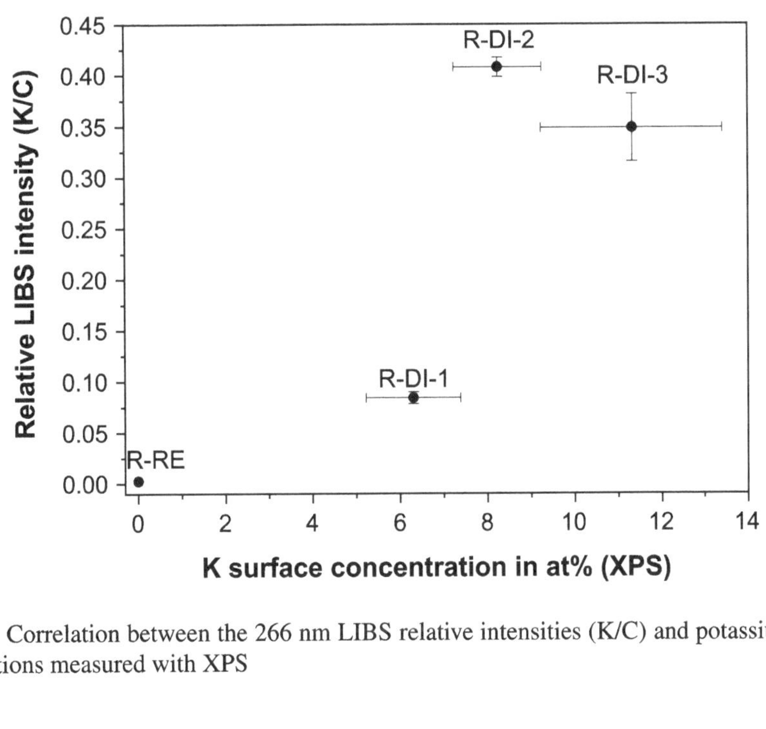

Fig. 3.62 Correlation between the 266 nm LIBS relative intensities (K/C) and potassium surface concentrations measured with XPS

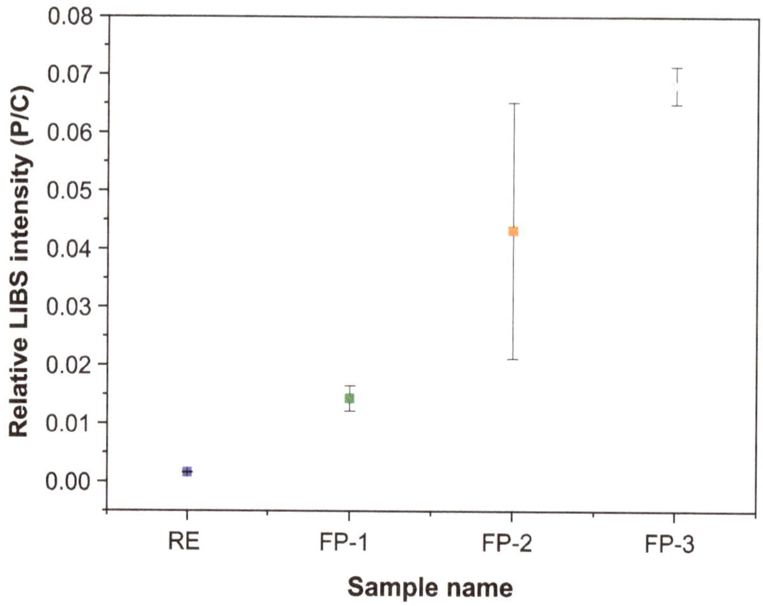

Fig. 3.63 1064 nm LIBS relative intensities (P/C) for distinct degrees of CFRP surface contamination with a phosphorous-containing hydraulic oil

DI-3 was not achieved. The comparatively large standard deviations of the XPS results for the surface concentrations (potassium) on DI contaminated samples indicate that the DI is non-uniformly distributed on the CFRP surfaces. Hence, depending on the area investigated with LIBS, different intensity ratios might be the result.

Fingerprint (P-FP) contamination scenario:

For this scenario, the same LIBS settings as elaborated for the DI detection were used. In this case, the fingerprints comprised a phosphorous-containing hydraulic oil. FP detection was successfully performed with both the 1064 and 266 nm plasma excitation wavelengths; see Fig. 3.63 (1064 nm) and Fig. 3.64 (266 nm). The three different contamination levels could be distinguished. However, the confidence interval for sample FP-2 was quite large when measured with the 1064 nm laser. A correlation with the XPS results was not achieved in this case. Comparing both measurements, we infer that the 266 nm laser excitation wavelength is again more surface sensitive, and thus gives larger P/C ratios (compared to the release agent (RA) scenario in the production user case).

Thermal degradation (TD) scenario:

Since thermally degraded surfaces do not contain a chemical element that exclusively and specifically marks the treated samples, a clear detection of CFRP samples that had undergone a TD impact was not achieved with LIBS in the current setup. Using a multivariate approach, there was no clear differentiation between the sample sets, and the prediction of unknown sample states failed. Improvements may be reached

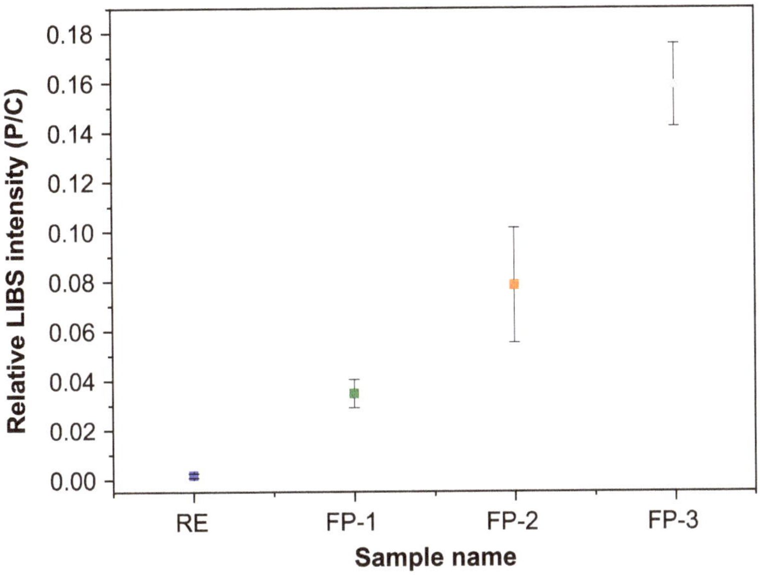

Fig. 3.64 266 nm LIBS relative intensities (P/C) for distinct degrees of CFRP surface contamination with a phosphorous-containing hydraulic oil

by using a setup dedicated to oxygen detection, which was not the focus of our current LIBS setup.

Summary of the LIBS results for coupon level samples:

Table 3.9 summarizes the LIBS results for the coupon level samples. For the production user case, the detection of the contaminant was possible for the RA and FP scenarios. A clear differentiation of the three contamination levels was possible for

Table 3.9 Categorizing summary of the LIBS results for the coupon level CFRP samples of distinct contamination scenarios in the production and repair user cases

		Detection of contaminants	Differentiation of cont. levels
production	MO	✖	✖
	RA	✚	✚
	FP	✚	▬
repair	DI	✚	✚
	FP	✚	✚
	TD	✖	✖

✚ Clear

▬ Partially/Improvement nescessary

✖ No

the RA scenario. Concerning the repair scenarios, the detection as well as the differentiation of the levels of contamination were successfully demonstrated for the FP and DI scenarios.

3.5.2.2 LIBS Results on Pilot Level Samples

The LIBS measurements on pilot level CFRP specimens from the production user case were performed on curved CFRP surfaces after peel ply removal. The laser wavelength of 1064 nm was used with a 180 mJ laser pulse energy. The inspected area was 4 mm × 4 mm with 1600 LIBS single shots within this area.

Combined contamination of RA and FP:

Two contaminants were applied to the samples: a release agent (RA) and a fingerprint with artificial sweat (FP). On the coupon level samples, these two contaminations could be detected individually using LIBS. For the combined FP and RA contamination, Fig. 3.65 shows the evaluation of the silicon emission line that is specific for the silicon-containing RA. The Si/C ratio increases with increasing RA concentration on the CFRP surface. We clearly detected the low RA concentration (compared to a clean reference sample) and also distinguished it from the higher RA concentration. Figure 3.66 shows the good correlation of the LIBS results (giving the Si/C signal intensity ratio) with the XPS results (giving the silicon surface concentration).

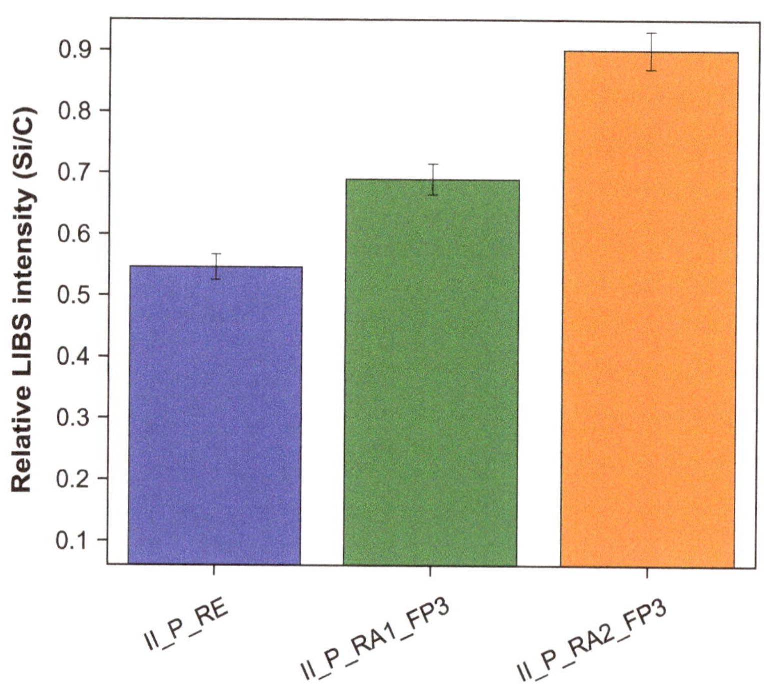

Fig. 3.65 Evaluation of the Si/C LIBS signal for the pilot level CFRP samples with combined release agent (RA) and fingerprint (FP) contaminations

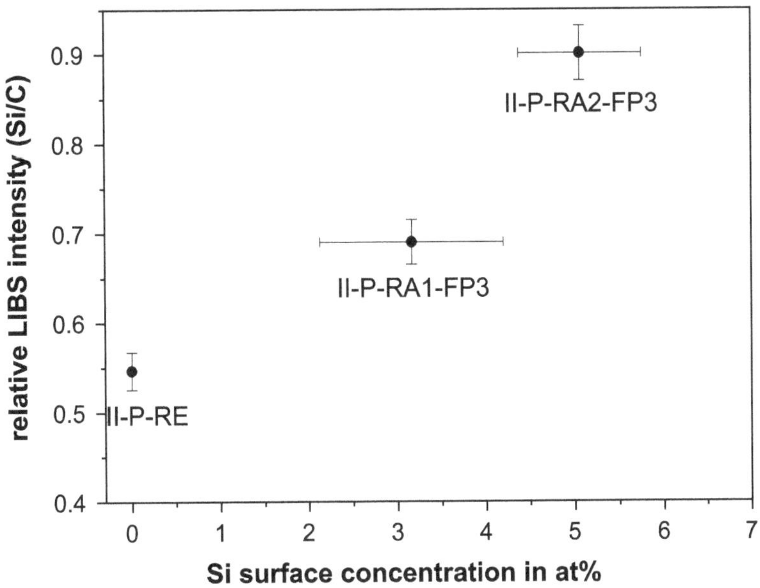

Fig. 3.66 Evaluation of the Si/C LIBS signal for the pilot level CFRP samples correlated to the XPS results

Based on the visual inspection, we inferred that the FP contamination did not spread over the entire sample surface (as it is the case for the release agent) but is located somewhere in the middle of the samples. In order to detect this contamination, an area of 4 cm × 2.4 cm containing 240 LIBS measurement spots was inspected for each sample. We did not obtain a clear FP detection (as for the coupon level samples). In Fig. 3.67, the Na/C LIBS mapping result for the reference sample is shown, and in Fig. 3.68, we present the corresponding mapping for the contaminated sample II-P-RA1-FP3. There is an overall increased sodium concentration (characteristic of the artificial sweat in the fingerprinted surface regions) for the contaminated sample (in the more orange and red areas), but the detection of a clear fingerprint shape was not possible.

Within the repair user case based on pilot level samples, the LIBS measurements were performed on the CFRP surfaces after scarfing. The laser wavelength of 1064 nm was used with a 180 mJ laser pulse energy. The inspected area was 0.4 cm × 0.4 cm, with 1600 LIBS single shots within this area.

Combined DI and TD contamination scenario:

The scarfed CFRP specimens underwent two contaminations and showed the surface modifications resulting from contact with de-icing (DI) fluid and exposure to thermal impact (TD). On coupon level samples, LIBS was a suitable method for the detection of DI, but not for TD. For the pilot samples, we, therefore, focused our measurements on the detection of potassium-containing DI.

The LIBS results for K/C are shown for all measured samples in Fig. 3.69. We obtained a clear detection of DI in comparison to the reference CFRP sample. Also,

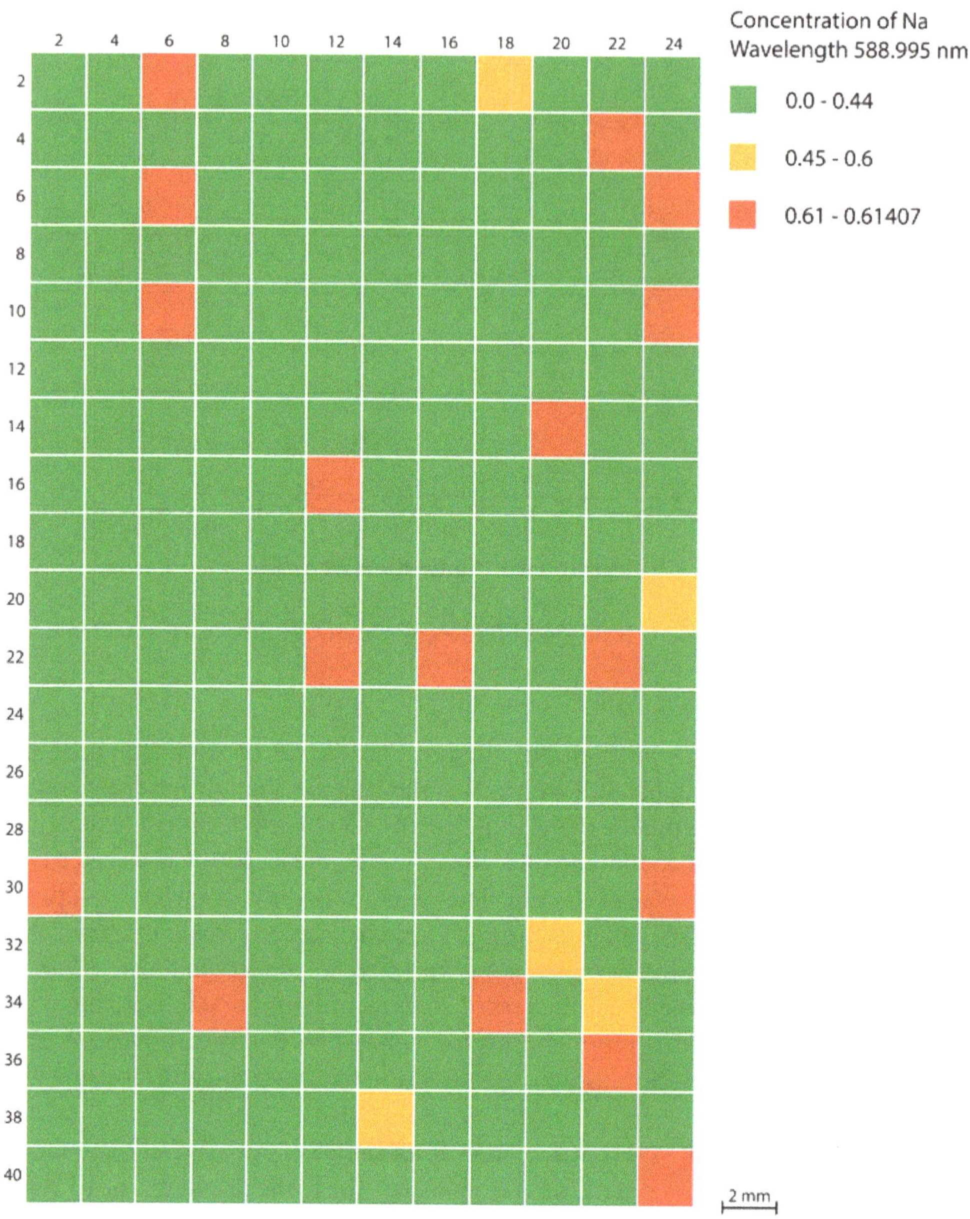

Fig. 3.67 LIBS map for the Na/C signal intensity ratio on a pilot level CFRP reference sample

the two levels of combined contamination were differentiated. The LIBS results correlated well with the XPS results for potassium concentration on the surface, as shown in Fig. 3.70.

Summary of results for the pilot level CFRP specimens:

Concerning the measurement of the pilot samples for the respective production user case, LIBS enabled detection and even a differentiation of the different levels of release agent. The fingerprint composed of artificial sweat was detected, but the

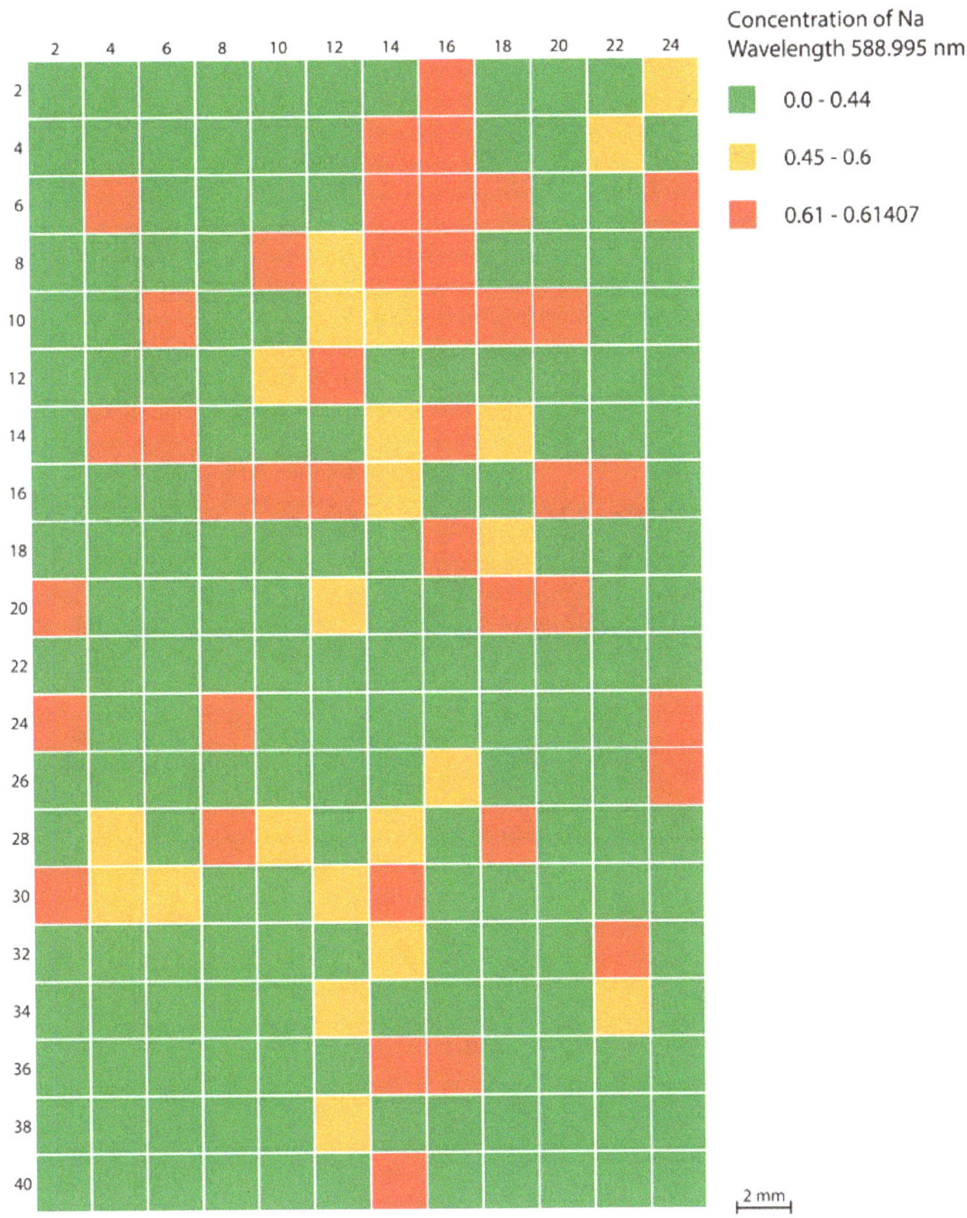

Fig. 3.68 LIBS map for the Na/C signal on a pilot level CFRP sample II-P-RA1-FP3

respective ENDT procedure might even be improved. The detection was not as clear as for the coupon level samples. We suggest that one reason for this outcome might be the rough surface structure (the peel ply was removed without subsequent abrasion treatment) of the examined samples.

Within the repair user case comprising pilot level CFRP specimens, LIB investigations were performed on scarfed samples. The geometry was easily accessed and handled by the LIBS system. The deposited de-icing fluid was neither detected nor

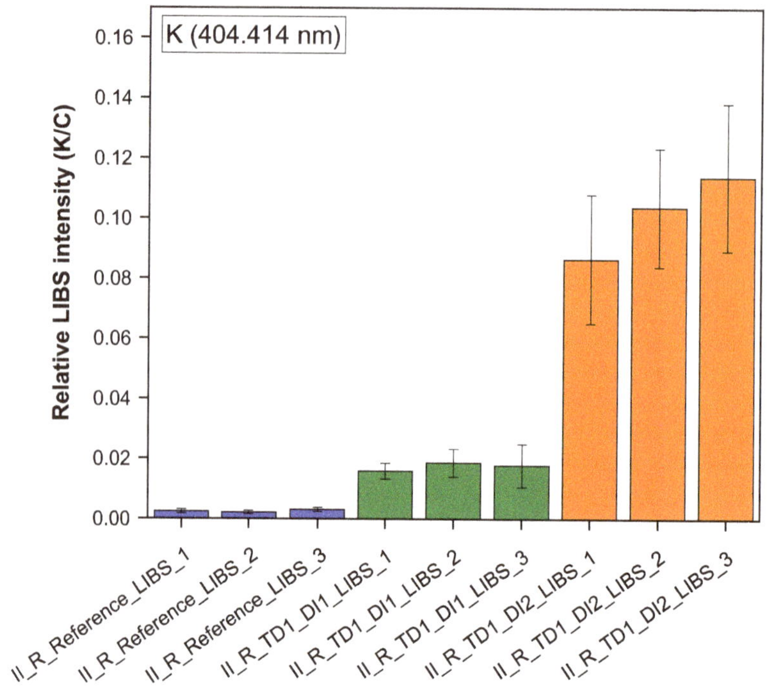

Fig. 3.69 Evaluation of the K/C LIBS signal for the pilot level CFRP samples in the repair user case

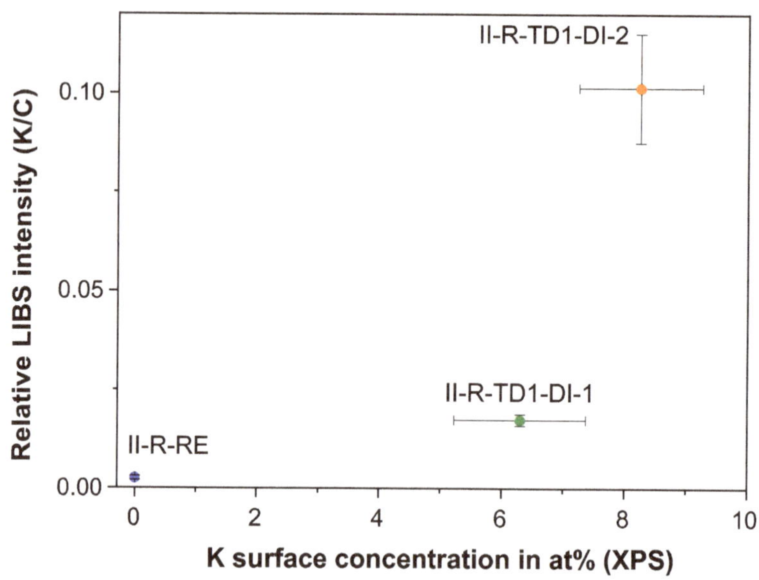

Fig. 3.70 Correlation of the K/C LIBS signal for the pilot samples in the repair user case with the XPS results for the potassium surface concentration

quantified. Thermal degradation was not detected (as when inspecting coupon level samples) because no element contrast was present in this scenario.

3.5.3 Performance in Inline Surface Quality Assurance

In the following, we summarize the findings revealing the performance of laser-induced breakdown spectroscopy (LIBS) as an ENDT technique for the surface quality assessment of CFRP composite specimens in distinct user cases. These comprised first a production and a repair user case, both based on coupon level samples, then pilot level CFRP samples with a more complex shape applied in a production and a repair user case, and finally user cases relying on technologically realistic CFRP parts.

In a first and trendsetting approach within the ComBoNDT research project, we applied LIBS to map CFRP test coupons prepared by intentionally depositing distinct contaminants within several contamination scenarios. Some of these scenarios comprised contaminants containing chemical components that feature elements that are not found on clean CFRP substrates and thus, may be used as tracer species for LIBS-based surface assessment procedures. We advanced these spectroscopic procedures to achieve technologically relevant detection limits and to accomplish fast data acquisition and evaluation that permit surface investigations by applying a high area density of LIBS spots. In this way, even locally applied contaminants relevant for aeronautical environments were detected.

3.6 Fourier-Transform Infrared Spectroscopy (FTIR)

In this section, we introduce Fourier-transform infrared spectroscopy (FTIR) as a tool for surface quality assessment and detail how its performance was enhanced in the ComBoNDT project for the in-process monitoring of CFRP adherends.

3.6.1 Principle and Instrumentation

In FTIR spectroscopy, molecules close to the surface are excited by infrared light. This excitement results in a partial absorption of the infrared radiation by the molecules, as depicted in Fig. 3.71. The absorbed infrared radiation is missing in the received infrared spectra, providing information about the chemical composition of the surface. The measured FTIR spectra have complex structures that can be interpreted by a partial least squares (PLS) algorithm that correlates the material properties given in the calibration with significant features in the FTIR spectrum recorded on a substrate. In the second step, the calibration can be used to predict

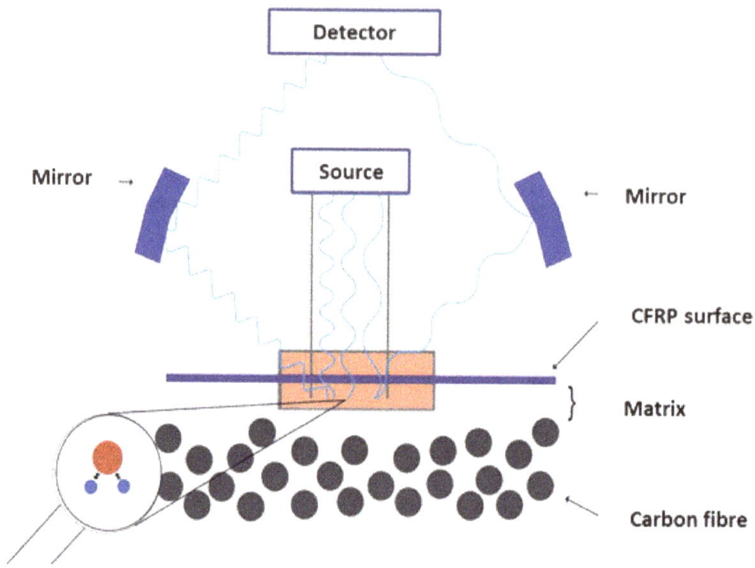

Fig. 3.71 Working principle of diffuse reflection FTIR spectroscopy

the material properties from an FTIR spectrum made on a substrate with unknown properties. The advantage of this evaluation algorithm is that the FTIR spectra can be interpreted quantitatively.

The measurements were conducted with an Exoscan 4100 portable handheld FTIR spectroscope. The test parameter was set to 64 scans at a resolution of 8 cm^{-1}. All measurements were performed in diffuse reflection.

3.6.2 FTIR Results

The following subchapter summarizes the results obtained for the three sample levels. Since no sensitivity toward release agent or fingerprint contamination was found during this experiment, a successful evaluation of these samples was not possible; therefore, the results are not described in this chapter.

3.6.2.1 FTIR Results on Coupon Level Samples

The coupon level samples were used to create a PLS algorithm model. For this purpose, FTIR spectra recorded on surfaces with different contaminations were used to build a calibration model. The presented results show the validations of these models with additionally recorded FTIR spectra. Figure 3.72a depicts the validation of a thermally degraded CFRP surface. The results show a satisfactory prediction

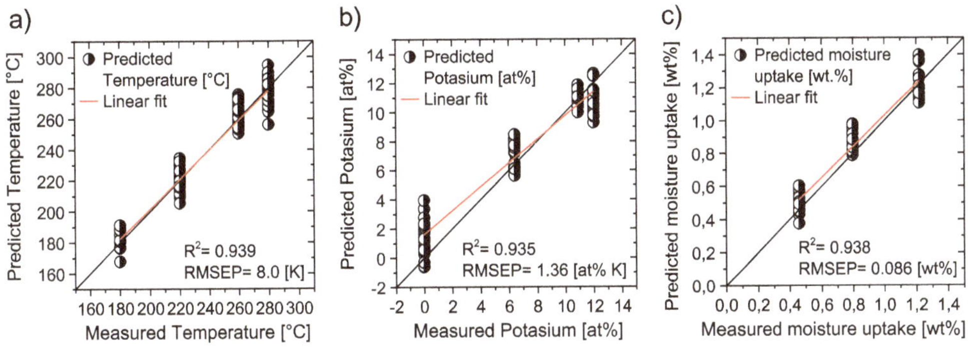

Fig. 3.72 **a** Prediction of the temperature during thermal degradation based on the PLS algorithm from the FTIR spectra recorded on a CFRP surface; **b** prediction of the amount of potassium on the surface based on the PLS algorithm from the FTIR spectra recorded on a CFRP surface contaminated with de-icing fluid; **c** prediction of the moisture uptake on the surface based on the PLS algorithm from the FTIR spectra recorded on a CFRP surface after moisture conditioning

accuracy with a root mean square error of prediction (RMSEP) of about 8 °K and a coefficient of determination (R^2) of about 0.94.

For the prediction of the amount of residual de-icing fluid, the amount of potassium (K) from XPS measurements was used as a calibration reference. The prediction quality is quite satisfactory, with a root mean square error of prediction (RMSEP) of 1.4 at.% K. The uptake of moisture is a well-known property of CFRP material. In Fig. 3.72c, the prediction of moisture uptake correlates with an accuracy of 0.086 wt% (RMSEP) and an R^2 of 0.938. This prediction accuracy is in good accordance with similar investigations. In [32], a method was described that uses additional conditioning of the surface with an intermediate heating step. This investigation shows at an early stage that it is possible in principle to also measure non-homogenously saturated samples.

3.6.2.2 FTIR Results on Pilot Level Samples

The measurements for a thermally degraded scarfed surface are depicted in Fig. 3.73a. Both the reference and the thermally damaged sample showed a high scattering of the predicted temperature. It is assumed that this phenomenon was caused by the sample preparation. The samples on the coupon level were scarfed in a gentle manner to remove only the resin above the top fiber layer. In contrast, the pilot level samples were scarfed with an elevated pressure to reach an appropriate scarfing depth for the repair patches. Such friction leads to high temperatures on the surface. The fact that the information depth of FTIR spectroscopy is limited to near-surface areas can explain the predicted temperatures and their scattering. Further development of the method could include a defined surface preparation after scarfing in order to prevent a high-temperature impact.

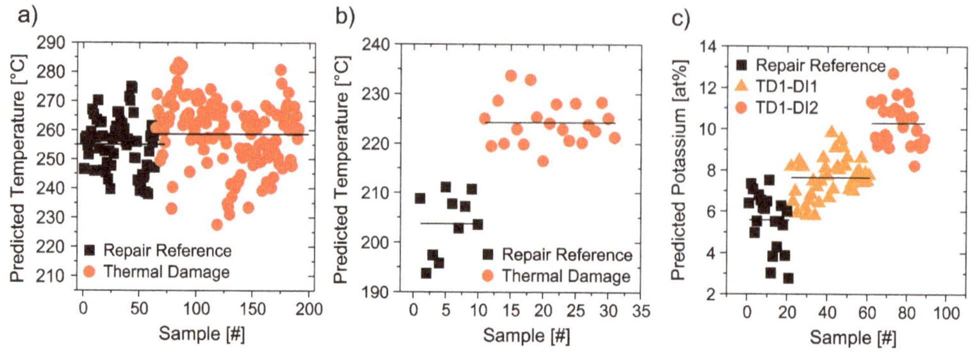

Fig. 3.73 **a** Prediction of the temperature during thermal degradation based on the PLS algorithm from the FTIR spectra recorded on thermally degraded and de-icing fluid contaminated scarfed CFRP surfaces; **b** prediction of the temperature during thermal degradation based on the PLS algorithm for thermal-oxidative degradation; **c** prediction of the amount of potassium on the surface based on the PLS algorithm from the FTIR spectra recorded on scarfed CFRP surfaces with thermal degradation and additional de-icing fluid contamination

In order to check the repeatability of the FTIR method, un-scarfed areas of the sample were also investigated. For this evaluation, an in-house PLS model was used that was available from previous investigations on thermal-oxidatively degraded surfaces. In Fig. 3.73b, the evaluation shows a quite significant differentiation between the reference and thermally degraded surfaces. The evaluation of the measured amount of de-icing fluid is displayed in Fig. 3.73c. It was observed that contamination from de-icing fluid was detected on all samples. Furthermore, a clear increase in the amount of predicted potassium could be detected between the reference sample and the contamination levels I and II. For the reference sample, a high scattering was obtained.

3.6.3 Performance in Inline Surface Quality Assurance

During the different measurement series, FTIR spectroscopy demonstrated high sensitivity in detecting and quantifying thermally damaged parts, the amount of moisture uptake inside an assembly, and de-icing fluid contaminations on CFRP surfaces.

Moisture uptake could be correlated successfully with an error of 0.09 wt%. This sensitivity is sufficient to set a reliable process window to prevent bondline failure due to unknown moisture uptake.

Coupon level investigations proved the capability of determining thermally damaged samples with an error of 8 °K. This is a very good result for sanded CFRP surfaces. However, this accuracy was not achieved for the pilot samples. At present, it is assumed that the heat friction during the scarfing process leads to high temperatures and causes thermal damage on the sample surface with higher local variations.

On the coupon sample level, the surface contamination with de-icing fluid was well predicted with an error of only 1.4 at.% K. These results were subsequently validated with the contaminated pilot samples, with appropriate separation of lower (DI-I) and higher (DI-II) levels of de-icing concentration. For measurements on the demonstrator part, the de-icing fluid was found on the contaminated area as well as on the reference side (in smaller concentrations). It is assumed that during the drying step in the oven, the de-icing fluid was spread across the whole surface of the sample.

With this newly gained knowledge and the identified tasks, the process of FTIR sampling can be significantly improved. Hence, FTIR spectroscopy has made another step toward finally being employed on the shop floor.

3.7 Vibrometry Inspection

In this section, we introduce scanning laser Doppler vibrometry (SLDV) as a tool for surface quality assessment and detail how its performance was enhanced in the ComBoNDT project for the in-process monitoring of CFRP adherends.

3.7.1 Principle and Instrumentation

A scanning laser Doppler vibrometer is a non-contact measuring device that makes use of the Doppler effect to register the vibration velocity. A beam of laser light is focused on a point on the measured surface, see Fig. 3.74. The light is reflected and due to the Doppler effect, its frequency is shifted proportionally to the velocity of the measured point. A built-in SLDV interferometer is used to estimate this shift, thus measuring the point velocity. A set of motors and mirrors in the scanning head enables laser beam steering and, together with developed software, scanning along a defined grid of points. Possibilities for SLDV application in research related to guided wave measurements have been extensively studied over recent years. In [33], a combination of experimental analysis using laser vibrometry and numerical analysis for the thin aluminum plate is presented. Ruzzene [34] proposed a technique for full wavefield analysis in the wavenumber/frequency domain for damage detection. This is a filtering technique that improves damage localization results. The author performed numerical and experimental analyses of simple aluminum plates with crack and disbonded tongue and groove joints. In [35], scanning laser vibrometry and imaging techniques were utilized for the detection of hidden delaminations located in a multi-layer composite. The authors analyzed the wave interactions with delamination and utilized several image processing techniques, such as Laplacian filtering. It should be noted that the mentioned work presented a completely non-contact system for elastic wave generation and sensing. Elastic wave generation was performed using a continuous wave (CW) laser source in connection with a photodiode that excited a piezoelectric transducer. Elastic wave sensing was performed

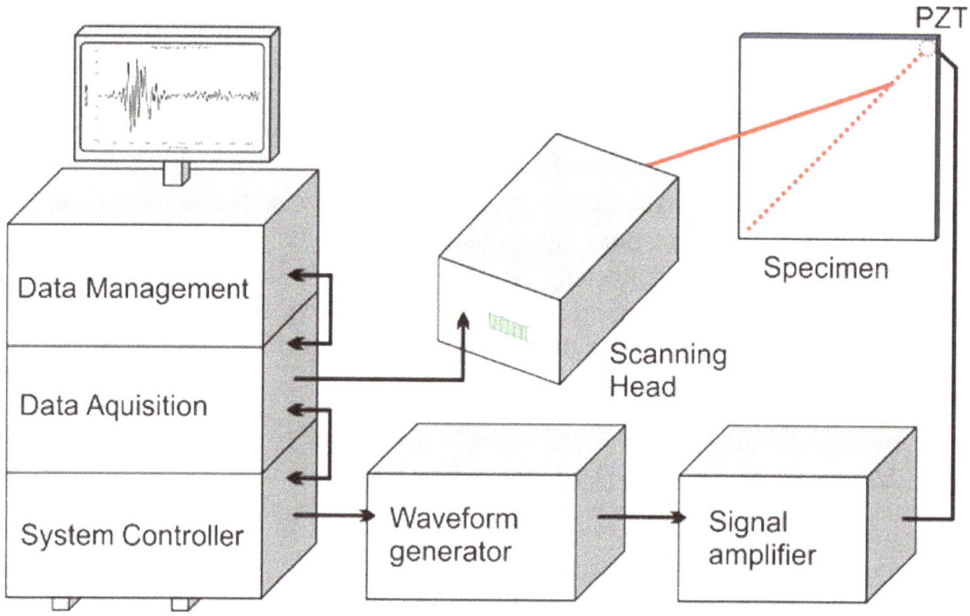

Fig. 3.74 The experimental setup

using a conventional scanning laser vibrometer. Non-contact elastic wave generation based on a thermoelastic effect was utilized in [36]. The authors utilized broadband excitation based on a low-power Q-switched laser. The research presented here used a piezoelectric transducer (Sonox P502) with a diameter of 10 mm, which was attached to the surface to excite guided waves in the samples. An excitation signal in the form of a 5 μs Hanning was generated by an arbitrary waveform generator and amplified to ±180 V by a signal amplifier. The excitation was synchronized with the SLDV system.

3.7.2 *Vibrometry Inspection Results*

3.7.2.1 Vibrometry Inspection Results on Coupon Level Samples

To excite guided waves in the coupon samples, the piezoelectric transducer was attached to the back surface of the specimen in the top-right corner, as depicted in Fig. 3.75. Measurements were made along a diagonal line of 592 equally spaced points. At each point, 100 time samples were registered with a 512 kHz sampling frequency (192 ms total registered time). Each measurement was repeated 10 times and used for averaging to improve the signal-to-noise ratio. The specimen was laid flat on foam to minimize any propagating wave distortion.

Fig. 3.75 The specimen

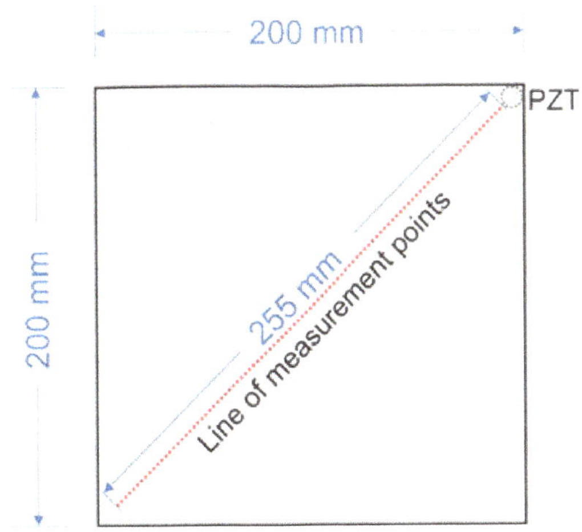

An example of a registered signal, presented in Fig. 3.76, is in the form of a time response at one point situated on a diagonal line, 50 mm from the top-right corner of the specimen.

The complete registered data are presented in the form of a waterfall plot in Fig. 3.77.

Due to the high wave amplitude diversification, the same results are presented in the logarithmic amplitude scale in Fig. 3.78. Three wave modes are visible, one dispersive and two nondispersive (parallel lines).

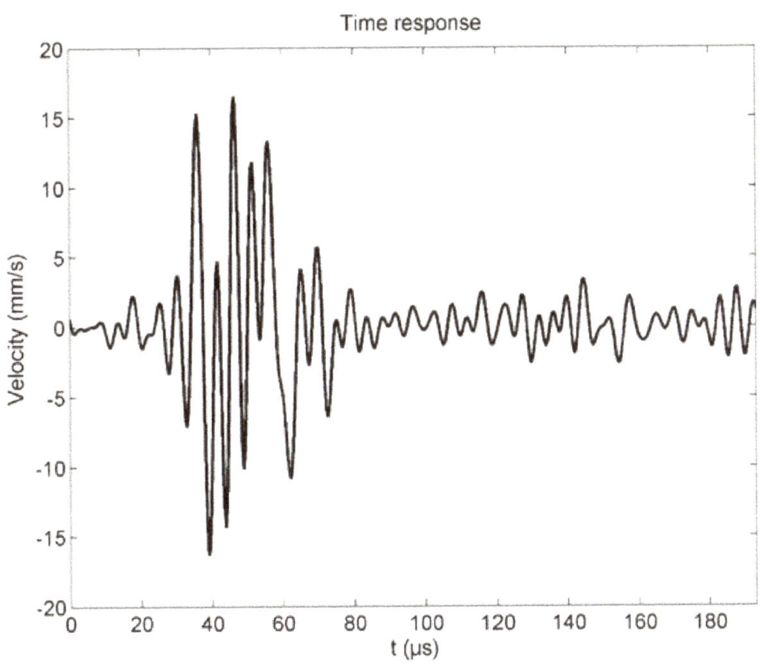

Fig. 3.76 Time response at one point at d = 50 mm

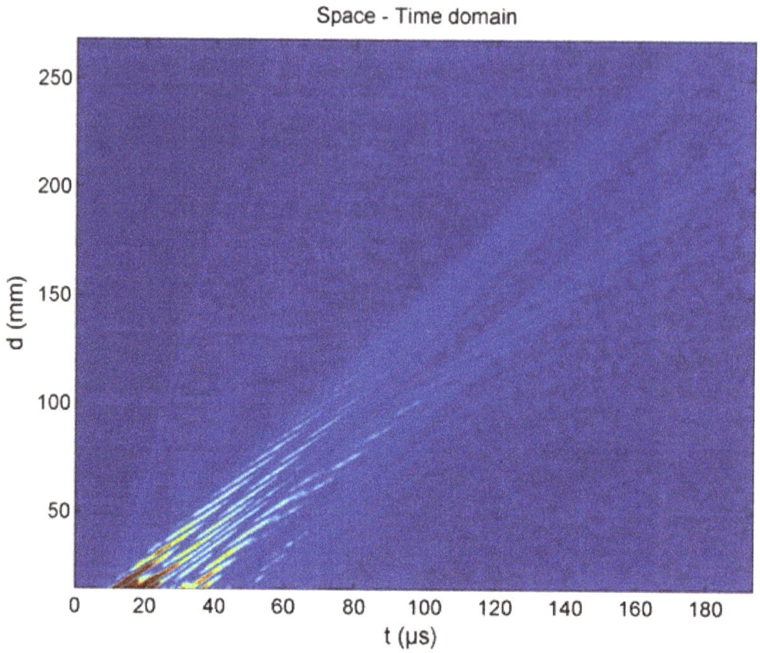

Fig. 3.77 Waterfall plot

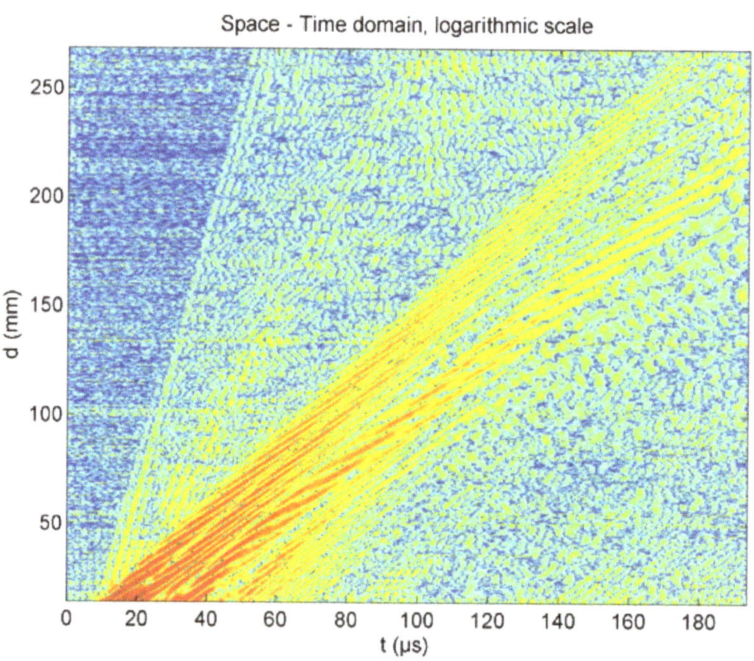

Fig. 3.78 Waterfall plot in the logarithmic scale

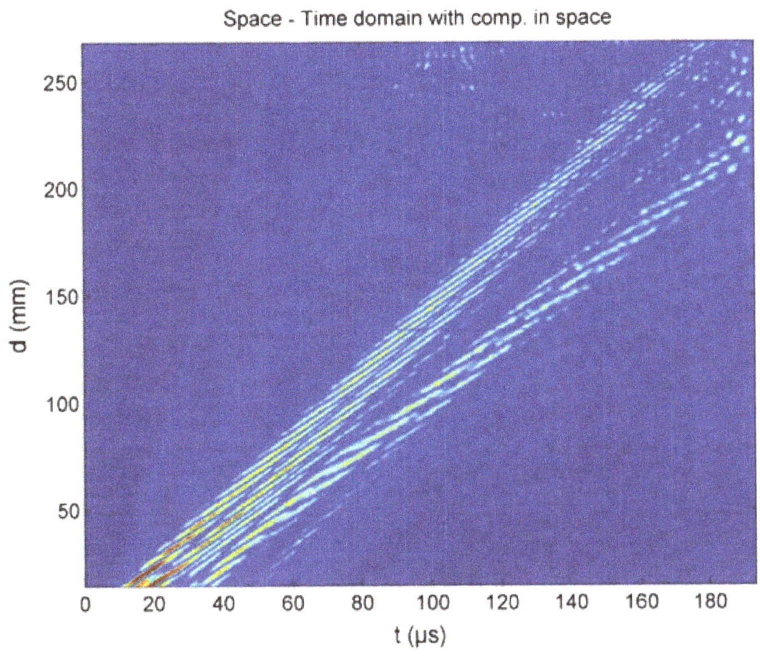

Fig. 3.79 Waterfall plot with space-amplitude compensation

Guided Wave Amplitude Compensation in Space

As the waves propagate along the specimen due to the damping phenomenon and the geometrical wave spread, the amplitude decreases dramatically with distance traveled. To compensate for this effect, an amplitude compensation procedure was proposed.

The average energy of the propagating waves at every measured point was determined as follows:

$$E(d) = \frac{1}{T}\sum_{t=0}^{T-1}\sqrt{w[d,t]^2}, \tag{3.1}$$

where T is the total number of measured time samples. Subsequently, the registered signal at each point was divided by its average energy:

$$\tilde{w}(d,t) = \frac{w(d,t)}{E(d)}. \tag{3.2}$$

The results, taking into account the amplitude compensation in space, are presented in the form of a waterfall plot in Fig. 3.79.

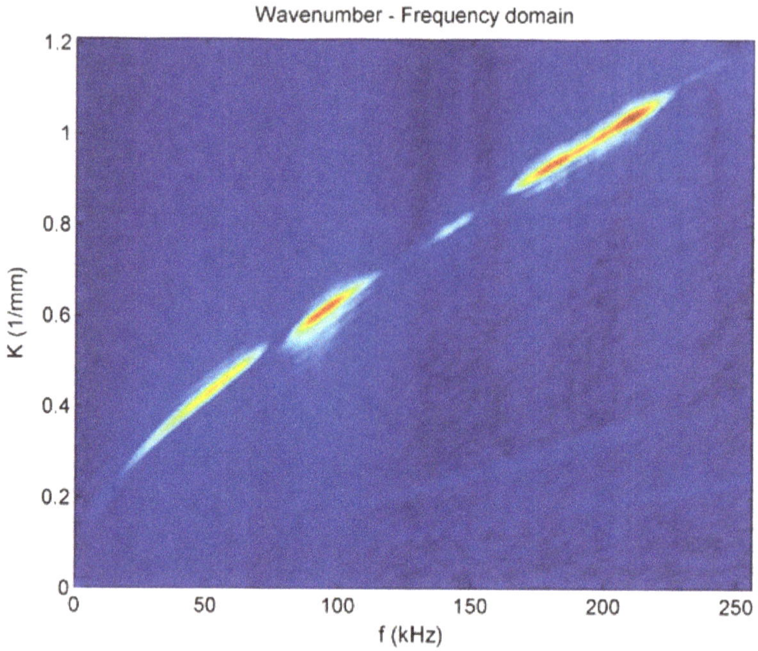

Fig. 3.80 Wavenumber–frequency relationship

Two-Dimensional Fourier Transform

To transform the registered data from the space–time domain into the wavenumber–frequency domain, the two-dimensional Fourier transform (2D FT) was used:

$$W(k, f) = \int\int_{-\infty}^{+\infty} w(d, t)e^{-j2\pi(kd+ft)}\mathrm{d}d\mathrm{d}t. \tag{3.3}$$

The discrete form of the two-dimensional Fourier transform may be noted as follows:

$$W[k, f] = \frac{1}{\sqrt{DT}} \sum_{d=0}^{D-1} \sum_{t=0}^{T-1} w[d, t]e^{-j2\pi\left(\frac{kd}{D}+\frac{ft}{T}\right)}, \tag{3.4}$$

where D and T are the total numbers of space and time samples, respectively.

Examples of the results in the wavenumber–frequency domain are presented in Fig. 3.80.

The same operation was repeated for the signal with the compensated amplitudes in space, leading to improved results, which are presented in Fig. 3.81. Such a curve is typical for a sample made of the given material with the given thickness. Next, a thresholding procedure was applied in order to create a binary curve. For the binary data, linear fitting was conducted. The reciprocity of the value of the slope obtained

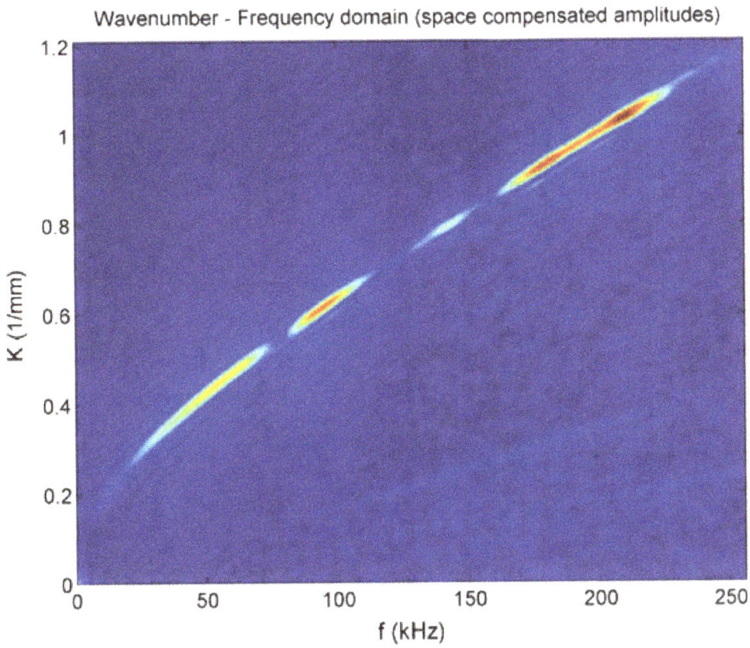

Fig. 3.81 Wavenumber–frequency relationship with space–amplitude compensation

from the fitting gives a linear approximation of the elastic wave group velocity *c*. This velocity was taken for the comparison of the coupon samples.

Two groups of samples were investigated in the coupon level samples; the sample set investigated for the production user case comprised three reference samples (RE), nine samples contaminated with release agent (RA), nine samples contaminated with fingerprints containing artificial sweat (FP), nine samples containing moisture (MO), and six samples with a mixed contamination of release agent and fingerprints. Because moisture samples were investigated in this set, it was important to compare them with very dry samples. The reference samples were measured for the first time after arrival from Fraunhofer IFAM (Bremen, Germany) and the wave velocity was estimated. Next, the samples were dried in an oven with air circulation. It was observed that the mass drop for the I-P-RE-1, I-P-RE-2, and I-P-RE-3 samples was 0.19%, 0.63%, and 0.18%, respectively. The second measurement was made after drying. The estimated wave velocity values for the three samples before and after drying are plotted in Fig. 3.82. A significant drop in velocity was observed for the second sample (I-P-RE-2). The observed relative change in velocities due to drying was 1.80%, −15.63%, and 0.73% for the I-P-RE-1, I-P-RE-2, and I-P-RE-3 samples, respectively. For the next analyses, the dried reference samples were taken.

Figure 3.83 depicts the results obtained for the feature velocity when investigating samples with single contaminations in the production user case based on CFRP coupons. No correlation between the contamination level and wave velocity is observed for the release agent contamination (RA). An exceptionally high value of velocity is observed for the I-P-RA-2-2 sample, while the I-P-RA-2-1 sample has the

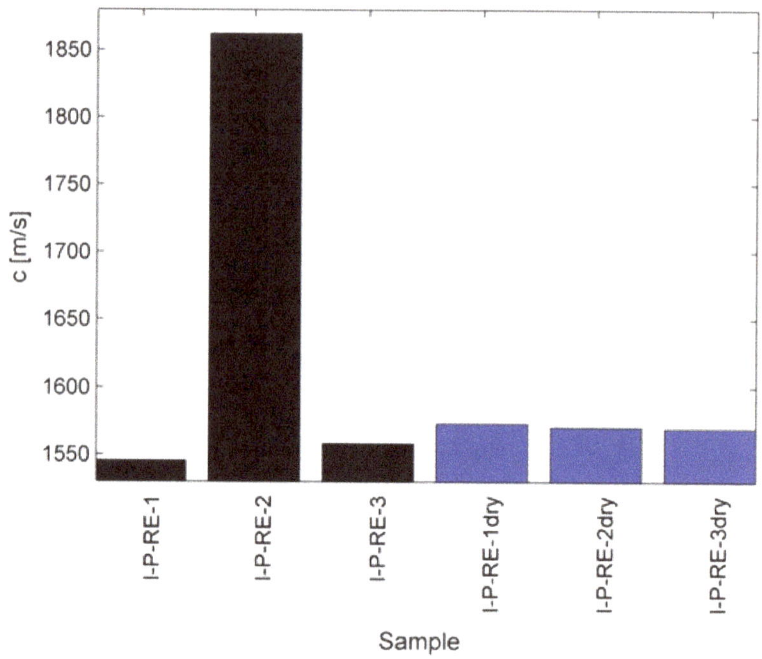

Fig. 3.82 Vibrometry results; comparison of the reference (RE) CFRP coupon samples of the production user case

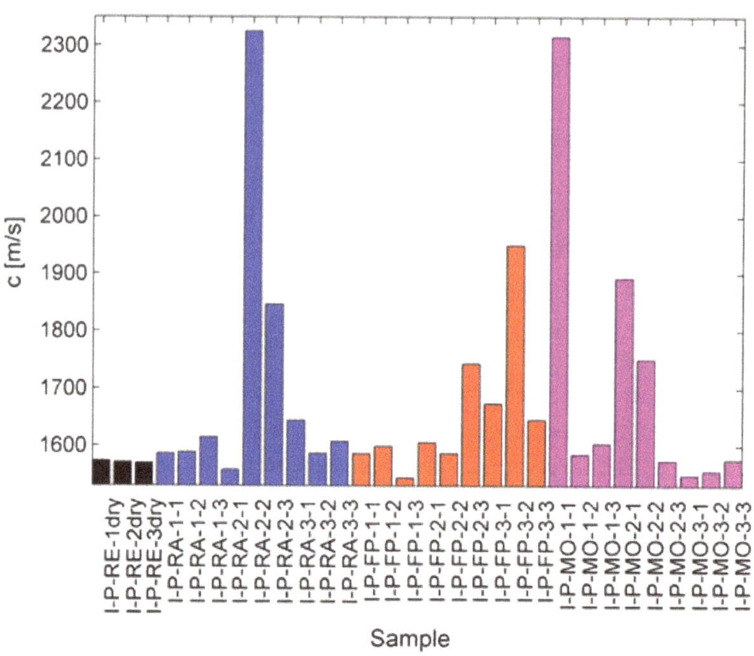

Fig. 3.83 Vibrometry results for the feature c (velocity of wave propagation) obtained during the surface quality assessment of CFRP coupons within the production user case; each sample had undergone one single modification. More details are given in the text

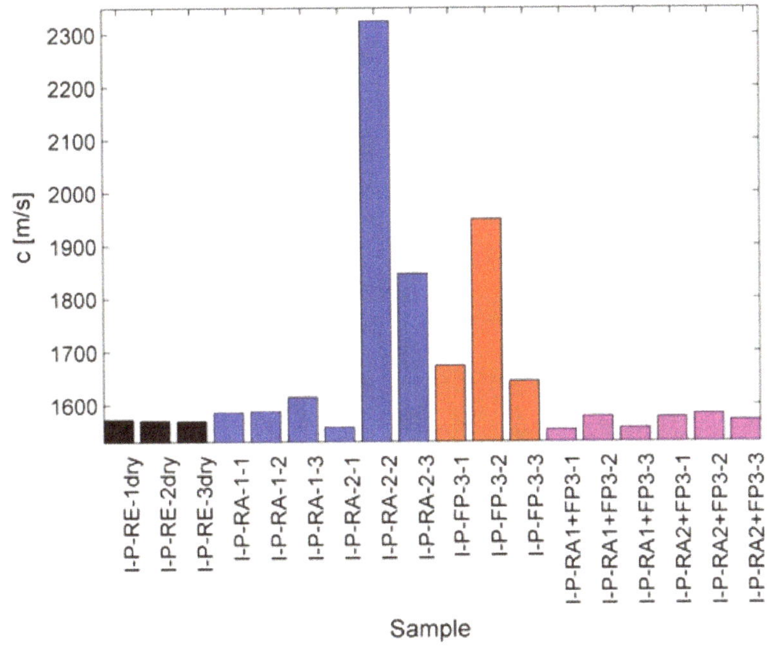

Fig. 3.84 Vibrometry results obtained with CFRP coupons in the production user case; the findings for samples prepared by applying combined contaminations are compared to the ones for the reference (RE) samples and the samples with the respective single contaminations

lowest value among all the RA samples, even though it has the same contamination level. In the case of the fingerprint contamination (FP), there is a slight increase in the wave velocity going from the lowest level (I-P-FP-1-x) to the highest level (I-P-FP-3-x). The numbering of the moisture samples cannot be treated as being representative of contamination content because each of the samples absorbed different amounts of moisture during conditioning. The sample with the highest velocity (I-P-MO-1-1) gained 0.59% of mass, the same amount as the I-P-MO-3-1 sample, which has the lowest velocity. The highest mass increase (0.71%) is observed for the I-P-MO-2-2 sample, but this is not distinguished by an extreme value of the velocity. The moisture case is further studied in a separate subsection.

In the next step, the samples with combined contaminations were studied. The combined case comprised both fingerprint and release agent contaminations, so the comparison was made not only with the reference samples but also with the RA and FP samples with single contaminants at the respective levels. The results are depicted in Fig. 3.84. It should be noted that the wave velocity values for the samples with combined contaminations are on the level of the dry reference samples. The velocity for single contamination samples was higher in most of the cases.

The CFRP coupon sample set investigated for the repair user case comprised three reference (RE) samples, nine thermally treated (TD) samples, nine samples contaminated with de-icing fluid (DI), nine samples contaminated with oily fingerprints (FP), and six samples with mixed modification obtained by a combination of thermal impact and subsequent contamination with de-icing fluid. In comparison to

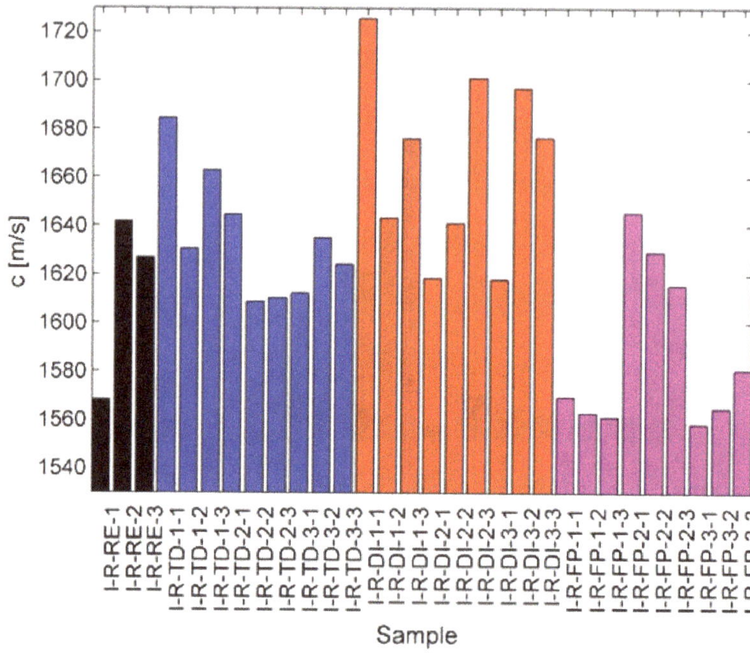

Fig. 3.85 Vibrometry results for the CFRP coupons from the repair user case; findings related to the feature c (velocity) are presented for all the investigated samples with a single modification

the findings within the production user case, the reference samples differ more from each other, see Fig. 3.85. Especially for the first sample (I-R-RE-1), the value of velocity is significantly low. The non-uniformity of the estimated velocities could be related to the surface preparation of the samples. The samples for the repair user case were ground down to the fibers. For the velocity values for the samples with modifications, there is no correlation with the de-icing fluid contamination level. Relatively good repeatability is observed for the fingerprint contamination. The samples with the same level of contamination are characterized by similar values of velocity; however, they cannot be distinguished from the reference samples. In the case of the thermal treatment, the highest value of velocity was observed for the lowest temperature of treatment, namely 220 °C (I-R-TD-1-x).

In the next step, the samples with combined modifications were studied. The combined case comprised thermal degradation and de-icing fluid contamination, so the comparison was made not only with the reference samples but also with the respective TD and DI samples. Generally, a higher level of combined modification results in an increase in the velocity value, see Fig. 3.86. However, only two samples (I-R-TD1+DI-2-2 and 2-3) are characterized by a higher velocity than the reference samples. Moreover, the combined modifications cannot be distinguished from the respective single modifications (DI and TD).

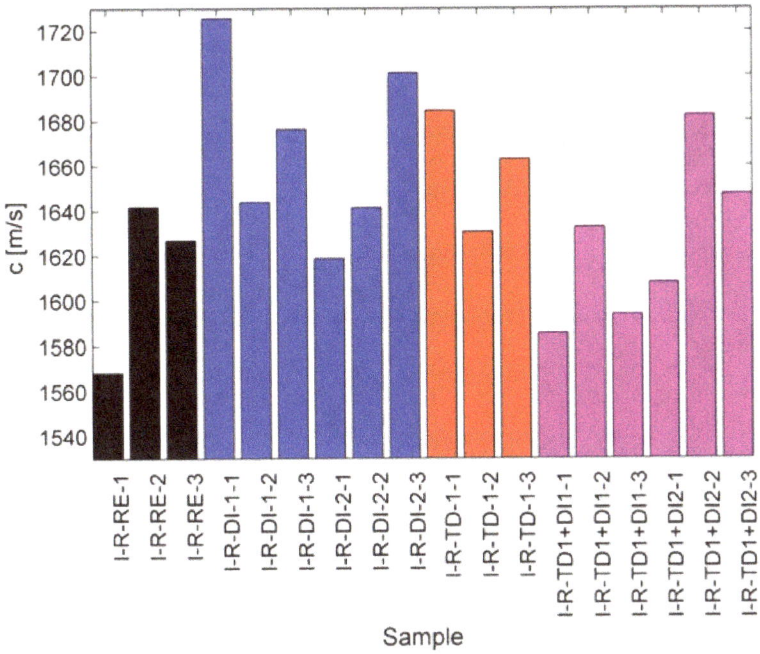

Fig. 3.86 Vibrometry results for the CFRP coupons from the repair user case; findings for the combined modifications are compared against those obtained for the reference (RE) samples and for coupons with the respective single modifications that contribute to the combined modification

Focus on the Moisture Uptake

After inspecting all coupon samples, it was decided to further investigate the samples with moisture contamination. This was because a significant influence of the drying of the I-PRE-x samples was observed. Additional analysis was conducted, and the undried reference samples were also treated as being contaminated with moisture. Because the moisture absorption of each sample was different, it was decided not to use the sample symbols but rather to represent moisture intake as the percentage of mass increase. The new results are plotted in Fig. 3.87. This time, there were 12 cases with moisture because the undried reference samples were included in the set. The wave velocity value is on the same level for a considerable number of samples, starting at 0% moisture content and ending at about 0.7%.

A correlation of the velocity with moisture content was not observed. This scattering of the results could be related to the fact that the considered samples were not saturated with moisture. This could explain the fact that we observed distinct results for samples with similar mass increases. The moisture content on the wave path may vary between samples with the same mass increase. Moreover, the moisture content on the wave path may also vary from point to point for a single sample. Also, according to the numerical investigation, the moisture increase influences the wave amplitude far more than it does the wave velocity value. Therefore, it was decided to analyze the vibration energy from point to point. At each measurement point i, the RMS value was calculated:

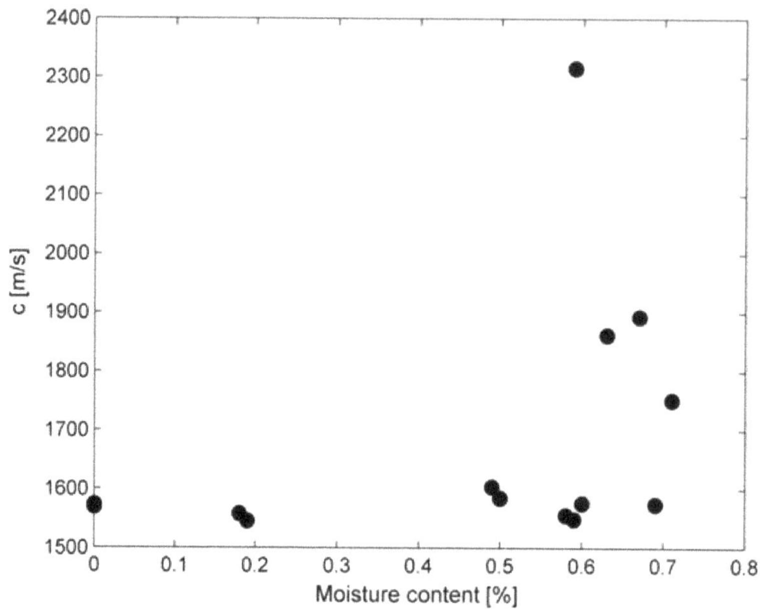

Fig. 3.87 Estimated wave velocity plotted against mass increase due to moisture

$$RMS_i = \sqrt{\frac{1}{N}\sum_{n=1}^{N} X_{i,n}^2}, \tag{3.5}$$

where $X_{i,n}$ is the n-th time sample of a signal X measured at point i. The indices that are assumed to give the average state of the surface are the mean value of RMS_i, defined as

$$E = \frac{1}{M}\sum_{i=1}^{M} RMS_{i,} \tag{3.6}$$

where M is the number of measurement points, and the three-sigma is defined as three times the standard deviation:

$$3\sigma = 3 \times \sqrt{\frac{1}{M-1}\sum_{i=1}^{M}(RMS_i - E)^2}. \tag{3.7}$$

The three-sigma value consists of 99.7% of all data, meaning this encompasses almost all the calculated *RMS* values. The results are depicted in Fig. 3.88. Again, the samples with moisture contamination cannot be distinguished from the dry samples.

The problem in identifying the samples with moisture contamination could lie in the fact that only a linear measurement is done along the diagonal of the samples. If the samples are not saturated, one cannot be certain where the moisture is located in the sample because only the mass of the sample is taken as the indicator of the

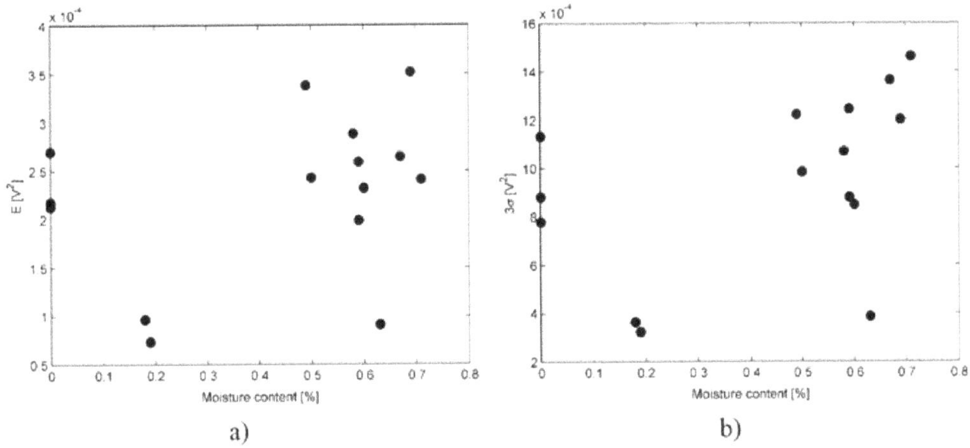

Fig. 3.88 Results of the RMS analysis for the moisture contaminated samples; **a** E index, **b** three-sigma index

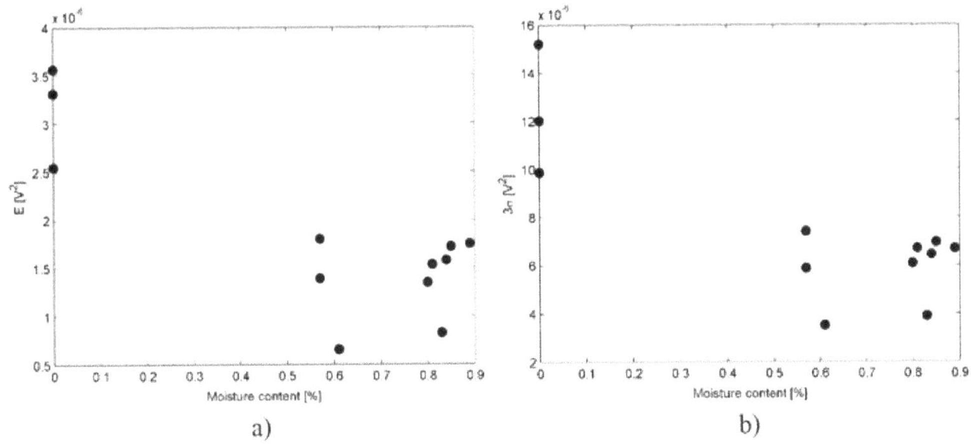

Fig. 3.89 Result of the whole area analysis for moisture contaminated samples; **a** E index, **b** three-sigma index

moisture content. For this reason, a new set of measurements was conducted. The time signals were registered in a dense grid of measurement points defined at the whole sample surface. Again, the *E* and three-sigma indexes were calculated. It should be noted that these measurements were taken after some time, so the moisture content was slightly different than the results depicted in Fig. 3.88. The new results for the whole area scan are presented in Fig. 3.89. It can be observed that the dry reference samples (zero moisture content) differ slightly from each other. This could be caused by slight differences in surface quality that influence the reflection of the vibrometer laser beam. However, the more important observation is that the reference sample results clearly differ from those of the samples with moisture.

3.7.2.2 Vibrometry Inspection Results for Pilot Level Samples

In the next step, pilot samples were investigated using whole area analysis. The production samples contaminated with both release agent (RA) and fingerprints (FP) had a curved shape, so they were measured from both sides. The results for the convex side are presented in Fig. 3.90. The values for the reference samples do not differ significantly. The E and three-sigma values for contaminated samples are dispersed. The II-P-RA1+FP3-2 and II-P-RA2+FP3-2 samples have similar values as the reference samples. Looking at the results for the concave side, see Fig. 3.91, it can be noted that the scale is 10 times larger. This could be related to the different

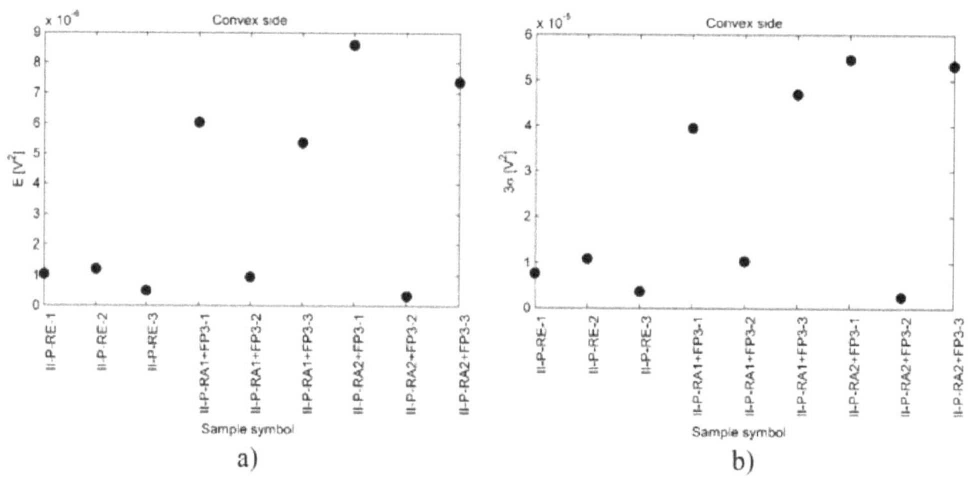

Fig. 3.90 Results of the full-field analysis for the production pilot samples with combined RA+FP contamination; convex side; **a** E index, **b** three-sigma index

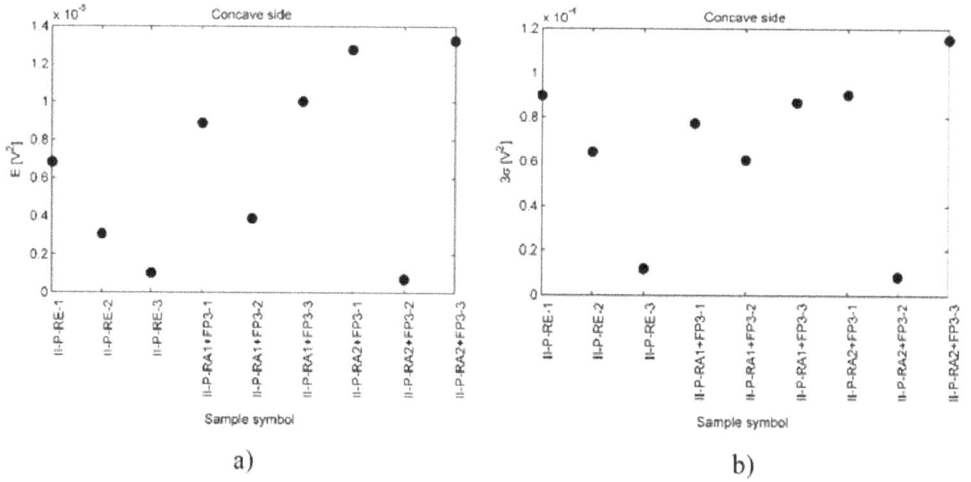

Fig. 3.91 Results of the full-field analysis for the production pilot samples with combined RA+FP contamination; concave side; **a** E index, **b** three–sigma index

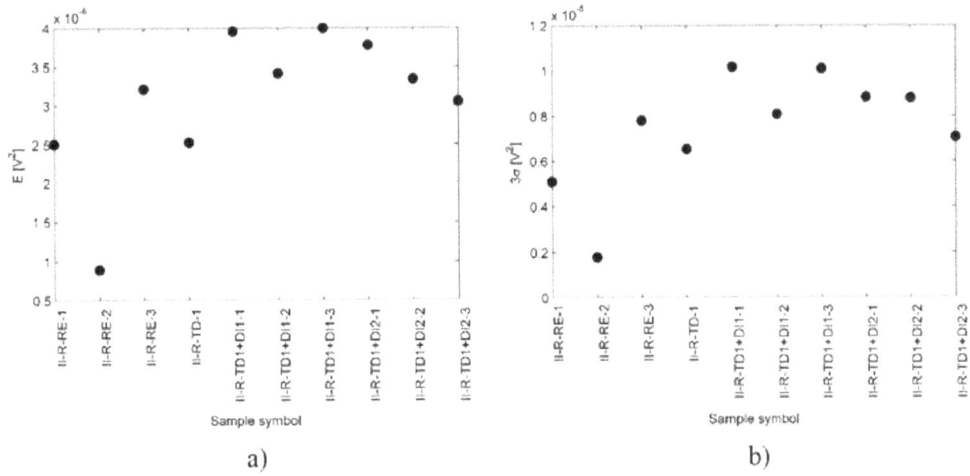

Fig. 3.92 Results of the full-field analysis for the repair pilot samples with combined TD+DI modification; **a** E index, **b** three-sigma index

surface quality and shape of this side. The behavior of the reference samples is no longer uniform. The difference between the reference and contaminated samples is also not evident.

The pilot samples for the repair user case comprised a mixed modification by thermal degradation (TD) and de-icing fluid contamination (DI). These samples were scarfed samples with a smoothly decreasing thickness. They were measured only from the side with decreasing thickness. The results are depicted in Fig. 3.92. Apart from the three reference samples, one sample that was only treated thermally was measured (TD1) for comparison with samples with mixed modification. The reference samples differ from each other. The TD1 sample result is within the interval given by the reference sample measurements. There is no clear separation between the TD1+DIx results and the values for the reference samples.

3.7.3 Final Remarks

In the research reported in this subchapter, the laser vibrometry measurements were first conducted at points defined along the samples' diagonal. A signal processing method was applied that allowed the wavenumber to be extracted as a function of frequency. Based on this relationship, the wave group velocity was estimated. The most promising results were observed for the moisture contamination, although the samples were not fully saturated with moisture. The used approach was local, so only a limited part of the sample was measured (diagonal measurement). However, as the waves propagated in the whole sample, the change in the wave field caused by the surface modification may have been too weak to be noticed after the wave had traveled a considerable distance (from the transducer to the edges and then to the measurement point). Considering that the presence of degradation or contamination can have a very slight influence on the wave, a new approach was proposed based on full-field measurements. The time signals were registered in a dense grid of measurement points defined at the sample surface. It was decided to choose an index that could give information about the average state of the surface because it was assumed that the real distribution of the contaminant/degradation is unknown. It was observed that the samples contaminated with moisture clearly differed from the reference samples. However, a sensitivity to the moisture level was not observed in the results. The pilot samples for the production user case were measured from both sides due to their curvature, and the results obtained for the reference samples on the convex side were comparable; however, this was not observed for the concave side. Regarding the contaminated samples, it cannot be clearly stated that they differ from the referential samples. In the case of the pilot samples from the repair user case, the reference samples differ from each other. There is no clear separation between the results for the modified samples and the values for the reference samples. The same approach applied to the realistic part also showed no sensitivity to the contamination of the surface. This could be associated with the type of contamination used.

3.8 Laser-Induced Fluorescence (LIF)

In this section, we introduce laser-induced fluorescence (LIF) as a tool for surface quality assessment and detail how its performance was enhanced in the ComBoNDT project for the in-process monitoring of CFRP adherends.

3.8.1 Principle and Instrumentation

LIF is a technique based on the analysis of the spontaneous emission of atoms or molecules excited with a laser, whereby the analyzed material can be in the gas, liquid, or solid-state phase. Typical instrumentation for LIF analysis consists of a laser to illuminate the investigated material and a detection system to record the fluorescence. Depending on the chemical composition of the analyzed material, the laser can be tuned to match the wavelength to the absorption lines or bands of specified atoms or molecules, producing an electronically excited state that can radiate. The fluorescence emission can be detected using bandpass filters and sensitive detectors, i.e., photomultipliers in the case of specified spectral band analysis or monochromators equipped with sensitive CCD detectors for full spectra recording. Since the fluorescence signal is much weaker than the excitation laser radiation, a laser cutoff filter is recommended in the detection system. The LIF analysis can be provided in two regimes: (i) the time-integrated mode, in which the spectra are integrated over a long time—in this mode both CW and pulsed lasers can be used for sample excitation;

(ii) the time-resolved mode, in which the fluorescence decay is taken into account—this mode requires a short pulse duration of the excitation laser. The fluorescence spectrum is usually represented as a combination of vibrational, rotational, and fine structures, depending on the chemical composition and spectral range. In the case of composite materials characterized by complex chemical composition, the fluorescence bands usually overlap, but subtle changes in the structure of the material can still be seen as changes in the intensity or profile of the fluorescence spectrum.

The main advantages of the LIF technique are its sensitivity, spatial and temporal resolution as well as the non-invasive nature of the analysis. Measurements can be provided for the individual points on the material, or large areas of the surface can be scanned with a laser to visualize the local structure or composition changes [37]. The analysis does not require sampling, and the measurement can be performed on the tested object. The detection of kerosene and hydraulic fluid on CFRP surfaces has been previously reported [38]. Good results were obtained for 266 nm excitation. In another work [39], three wavelengths were studied (266, 355, and 532 nm). The detection of hydraulic fluid, release agent, and moisture contaminations as well as thermal treatment has also been presented. It was shown that at 532 nm, the thermal treatment can be easily distinguished from the remaining cases. Moreover, it was shown that samples treated at 190, 200, and 210 °C are characterized by increasing LIF intensity and can be distinguished from each other and from reference samples.

For the results presented in the next sections of this chapter, the sample excitation was provided by a CW DPSS Nd:YAG laser operating at 532 nm (Spectra Physics). The laser power was set to 0.2 W and the laser intensity was $1\ \mathrm{W/cm^2}$ (5 mm laser spot diameter). The spectra of the laser-induced fluorescence (LIF) were recorded in the time-integration mode. The emission spectra were dispersed by a SR-303i 0.3 m spectrograph equipped with gratings of 600 and 150 grooves/mm and coupled to a time-gated ICCD camera DH 740 (Andor Tech). Spectra were acquired in the range of 300–800 nm with resolutions of 0.3 nm or 1.2 nm. An illustration of the setup is presented in Fig. 3.93.

3.8.2 *LIF Results*

3.8.2.1 LIF Results on Coupon Level Samples

In the set of coupon level samples, two groups of samples were investigated. The first examined production-related contaminations of the samples, and the second focused on repair-related modifications/contaminations of the samples. The sample set investigated for the production user case comprised three reference samples, nine samples contaminated with a release agent, nine samples contaminated with fingerprints (artificial sweat), and nine samples contaminated with moisture. The sample set investigated for the repair user case comprised three reference samples, nine thermally degraded samples, nine samples contaminated with de-icing fluid, nine samples contaminated with oily fingerprints (containing Skydrol hydraulic oil), and

Fig. 3.93 LIF equipment used for the investigations

six samples with mixed modifications obtained by combining thermal degradation and contamination with de-icing fluid.

Firstly, the reference samples from both user cases were compared. The fluorescence spectra were measured and then the area under the curve (LIF intensity) was taken for comparison, see Fig. 3.94. The production and repair user case samples differed significantly. The fluorescence intensity from the surface of the production samples was four times stronger than for the repair user case, see Fig. 3.95. The reason for this could stem from the surface preparation. The repair user case samples were ground down to the fibers in order to simulate the preparation for bonding repair, while the production user case samples were only slightly ground. Due to this difference, the modified samples from the repair user case should be compared with the respective reference samples. An analogous approach should be used for the production samples.

The results for the release agent contamination are depicted in Fig. 3.95. All the contaminated samples have a similar intensity to the reference samples. In the case of fingerprint contamination, the behavior is similar. The results for the contaminated samples are similar to the reference cases within the boundaries defined by the standard deviation, see Fig. 3.96. There is no correlation between the LIF intensity and the release agent or fingerprint contamination level.

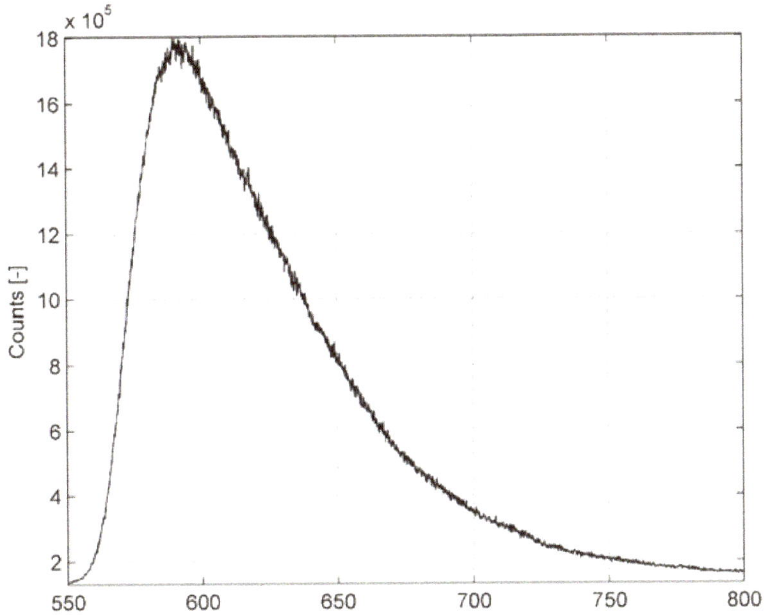

Fig. 3.94 An example of registered LIF spectra

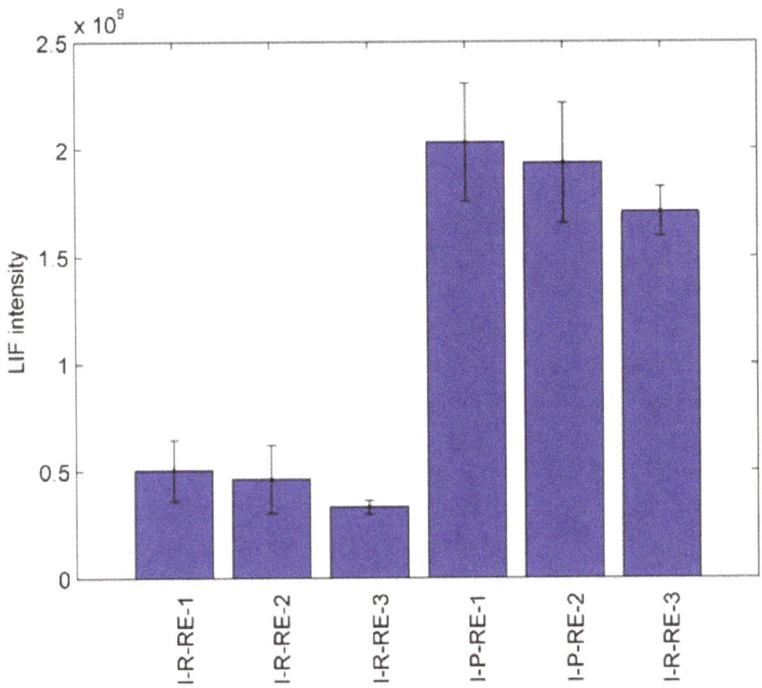

Fig. 3.95 LIF technique: Results for the three reference (RE) samples from each of the two user cases, namely a production (P) and a repair (R) aeronautical user case, involving flat CFRP coupons

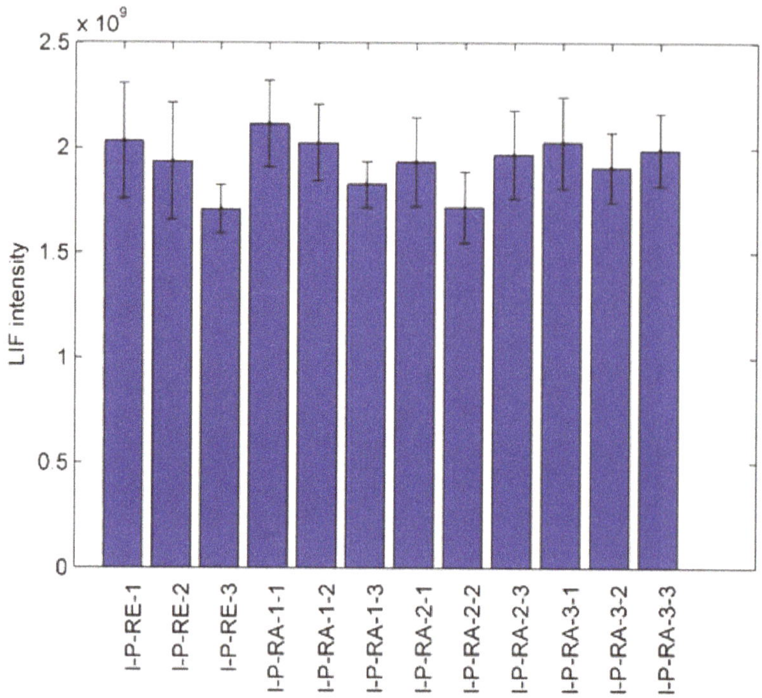

Fig. 3.96 LIF results for the production (P) user case as obtained for each of the three CFRP coupons (1, 2, and 3) that had been intentionally treated following a release agent contamination (RA) scenario at three distinct contamination levels (1, 2, and 3)

The first results for the repair user case are presented in Fig. 3.97 for the de-icing fluid contamination. These results are characterized by high dispersion (wide standard deviation bars). Some samples (IR-DI-1-1, 1-3, 2-1, 3-3) are characterized by significantly low-intensity values, but these do not correspond to the same contamination levels. The LIF intensity for oily fingerprint contamination, see Fig. 3.98 increases with increasing contamination. However, considering the intensity values for the reference samples, it becomes clear that these are located between the results for the two highest contaminations levels I-R-FP2-x and I-R-FP3-x. Considering the results for thermal degradation, a high response is observed at the first level (I-R-TD-1-x), see Fig. 3.99. This case was achieved by keeping the samples in an oven at 220 °C. The samples I-R-TD-2-x and I-R-TD-3-x were treated at higher temperatures, and there is no correlation between the intensity and the used temperature. Moreover, the response is very uniform. All six samples have a comparable level of mean intensity, and the standard deviation is low.

In a previously published work [39], there was a good detection of the thermal damage scenario with the LIF intensity measurement with 532 nm excitation. Different levels of thermal treatment correlated with the intensity level. The intensity value increased with the increasing temperature of the treatment. However, here we see different behavior. The first important difference is that the new samples (I-R-TD) were ground down to the fibers, whereby the previously investigated samples were measured just after the thermal treatment without any surface modification

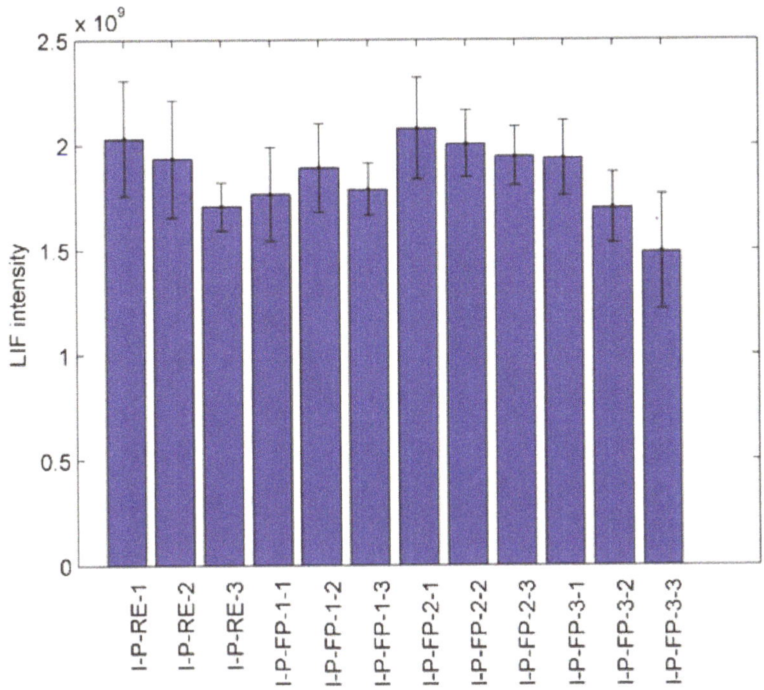

Fig. 3.97 LIF results for the production (P) user case as obtained for each of the three CFRP coupons (1, 2, and 3) that had been intentionally treated following a fingerprint contamination (FP) scenario at three distinct contamination levels (1, 2, and 3)

[39]. It is possible that the thermal treatment only influenced the surface layer, which was removed by the grinding, and thus the intensity increase could no longer be observed. The second important difference is that the I-R-TD-2-x and 3-x samples were treated at higher temperatures than before. Crucially, the other side of the sample was not prepared/modified in any way. It was decided to measure this second side and compare the results with those previously obtained, see Fig. 3.99. The new result is presented in Fig. 3.100. Again, the highest intensity is observed for the I-R-TD-1-x samples and, as before, the I-R-TD-2-x and I-R-TD-3-x samples have intensity values on the level of the reference samples. There is also a clear difference in the vertical scale between Figs. 3.99 and 3.100. This could be related to the considerable time interval between these two measurements. Thus, the obtained results suggest that treatments at 260 and 280 °C have different influences on the sample and that the LIF intensity is not influenced.

With the knowledge that the 220 °C (I-R-TD-1-x) scenario can be detected and distinguished, the combined modification scenario for the remaining cases was investigated. This set was treated at the same temperature and subsequently contaminated with de-icing fluid. Figure 3.101 presents the results for this new set of samples with mixed modification (I-R-TD1+DI) in comparison with the clear reference samples (I-R-RE) and samples that had only undergone thermal treatment at 220 °C (I-R-TD1-x). The first observation is that the combined modification was detected. It clearly differs from the reference. The higher level of de-icing fluid (DI-2) contamination

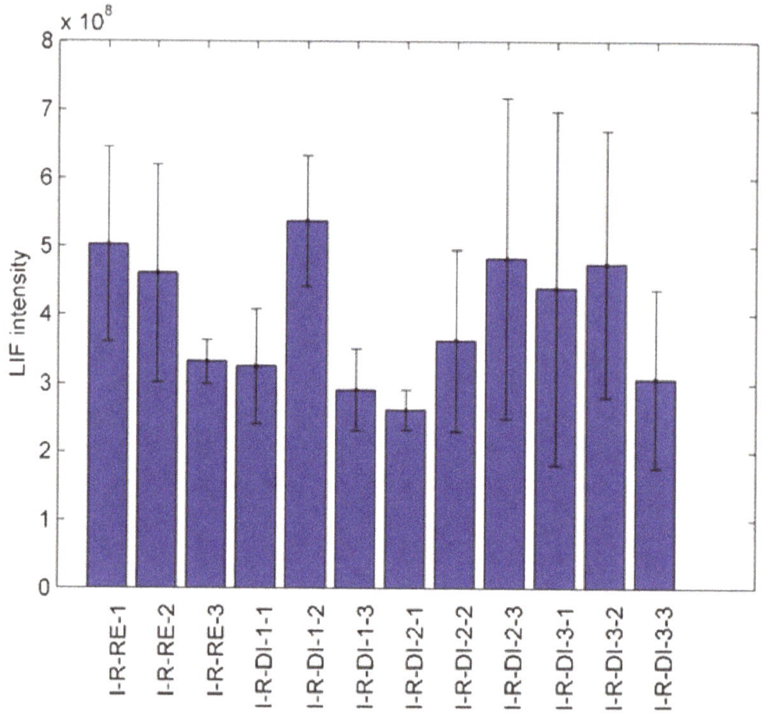

Fig. 3.98 LIF results for the repair (R) user case as obtained for each of the three CFRP coupons (1, 2, and 3) that had been intentionally treated following a de-icing fluid contamination (DI) scenario at three distinct contamination levels (1, 2, and 3)

is not related to any increase or decrease in relation to the lower level (DI-1). What we can observe is that the intensities for the samples with mixed modification have higher values than for the pure thermal treatment (I-R-TD-1-x). However, taking into consideration the standard deviations, all the results after modifications (I-R-TD-1-x and I-R-TD1+DI) lie within the boundaries of standard deviation for the I-R-TD-1-1 sample, see Fig. 3.102.

3.8.2.2 Conclusions

The conducted research showed that the thermal treatment at 220 °C could be clearly distinguished from the samples based on the intensity measurement conducted on both sides of the sample. Above 220 °C, the LIF intensity gives results comparable to the reference samples. No sensitivity to any of the higher thermal treatment levels was observed. The thermal treatment level could not be determined for 532 nm excitation. Taking into account the results presented in this paper as well as previously published results [39], it can be concluded that thermal degradation up to 220 °C can be distinguished. Moreover, it was shown that the detection of thermal degradation at 220 °C was possible even if the surface was subsequently contaminated with de-icing fluid. The other contaminations investigated in the production and repair user

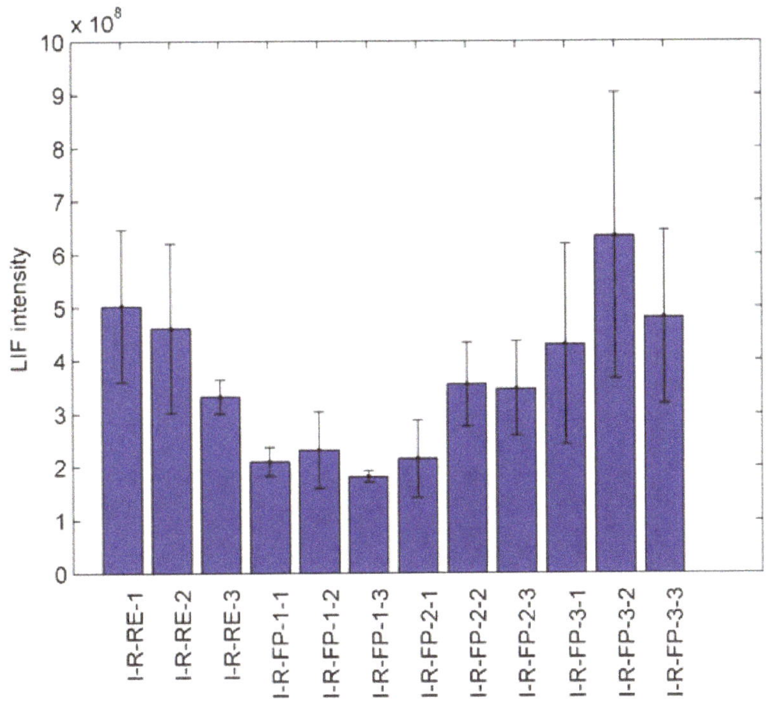

Fig. 3.99 LIF results for the repair (R) user case as obtained for each of the three CFRP coupons (1, 2, and 3) that had been intentionally treated following a fingerprint (with hydraulic oil) contamination (FP) scenario at three distinct contamination levels (1, 2, and 3)

cases of the ComBoNDT research project did not influence the LIF intensity in a way that the level of contamination could be distinguished. The results obtained for the Skydrol (FP) and release agent (RA) contaminations confirm previously published observations, namely that the LIF intensity observed for 532 nm excitation is not correlated to the amount of surface contamination. This was also observed for the realistic aircraft part, see Fig. 3.103.

3.9 Conclusion

We identified, defined, and intentionally implemented pre-bond contamination on CFRP surfaces in gradational levels. The respective CFRP parts were introduced as adherends in a qualified bonding process within two user cases involving different part geometries, namely flat coupons, curved pilot level specimens, and realistic or real CFRP parts, thereby contributing to the findings reported in Chap. 5. In this chapter, we described the findings obtained during the surface quality assessment with the ENDT methods that were advanced in the ComBoNDT research project for production and repair user cases relevant in aeronautical applications.

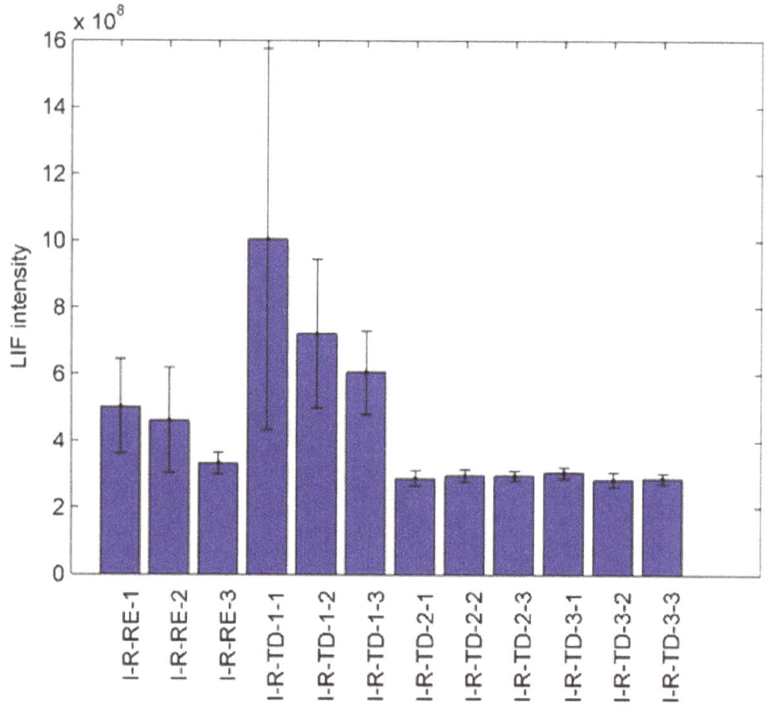

Fig. 3.100 LIF results for the repair (R) user case as obtained for each of the three CFRP coupons (1, 2, and 3) that had been intentionally treated following a thermal impact (TD) scenario at three distinct temperature levels (1, 2, and 3)

The in-process monitoring of CFRP composite surfaces was facilitated through the use of advanced setups and approaches based on the aerosol wetting test (AWT) performed with an enhanced bonNDTinspect® device by the consortium partner Automation W+R, optically stimulated electron emission (OSEE) performed by Fraunhofer IFAM, electronic nose (e-nose) testing performed by ENEA and Airbus, laser-induced breakdown spectroscopy (LIBS) performed by Fraunhofer IFAM, Fourier-transform infrared spectroscopy (FTIR) performed by Airbus, laser-induced fluorescence (LIF) performed by IMPPAN, and vibrometry inspection performed by IMPPAN.

Advanced ENDT procedures comprising in-process surface inspection with:

- an enhanced setup relying on AWT allowed for a differentiation between the surface states of clean and intentionally contaminated parts with the potential to discriminate between distinct levels of contamination for several contamination scenarios. This technology can be integrated into inline applications without major constraints.
- an enhanced setup relying on OSEE enabled detection of even small amounts of filmy contaminations. Therefore, this technique is most suited for specialized applications that require very clean and homogeneous surfaces.
- two enhanced setups relying on e-nose provided insight in the approach to combine custom-made-of-the-shelf (Airsense, Airbus E-Nose) with "open source" from

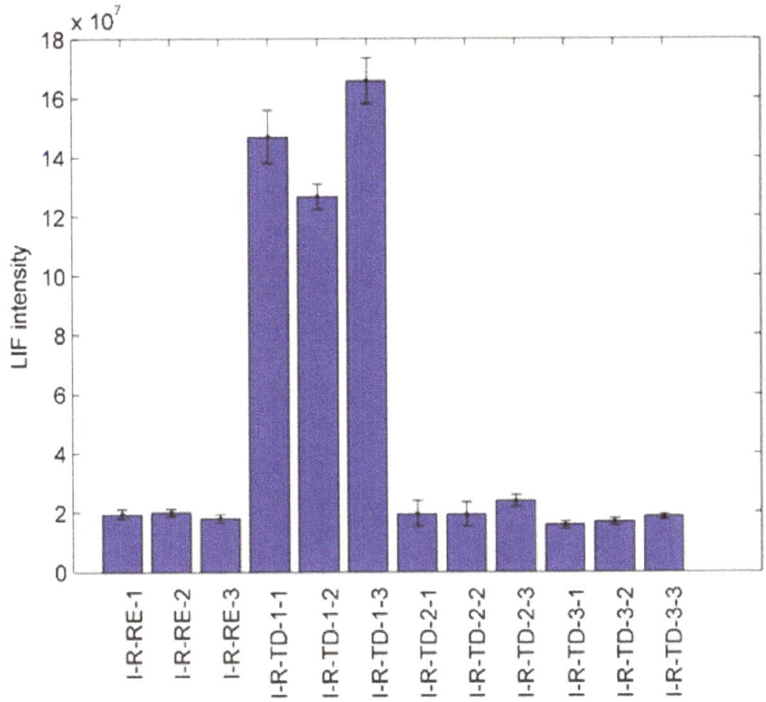

Fig. 3.101 The result for the repair user case for the thermal degradation of CFRP coupons; fluorescence intensity measured on the back face of the samples

scratch developed systems (ENEA). Considering smart sample taking and total system control seems to be the best path to further advance the sensor system performance of the e-nose method, which provides possibilities like a combined mode of operation to detect the chemical surface condition (clean or contaminated) at the same time as the moisture content.

- an enhanced setup relying on LIBS was used to map out contamination from coupon level sample to complex-shaped, technologically realistic CFRP parts. Using this approach, even locally applied contaminants can be detected on large areas in a realistic production environment.
- an enhanced setup relying on FTIR permitted detection and quantification of thermal degradation, the amount of moisture uptake inside an assembly, and de-icing fluid contaminations.
- an enhanced setup relying on LIF allowed for detection of thermal degradation at 220 °C.
- an enhanced setup relying on Vibrometry inspection facilitated moisture detection.

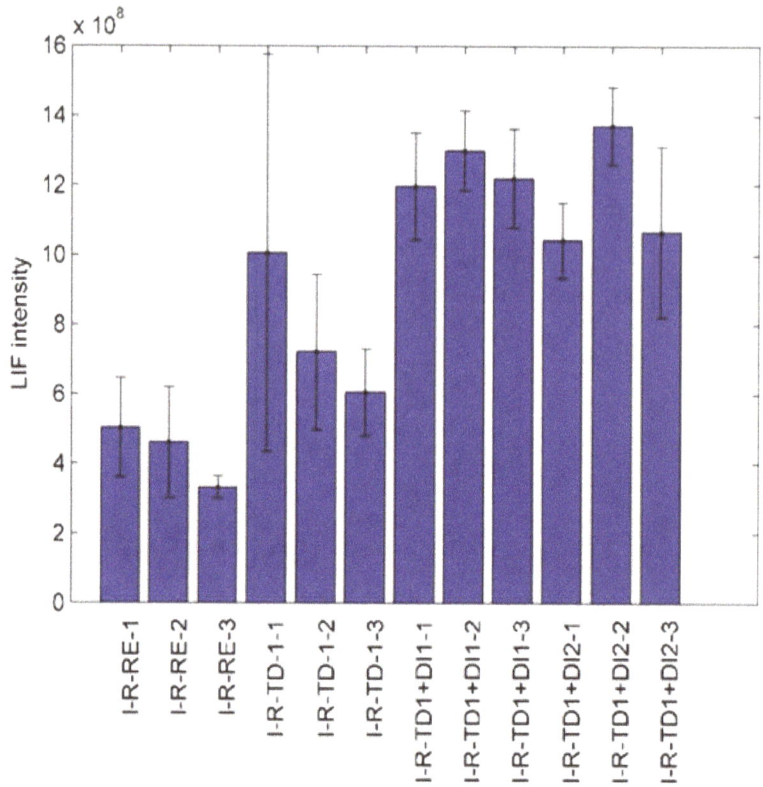

Fig. 3.102 LIF results for the combined modifications of the surface compared with the thermal treatment at the first level and referential results

In summary, our investigations revealed that in certain scenarios within user cases relevant for aeronautical applications, the ENDT methods of AWT, OSEE, e-nose, LIBS, FTIR, LIF, and vibrometry inspection are sensitive to impacts on CFRP specimen surfaces that would induce a bond strength reduction if these CFRP parts were to be used as adherends in a bonding process. The advanced ENDT methods can, therefore, be utilized to identify not-in-order (n.i.o.) adherend surfaces and—based on the findings for design-relevant mechanical adhesive joint properties reported in Chap. 2—also not-ready-to-bond adherend surface states. In the final research chapter of this book (Chap. 5), we underline this perception and prognosis with findings highlighting the performance of ENDT for the monitoring of quality-relevant operand features in adhesive bonding processes involving parts of real aerospace structures with stringers.

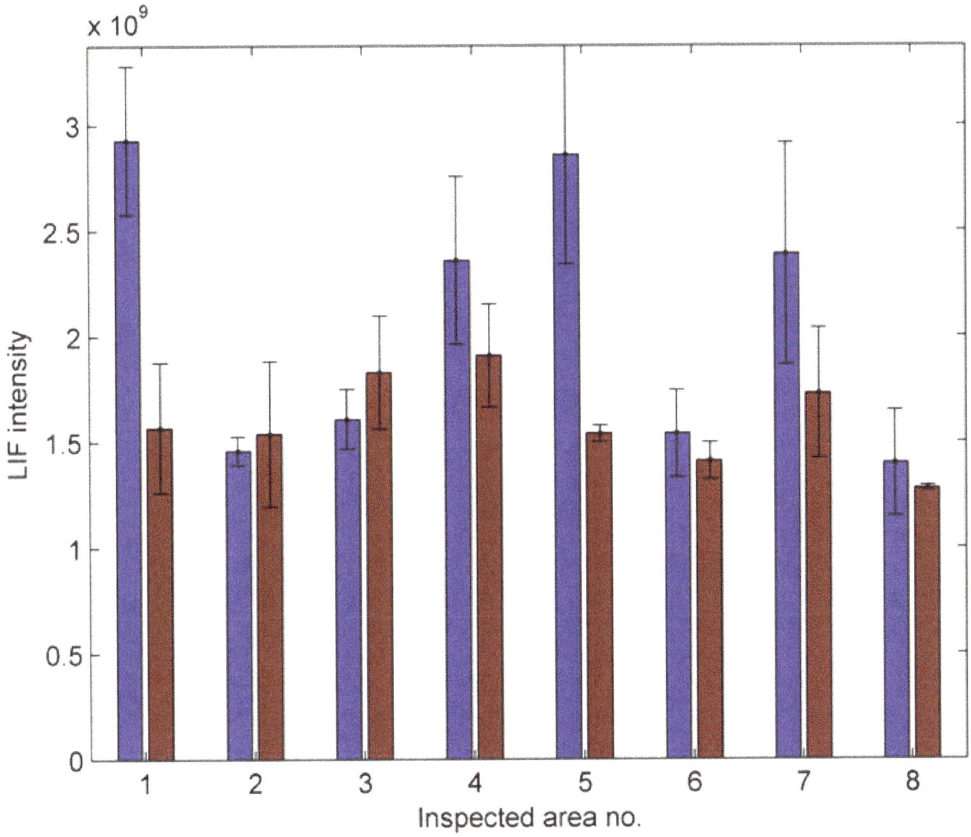

Fig. 3.103 LIF results of the realistic aircraft part inspection; blue bars correspond to clean areas; red bars correspond to contaminated areas; the numbers denote the inspected areas

References

1. Tornow C, Schlag M, Lima LCM et al (2015) Quality assurance concepts for adhesive bonding of composite aircraft structures—characterisation of adherent surfaces by extended NDT. J Adhes Sci Technol 29:2281–2294. https://doi.org/10.1080/01694243.2015.1055062
2. ComBoNDT "Quality assurance concepts for adhesive bonding of aircraft composite structures by advanced NDT" (2015–2018) Project funded from the European Union's Horizon 2020 research and innovation programme under grant agreement No 636494
3. Smith T (1983) Surface quality unit for inspection by nondestructive testing (SQUINT). In: 15th national SAMPE technical conference
4. Parker BM, Waghorne RM (1991) Testing epoxy composite surfaces for bondability. Surf Interface Anal 17:471–476
5. Tornow C, Schlag M, Cavalcanti W et al (2018) In-process quality assurance procedures for surface preparation in adhesive bonding of lightweight composite structures. In: 6th world congress on adhesion and related phenomena
6. Ledesma RI, Yost WT, Palmieri FL et al (2019) Optically stimulated electron emission analysis of surface contamination levels on epoxy composites and effect on failure mode of adhesively bonded specimens, p 70005
7. Mahmud KT, Chawla MK (1989) OSEE determination of fluorocarbon lubricant film thickness on magnetic disk media. Solid State Technol 32:135–136

8. Moutsompegka E, Tserpes K, Noeske M et al (2019) Experimental investigation of the effect of pre-bond contamination with fingerprints and ageing on the fracture toughness of composite bonded joints. Appl Compos Mater 26:1001–1019. https://doi.org/10.1007/s10443-019-09763-9
9. de Vito S, Massera E, Miglietta M et al (2016) Detection and quantification of composite surface contaminants with an e-nose for fast and reliable pre-bond quality assessment of aircraft components. Sens Actuators B Chem 222:1264–1273
10. Nakamoto T (ed) (2016) Essentials of machine olfaction and taste. Wiley, Singapore
11. Röck F, Barsan N, Weimar U (2008) Electronic nose: current status and future trends. Chem Rev 108:705–725
12. Fonollosa J, Fernández L, Gutiérrez-Gálvez A et al (2016) Calibration transfer and drift counteraction in chemical sensor arrays using direct standardization. Sens Actuators B Chem 236:1044–1053
13. Fernandez L, Guney S, Gutierrez-Galvez A et al (2016) Calibration transfer in temperature modulated gas sensor arrays. Sens Actuators B Chem 231:276–284
14. Reidt U, Helwig A, Plobner L et al (2017) Detection of microorganisms onboard the international space station using an electronic nose. Gravit Space Res 5:89–111
15. Katnam KB, Da Silva LFM, Young TM (2013) Bonded repair of composite aircraft structures: a review of scientific challenges and opportunities. Prog Aerosp Sci 61:26–42
16. Petrone G, Bruno M, Bocchetto F et al (2016) An innovative health monitoring system for aircraft landing gears. In: Proceedings of the European workshop on structural health monitoring (EWSHM 2016), pp 235–244
17. Markatos DN, Tserpes KI, Rau E et al (2013) The effects of manufacturing-induced and in-service related bonding quality reduction on the mode-I fracture toughness of composite bonded joints for aeronautical use. Compos B Eng 45:556–564
18. Moutsompegka E, Tserpes KI, Polydoropoulou P et al (2017) Experimental study of the effect of pre-bond contamination with de-icing fluid and ageing on the fracture toughness of composite bonded joints. Fatigue Fract Eng Mater Struct 40:1581–1591
19. Tserpes KI, Markatos DN, Brune K et al (2014) A detailed experimental study of the effects of pre-bond contamination with a hydraulic fluid, thermal degradation, and poor curing on fracture toughness of composite-bonded joints. J Adhes Sci Technol 28:1865–1880. https://doi.org/10.1080/01694243.2014.925387
20. Markatos DN, Tserpes KI, Rau E et al (2014) Degradation of mode-I fracture toughness of CFRP bonded joints due to release agent and moisture pre-bond contamination. J Adhes 90:156–173
21. Marco S, Gutierrez-Galvez A (2012) Signal and data processing for machine olfaction and chemical sensing: a review. IEEE Sens J 12:3189–3214
22. de Vito S, Esposito E, Salvato M et al (2018) Calibrating chemical multisensory devices for real world applications: an in-depth comparison of quantitative machine learning approaches. Sens Actuators B: Chem 255:1191–1210
23. Helwig A, Hackner A, Müller G et al (2018) Self-test procedures for gas sensors embedded in microreactor systems. Sensors 18. https://doi.org/10.3390/s18020453
24. Salvato M, de Vito S, Esposito E et al (2016) An holistic approach to e-nose response patterns analysis—an application to nondestructive tests. IEEE Sens J 16:2617–2626
25. Maier K, Helwig A, Müller G (2017) Room-temperature dosimeter-type gas sensors with periodic reset. Sens Actuators B: Chem 244:701–708. https://doi.org/10.1016/j.snb.2016.12.119
26. de Vito S, Miglietta ML, Massera E et al (2017) Electronic noses for composites surface contamination detection in aerospace industry. Sensors (Basel) 17. https://doi.org/10.3390/s17040754
27. Salvato M, de Vito S, Miglietta M et al (2017) Hydraulic oil fingerprint contamination detection for aircraft CFRP maintenance by electronic nose. In: 2017 ISOCS/IEEE international symposium on olfaction and electronic nose (ISOEN). IEEE, pp 1–3

28. AIRSENSE analytics GmbH portable electronic nose. https://airsense.com/en/products/portable-electronic-nose
29. Ricciardella F, Massera E, Polichetti T et al (2014) A calibrated graphene-based chemi-sensor for sub parts-per-million NO_2 detection operating at room temperature. Appl Phys Lett 104:183502
30. Pasquini C, Cortez J, Silva LMC et al (2007) Laser induced breakdown spectroscopy. J Braz Chem Soc 18:463–512
31. Ledesma R, Palmieri F, Connell J (2020) Laser induced breakdown spectroscopy of polymer matrix composites for real-time analysis of trace surface contaminants: a review. Int J Adhes Adhes 98:102528. https://doi.org/10.1016/j.ijadhadh.2019.102528
32. Heckner S (2017) Zerstörungsfreie Prüfmethoden zur Oberflächencharakterisierung von kohlenstofffaserverstärkten Kunststoffen. Dissertation, Technische Universität München
33. Ruzzene M (2004) Simulation and measurement of ultrasonic waves in elastic plates using laser vibrometry. In: AIP conference proceedings. AIP, pp 172–179
34. Ruzzene M (2007) Frequency–wavenumber domain filtering for improved damage visualization. Smart Mater Struct 16:2116–2129. https://doi.org/10.1088/0964-1726/16/6/014
35. Sohn H, Dutta D, Yang JY et al (2011) Delamination detection in composites through guided wave field image processing. Compos Sci Technol 71:1250–1256. https://doi.org/10.1016/j.compscitech.2011.04.011
36. Flynn EB, Lee, JR, Jarmer GJ et al (2012) Frequency-wavenumber processing of laser excited guided waves for imaging structural features and defects. In: Proceedings of the 6th European workshop on structural health monitoring, Dresden, Germany
37. Ludeker W, Gunther KP, Dahn HG (1995) Comparison of different detection set-ups for laser induced fluorescence monitoring of vegetation. EARSeL Adv Remote Sens 3
38. Wachinger G, Meer T, Sea B (2012) Detection of contaminants—a key aspect for composite adhesive repair bonding. In: ECCM15—15th European conference on composite materials
39. Malinowski PH, Sawczak M, Wandowski T et al (2014) Characterisation of CFRP surface contamination by laser induced fluorescence. In: Kundu T (ed) Health monitoring of structural and biological systems 2014. SPIE, 90640E

4

Bondline Quality Assessment

Paweł H. Malinowski, Tomasz Wandowski, Wiesław M. Ostachowicz, Maxime Sagnard, Laurent Berthe, Romain Ecault, Igor Solodov, Damien Segur, and Marc Kreutzbruck

Abstract We present the results of extended non-destructive testing (ENDT) methods for bond line quality assessment in adhesive joints. The results presented were derived for important application scenarios with regards to aircraft manufacturing and the in-service repair of composite structures. The electromechanical impedance (EMI), laser shock adhesion testing (LASAT), and nonlinear ultrasound scanning (NUS) were used on flat coupon samples, scarfed samples, and curved samples. The EMI method applied to the flat coupons showed some relation of the frequency shift to the level of contamination. For the curved samples, there was insufficient sensitivity to differentiate distinct levels of contamination, while for scarfed samples in most cases both detection and distinction were possible. The LASAT method gave good results for the coupon samples, which were also in accordance with the results of the G_{IC} and G_{IIC} tests. For coupon samples with multiple contaminations, we obtained results with varying significance. In the case of NUS, the measurements revealed an increase in nonlinearity affected by contamination at the interphase between the CFRP adherend and the adhesive layer for the majority of scenarios comprising single contamination of flat coupons and scarfed samples. The effect of multiple contaminations was a decrease in nonlinearity for the curved samples.

P. H. Malinowski (✉) · T. Wandowski · W. M. Ostachowicz
Institute of Fluid–Flow Machinery, Polish Academy of Sciences, Fiszera 14, Gdańsk, Poland
e-mail: pmalinowski@imp.gda.pl

M. Sagnard · L. Berthe
PIMM, CNRS-ENSAM ParisTech, 151 Bd de l'Hôpital, 75013 Paris, France

R. Ecault
Airbus Operations S.A.S., 316 route de Bayonne, B.P. D4101, 31060 Cedex 9 Toulouse, France

I. Solodov · M. Kreutzbruck
Institut für Kunststofftechnik, Universität of Stuttgart, Pfaffenwaldring 32, 70569 Stuttgart, Germany

D. Segur
CEA LIST, CEA Saclay, 91191 Gif-sur-Yvette, France

Keywords NDT · NDE · Adhesive bonds · Electromechanical impedance · Laser shock adhesion test · Nonlinear ultrasonics

4.1 Introduction

The merits of lightweight, high-strength composite materials characterized by unprecedented performance, energy efficiency, safety, and environmental compatibility are widely recognized and applied in many sectors of the economy. The role of technology in joining and consolidating composite parts and other materials is of primary importance to provide manufacturing and maintenance flexibility for high-performance composite components. In this context, the adequate non-destructive testing (NDT) and non-destructive evaluation (NDE) of interfacial adhesion, which determines the reliability of a structural bond, has long been considered a challenge for the NDT/E community [1].

This chapter presents methods for the assessment of the bond lines of aircraft composite structures, including a description thereof, as well as a discussion of the results obtained with each method. With regards to potentially suitable methods for the bond line quality assessment of aircraft composite structures, we have perceived a considerable advancement over the last decade. From a chronological standpoint, the research presented in this chapter is a continuation of the studies initiated within the framework of ENCOMB (extended non-destructive testing of composite bonds [2], an FP7 EU project. Originally, the idea behind ENCOMB was to screen assessment methods that could be utilized in standardized quality assurance procedures for bonded aerospace composite structures because, in essence, the NDT methods currently used in this field are neither sufficient nor reliable for the purpose of certifying bonded composite structures utilized in aerospace structures. This problem was initiated by the fact that more and more aerospace structures are being manufactured from fibre-reinforced composites, and aerospace engineers are looking for new methods to join these. The idea was to develop joining methods based on adhesive bonding, enabling a reduction of the aircraft weight through a reduction or even a total elimination of riveted joints, which are commonly used in aluminium-based aerospace structures. We would like to underline here that rivets are also common in aerospace composite structures, whereby they are joined in conjunction with an adhesive bond line in order to ensure the reliability of the joint in cases of bond line failure, an approach that is also referred to as "chicken rivets". Nowadays, according to the certification rules for aerospace structures, it is not possible to certify an aircraft composite structure that includes solely bonded joints (without additional rivets for safety). This is due to the continuing lack of reliable NDT methods and procedures for the certification of such structures.

The ENCOMB project screened NDT methods that are able to both assess the surface of composite structures prepared for adhesive bonding and evaluate weak bonds in composite structures. The process of adhesive bonding is applied to new composite parts at the manufacturing level or during a composite repair process.

Improper preparation of the surface of the adherend (the composite part prepared for adhesive bonding) is one of the potential sources of a weak bond. Such problems can be caused by chemical contamination by hydraulic fluids (Skydrol) or kerosene during the repair process of a composite aerospace structure. The term weak bond refers to a bond line with reduced strength properties. Another source of weak bonds could be problems in the curing process of the adhesive caused by incorrect parameters, such as temperature. In ENCOMB project, typical scenarios for aerospace structures were selected and investigated, whereby small, flat carbon fibre reinforced polymer (CFRP) samples were utilized individually for surface contamination detection and bonded for bond line assessment. In each sample, only one type of contamination scenario was investigated. Subsequently, the appropriate NDT method was verified for new application areas related to the assessment of bond lines. Hereby, the following methods were investigated in the ENCOMB project: active thermography (vibrothermography and with optical excitation), THz/GHz reflectometry, nonlinear ultrasound, ultrasonic frequency analysis, laser ultrasound, laser scanning vibrometry, and electromechanical impedance. Some preliminary adaptations of NDT methods were made, and these adapted methods are referred to as extended non-destructive testing methods (ENDT).

In this chapter, we present selected results of the research related to the more recent and advanced ComBoNDT (quality assurance concepts for adhesive bonding of aircraft composite structures by advanced NDT) [3] H2020 EU project. This project is partially a continuation of the ENCOMB project; however, only the most promising ENDT methods from the ENCOMB project were further investigated, whereby the following methods are presented in this chapter: electromechanical impedance (EMI), laser shock adhesion testing (LASAT) and nonlinear ultrasound scanning (NUS). The research used small, flat CFRP samples with single contaminations, but further extended the study to include samples with combined contaminations, as well as curved samples and samples with scarfed bonding. Finally, components of real aerospace structures with stringers were also investigated, and the respective findings are reported in the next chapter.

The results presented in this chapter were achieved from the research conducted into the applied ENDT methods. Looking ahead, we would like to highlight that, in certain scenarios, these methods are sensitive to impacts affecting a bond strength reduction, and therefore, could potentially be utilized to identify not in order (NIO) joints during a bond line assessment.

4.2 Electromechanical Impedance

In this section, we will describe the principle and instrumentation of the electromechanical impedance (EMI) method and detail the results recently obtained in the ComBoNDT project.

4.2.1 *Principle and Instrumentation*

The EMI technique utilizes measurements of the electrical parameters of a piezoelectric transducer bonded to the surface of the host structure. Due to the electromechanical coupling of the transducer and the structure, the mechanical characteristics of the structure have an influence on the electrical characteristics of the piezoelectric transducer. EMI is an active method. The transducer works both as an actuator by exciting the structure and as a sensor to sense the response in the defined frequency band. Data analysis for this method is performed in the frequency domain. Adhesive bonds have already been investigated with the EMI method by embedding a transducer into a lap joint [4, 5], these investigations were both numerical and experimental, but they have focused only on aluminium lap joints. Here, we investigate the EMI method in its application to the adhesive bonds of CFRP composites, whereby some of the initial results on this topic have been published in [6]. In the research reported here, the IM3570 laboratory impedance analyzer (Hioki) (Fig. 4.1a) was used and the focus was mostly on conductance spectra measured up to 5 MHz. The measurements were performed with a piezoelectric disc transducer that was bonded to the investigated sample with cyanoacrylate glue (Fig. 4.1b). The disc had a 10 mm diameter and a 0.5 mm thickness and was manufactured by CeramTec (SONOX P502 material). For the analysis, the root mean square value was calculated for the considered parts of the spectra, and the location and magnitude of the resonance peaks were also tracked.

4.2.2 *EMI Results*

4.2.2.1 EMI Results on the Coupon Level Samples

For the coupon samples, two groups of samples were investigated. The first group represented production-related contamination of the samples, while the second group represented repair-related modifications/contaminations of the samples. The sample set investigated for the production scenario comprised three reference samples, nine samples contaminated with a release agent, nine samples contaminated with moisture, nine samples contaminated with fingerprints (artificial sweat), and six samples with a combined contamination of release agent and fingerprints. The sensors were glued to the middle of the sample surface. After considering the registered spectra, some of the results were rejected (I-P-RE3, I-P-RA-1-2, I-P-RA-2-3, I-P-RA-3-2, I-P-MO-2-2, I-P-FP-1-2, and I-P-FP-2-2) due to the different shapes of the spectra. In a previously reported study referring to samples with moisture, release agent, and fingerprint contamination, the bandwidth corresponding to the maximum peak of conductance was inspected [7], whereby the average values of RMS on differential spectra were also analyzed and no clear sensitivity to the contamination level was observed. In the research reported here, the measured spectra were inspected and a local resonance was identified for the bonded samples in the bandwidth 4.3–4.7 MHz.

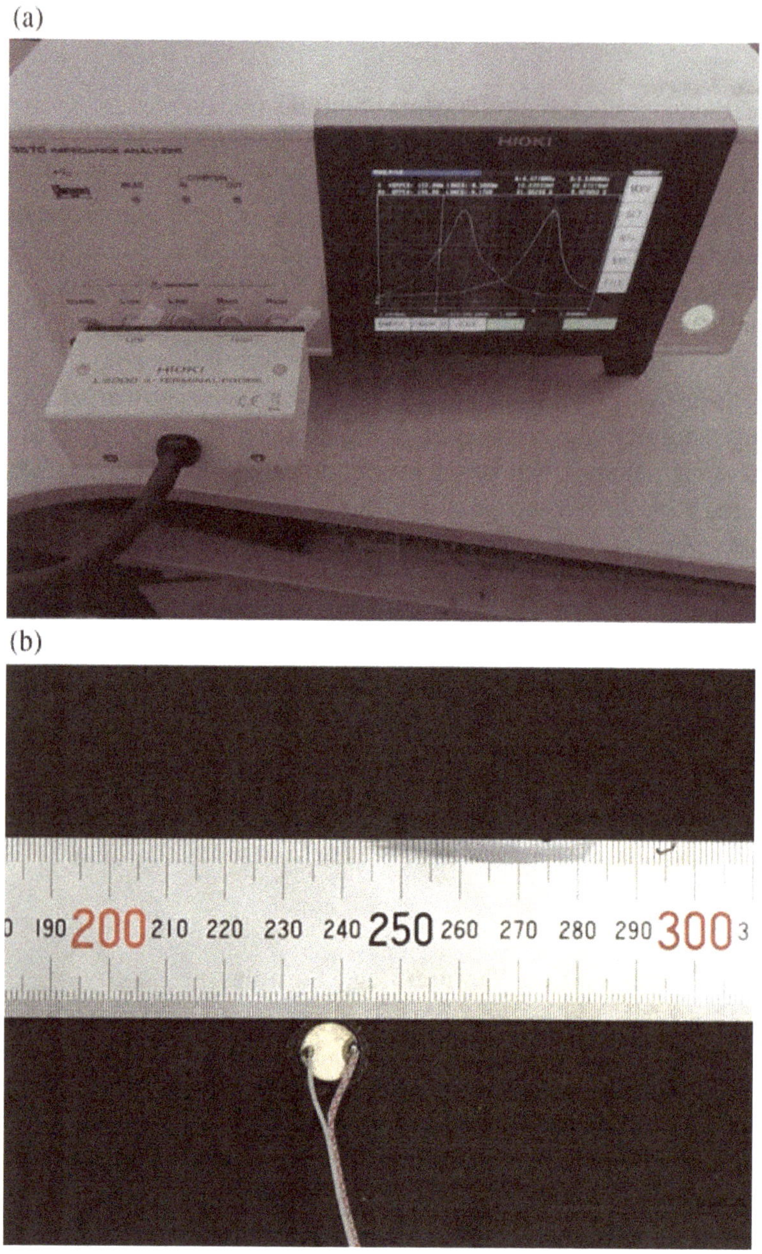

Fig. 4.1 Photographs of the experimental setup for the electromechanical impedance (EMI) measurements; **a** impedance analyzer and **b** surface-bonded piezoelectric sensor

Considering the difference in the root mean square (RMS) values, it can be noted that in this frequency region, an increase in the energy (RMS) value is observed because all the differential values are non-negative (Fig. 4.2). Next, the location of the resonance peak in the considered bandwidth was tracked for all samples. In most of the cases, the peak shifted left. The relative frequency shift between the two reference samples was only −0.027% (Fig. 4.3). Considering the absolute values, all the results

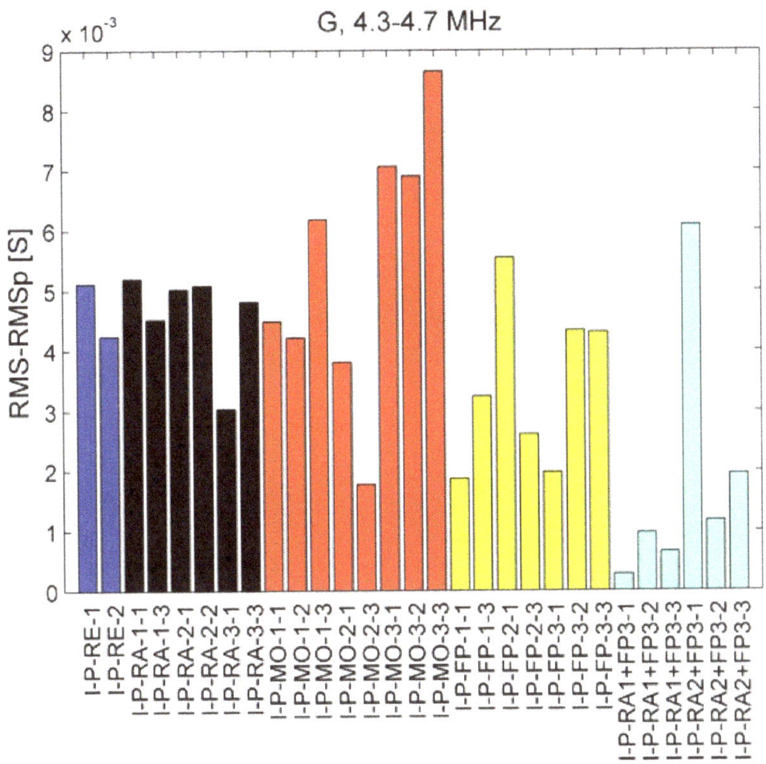

Fig. 4.2 Differences of the RMS values of EMI sensors bonded to flat coupon samples and free sensors for the production scenario

for the contaminated samples lie above this threshold. In general, the frequency shift increases with the increase of the moisture and fingerprint contamination level. However, there are some exceptions, such as very low values for the I-P-MO-1-2, I-P-FP-1-3, and I-P-RA2+FP3-3 samples and a high value for the I-P-FP-1-1 sample.

The sample set investigated for the repair scenario comprised three reference samples, nine thermally degraded samples, nine samples contaminated with de-icer, nine samples modified by a faulty curing of the adhesive, and six samples with a combined contamination of thermal degradation and contamination with de-icer. After considering the registered spectra, some of the results were rejected (I-R-RE-2, I-R-TD-1-2, I-R-TD-2-2, I-R-TD-3-3, I-R-FC-3-2, I-R-TD1+DI2-1, I-RTD1+DI2-2) due to the different shapes of the spectra. Moreover, previous results have shown that the I-R-TD-3-1 sample was delaminated [8]. The spectra were inspected in the 4.25–4.70 MHz region. A local resonance was observed, except for the sample with delamination. The presence of this local resonance is indicated by the increase in RMS values with respect to the free piezoelectric sensors (Fig. 4.4). The resonance location was tracked with respect to the I-R-RE-1 sample result. The relative frequency shift is shown in Fig. 4.5, and most of the samples were characterized by leftward peak shift. Considering the absolute value, all the samples are above the reference threshold defined by the I-R-RE-3 result, with the exception of the I-R-TD-2-3 sample. In the case of the thermal treatment (TD), the increase of the treatment temperature

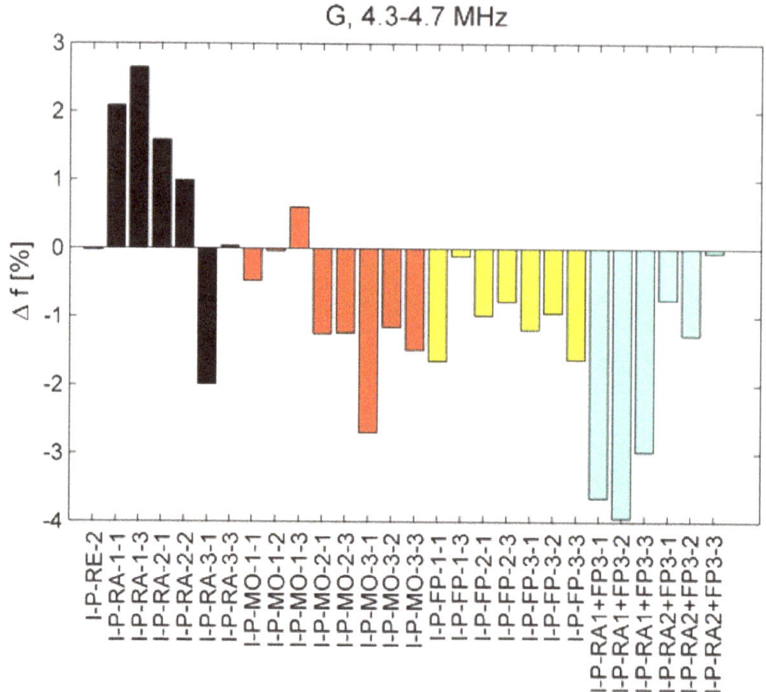

Fig. 4.3 Tracking of local EMI resonance in the 4.3–4.7 MHz range; frequency shift calculated in relation to the I-P-RE-1 sample

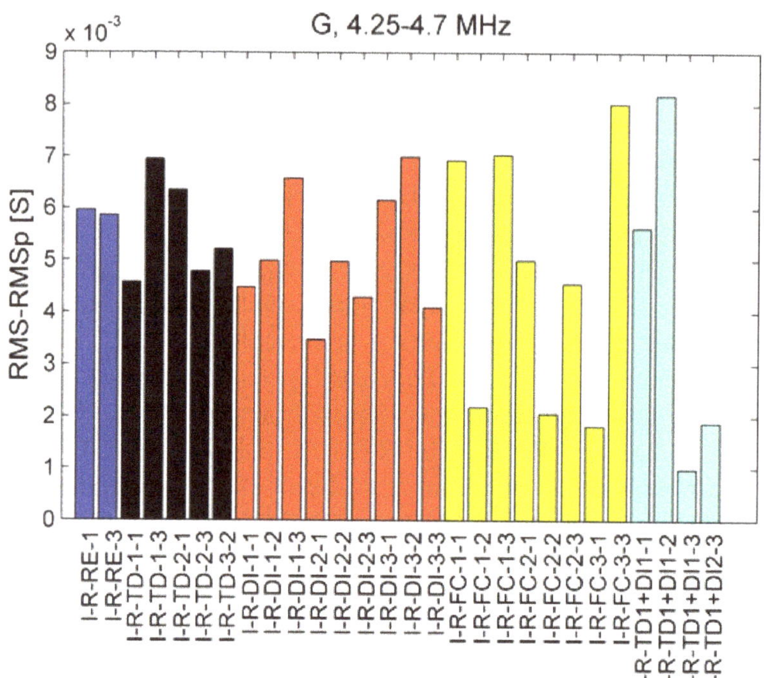

Fig. 4.4 Differences of the RMS values of EMI sensors bonded to flat coupon samples and free sensors for the repair scenario

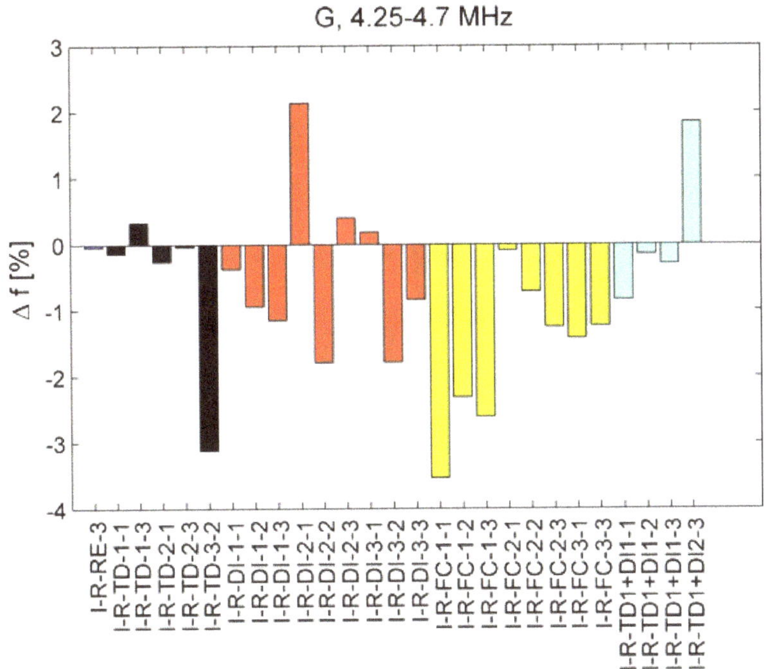

Fig. 4.5 EMI conductance peak shift in the 4.25–4.70 MHz range for the flat coupon samples (repair scenario)

results in an increase of the shift. In the case of the de-icer (DI) contamination, such dependence was not observed. The most intensive faulty curing was applied to the I-R-FC-1-x samples, and this was also observed for the results obtained with the EMI method. The results presented in Fig. 4.5 also show a sensitivity of the frequency shift to the mixed contamination level. However, these results should be treated with caution because only one sample (I-R-TD1+DI2-3) at the higher level of contamination was able to be considered.

4.2.2.2 EMI Results on Pilot Level Samples

In the case of the production scenario, the pilot samples were curved. The inspected conductance spectra were different, which should be attributed to the different sample geometry. No clear appearance of additional resonances was observed. Therefore, the highest peak in the spectrum was tracked. The frequency shift calculations were performed in relation to the II-P-RE-1 sample (Fig. 4.6). Four samples were characterized by a rightward shift, while three showed a leftward shift. Considering the absolute values of the shifts, the difference between the reference samples is 1.71%, so only three contaminated samples out of the six were above this threshold. The sensitivity to the contamination level was not clear in the considered case.

The pilot samples from the repair scenario had a scarfed bonding area. The spectra of free sensors and the spectra after bonding these sensors to the scarfed samples

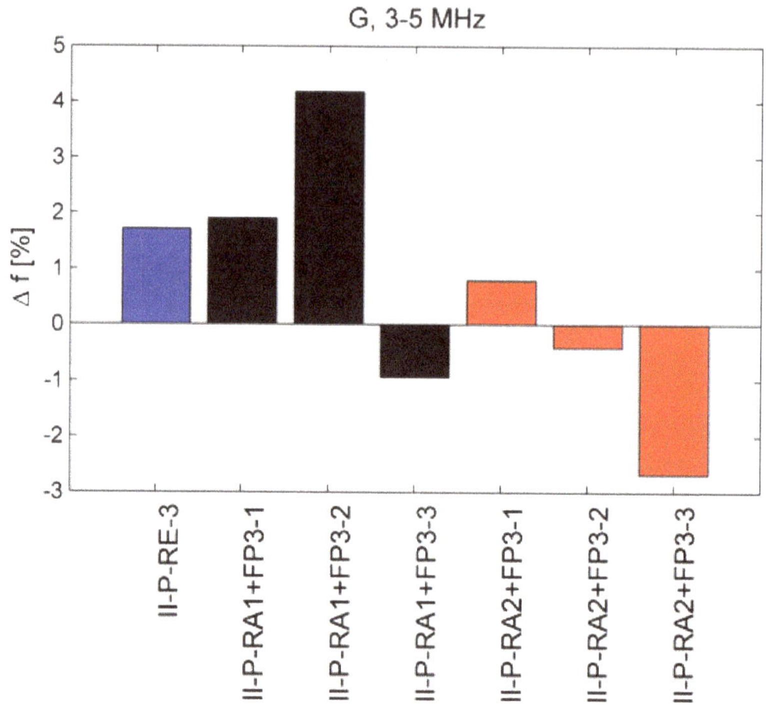

Fig. 4.6 EMI conductance local maximum shift in the range of 3–5 MHz for the curved samples

were inspected. The sample II-R-TD1+DI2-2 was rejected from the analysis due to the different shapes of the spectrum. For the rest of the samples, a local resonance was observed in the conductance range 3.95–4.15 MHz. In this frequency range, the root means square values were calculated for both the free sensors (RMSp) and the samples with bonded sensors (RMS). It was observed that all the differences (RMS-RMSp) are non-negative (Fig. 4.7). This indicates the influence of the sensor bonding to the inspected samples. This frequency region was further inspected, and the frequency shift was tracked in relation to the II-R-RE-1 sample. It can be observed that there were significant differences between the reference samples (Fig. 4.8a), which were higher than for the samples with contaminated adhesive bonds. The changes in the magnitude of this resonance in relation to the same reference sample are presented in Fig. 4.8b. The magnitude for most of the samples increased, while in two cases it dropped. In comparison with the reference samples, it can be noted that the absolute magnitude is higher for all the samples with contaminated bonds, except for the II-R-TD1+DI1-2 sample. While differences were detected, it cannot be stated that the contamination levels can be distinguished.

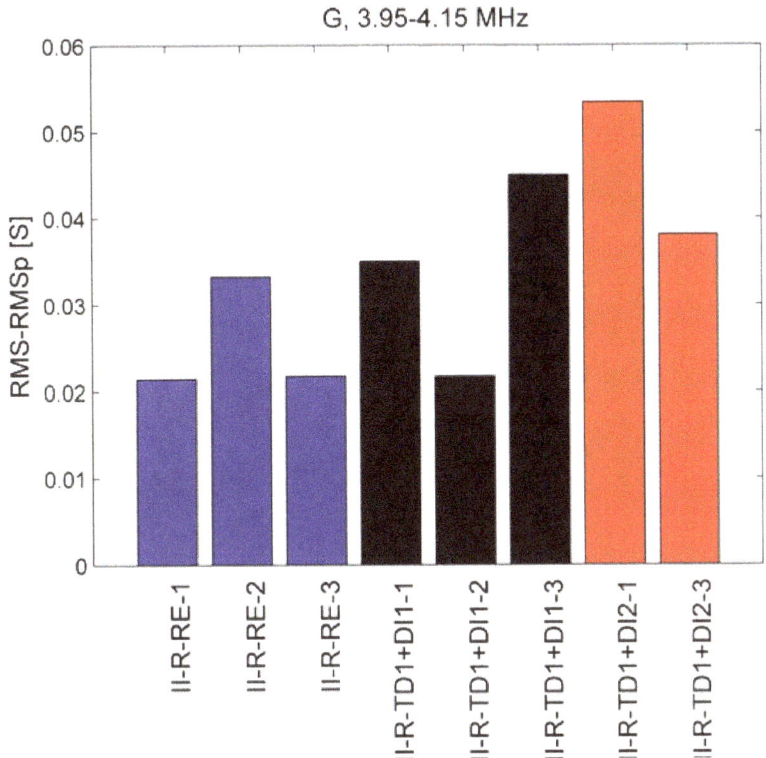

Fig. 4.7 Differences of RMS values of sensors bonded to scarfed samples and free sensors

4.3 Laser Shocks

In this section, we describe the principle and instrumentation of the Laser Shock Adhesion Test (LASAT) method and detail the results recently obtained in the ComBoNDT project.

4.3.1 Principle and Instrumentation

LASAT relies on the recombination of shock waves produced within the tested material to generate localized tensile stress. To produce the required shock, a high-power Nd-YAG laser is used. Its energy often varies between 1 and 40 J, and short pulses (for 3–20 ns) are necessary to generate the required power density (from 1 to 10 GW/cm^2). When the laser reaches the sample surface, a thin layer of the material is vaporized in dense plasma (within the pressure range of 1–5 GPa). Upon expanding, this plasma creates shock waves within the material Fig. 4.9. These shock waves (full lines in Fig. 4.10), followed later by release waves (dashed lines in Fig. 4.10), travel to the back face of the sample. Once the shock waves reach the back face, they

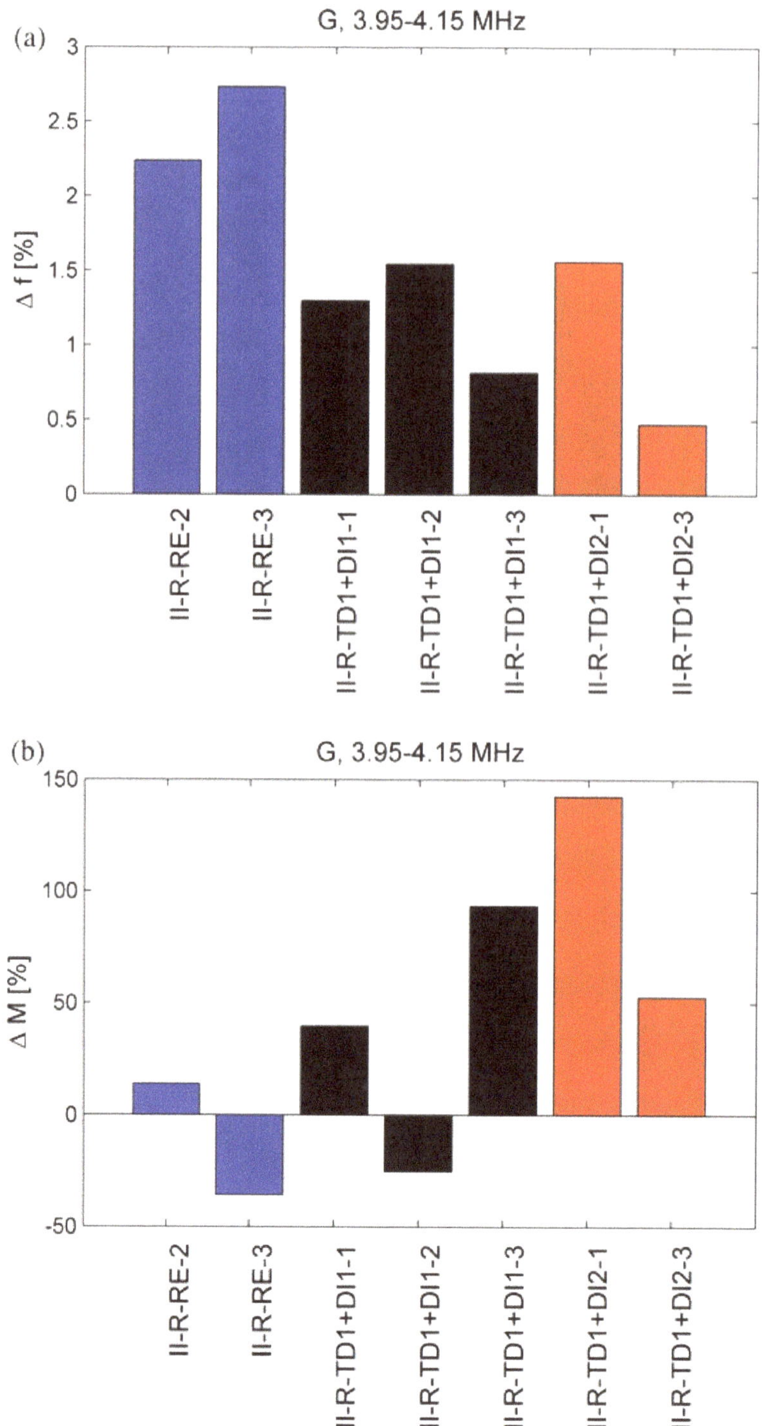

Fig. 4.8 **a** Conductance local maximum shift in the range of 3.95–4.15 MHz for the scarfed samples; the relative shift was calculated in relation to the II-R-RE-1 reference sample; **b** change of the conductance magnitude in the range of 3.95–4.15 MHz for the scarfed samples; the relative change was calculated in relation to the II-R-RE-1 reference sample

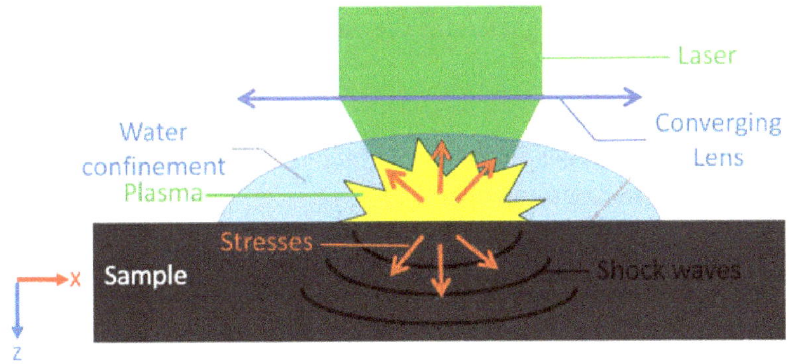

Fig. 4.9 Generation of shockwaves using a laser

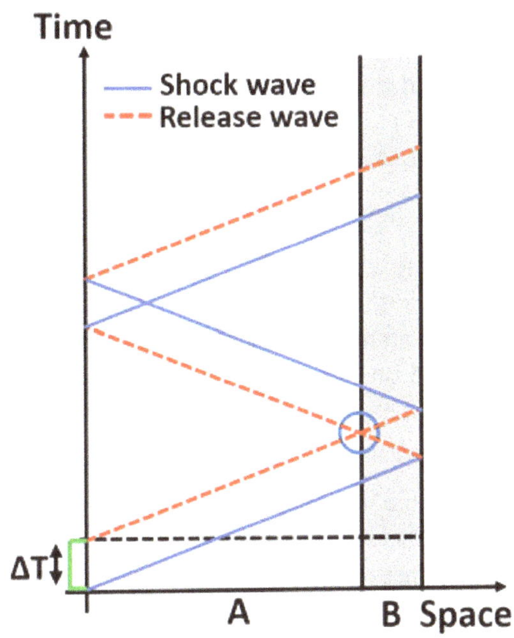

Fig. 4.10 Time/space diagram describing the shock wave behaviour within a material

are reflected as release waves. When the latter intersects the incident release wave, tensile stress is created (circle in Fig. 4.10).

Studies have shown that it is possible to increase the generated stress by using the water—as a confinement medium [9]. Hence, stresses can be multiplied by up to four times for the same laser intensity, allowing less powerful and more compact lasers to be used for this application.

It is possible to link the laser intensity applied to the surface to the pressure generated within the material [10]

$$P(\mathrm{GPa}) = 0.01\sqrt{\frac{\alpha}{\alpha+3}}\sqrt{I(\mathrm{GW/cm^2}}\sqrt{Z_{\mathrm{rel}}(\mathrm{kg\,m^{-2}\,s^{-1}})} \tag{4.1}$$

where

- α = part of the energy being used for the gas ionization
- I = laser intensity
- Z = the relative impedance

The relative impedance is a ratio between the material impedance of the sample (Z_{sample}) and the confinement impedance (Z_{conf})

$$Z_{rel} = 2 \times \frac{Z_{conf} \cdot Z_{sample}}{Z_{conf} + Z_{sample}} \tag{4.2}$$

Shock generation using a laser was first introduced by Vossen [11] to test the adherence of thin layers. Following these initial studies, Yuan [12] dealt in depth with the measurement of the interfacial stresses generated by this technique. More recently, Boustie [13] and Bolis [14], extended the laser shock technique to thicker metallic coatings. Gilath was among the first to test bond adhesion using laser shock generated tensile stresses [15]. Later, composite materials [16] and bonded composite materials [17] were also studied using the Laser Shock Adhesion Test.

Beginning in 2010, the ENCOMB project aimed at the development of new Non-Destructive Tests (NDT) that are capable of assessing the strength of bonded composite structures, and LASAT showed great promise in the detection of weak bonds and had the advantage of being contact-free. However, there was a major drawback: the rear part of the composite structure can be damaged within the range of laser intensity used to test the bond strength.

Based on the results of the project, three new configurations were proposed to control the tensile stresses inside the material by adjusting a temporal parameter of the laser:

- Variable laser pulse duration [18] (Fig. 4.11a);
- Double front face shocks [19] (Fig. 4.11b);
- Symmetrical laser shocks (Fig. 4.11c).

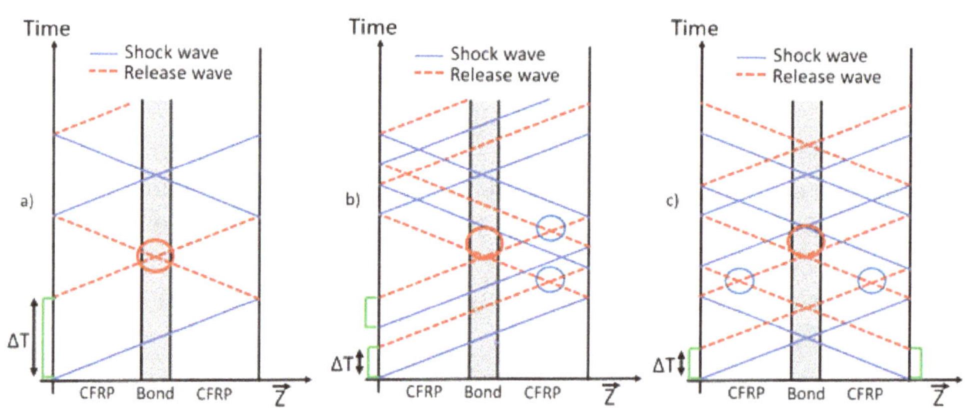

Fig. 4.11 LASAT optimisations: **a** variable laser pulse; **b** double shock; and **c** symmetrical shot

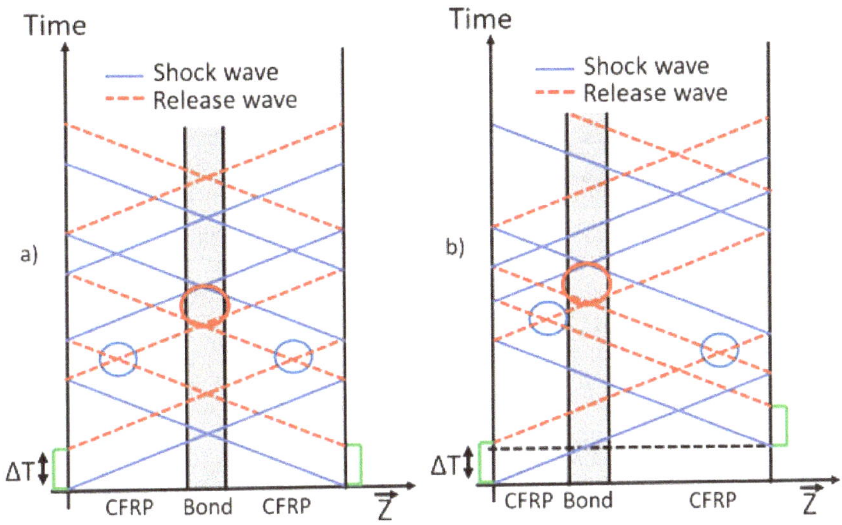

Fig. 4.12 Symmetrical LASAT: **a** without time delay and **b** with time delay

The symmetrical laser shock was implemented for ComBoNDT. The secondary tensile stress area, which is similar to that found in the single shot setting, is still noticeable (blue circles in Fig. 4.11c). However, in the symmetrical configuration, an even higher tensile stress area is located where both reflected shock waves cross (red circle in Fig. 4.11c). The position of this stress depends only on the time delay between both shots (Fig. 4.12).

For a given location ***A***, the time delay can be calculated using the following equation:

$$\Delta T = \sum_i t_{li} \times \frac{\rho_{li}}{Z_{li}} - \sum_j t_{rj} \times \frac{\rho_{rj}}{Z_{rj}} \tag{4.3}$$

where t_{li} represents the thickness of a layer situated to the left of ***A***, ρ_{li} is its density, and Z_{li} is its impedance. In the same way, each variable with the index letter "r" refers to the right part of ***A***.

4.3.1.1 Instrumentation and Method

Figure 4.13 describes the different steps of a LASAT procedure. It starts with the characterization in situ of the bond and the material properties of a reference sample. This initial step serves two purposes, namely finding the bond threshold and gathering the material data. Using these data, a series of numerical simulations are performed to optimize the laser parameters of intensity, time delay, and focal spot size. The sample is then prepared with an aluminium tape applied to both areas where the laser shot will occur. The tape serves as a sacrificial layer, preventing the sample from being ablated by the laser. It also ensures good repeatability of the shots. Indeed,

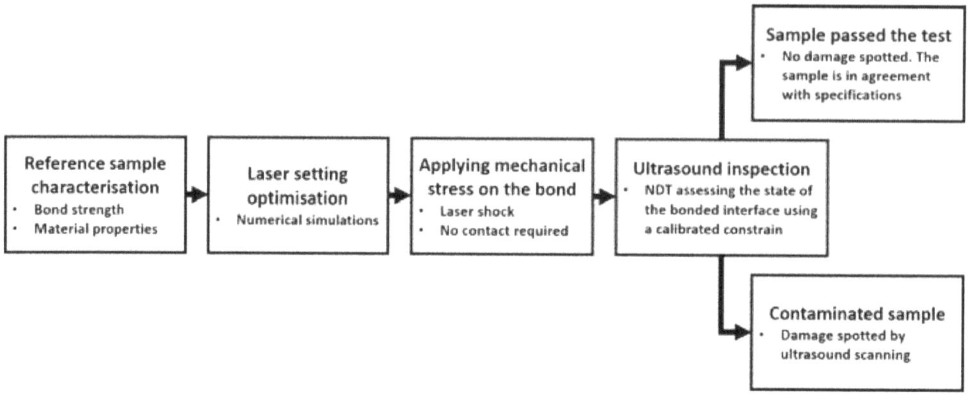

Fig. 4.13 Laser shock adhesion test methodology

the generated plasma significantly depends on the laser/matter interaction. By only using aluminium as a protective sacrificial layer, all parameters are fully controlled during the laser shot process.

The laser intensity is set to a percentage of the bond threshold found during the first step. In the aeronautical industry, a bond is considered faulty or weak if its adherence is below 80% of the nominal adherence defined in the specifications.

Once all the parameters are set, the laser shot occurs. If the bond is weak, the high tensile stress generated on the bond will create a small gap at the interface between the composite and the joint. This can be detected using ultrasound scanning, a reliable method that is often used to reveal these kinds of defects.

If no damage is spotted, then the sample is considered sound and has passed the test. Otherwise, the bond shows weakness and does not follow the specifications, meaning the component cannot be used and has to be replaced. Thus, this technology is only destructive when the tested part is not in accordance with the specifications. If the part is sound, it can be used as it is.

4.3.1.2 Description

LASAT is currently performed on the Hephaistos platform (Fig. 4.14). This facility consists of a high-power GAÏA laser (Fig. 4.14a), which has the particularity of consisting of two different lasers, with the beams being polarized from one another by 90°. Three different platforms are currently powered by the laser: one for laser shock peening studies, one for single shot LASAT, and one for symmetrical LASAT. The final one (Fig. 4.14b) uses a polarizer that, due to the two different polarisations, splits the main beam into two different laser beams (beam A and beam B). Using mirrors and lenses, each ray is then carried to the prepared sample (Fig. 4.14c).

In order to ensure a good control over the pressure spatial repartition, Diffractive Optic Elements (DOE) are added to both beam paths. These optics smooth the focal spot, creating a homogeneous focal spot on the sample (Fig. 4.15).

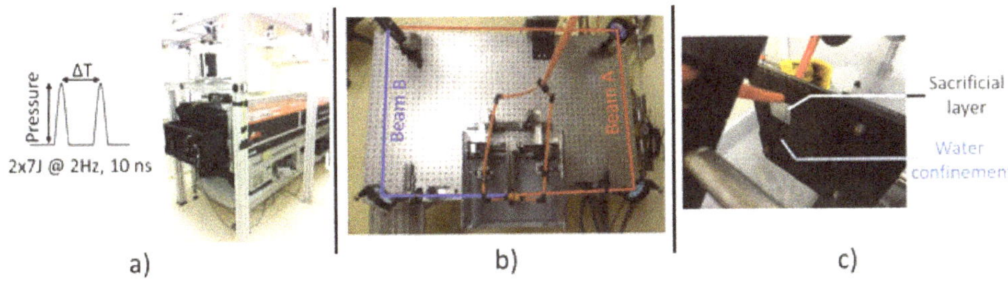

Fig. 4.14 Hephaistos facility: **a** GAÏA laser from Thales; **b** symmetrical experimental setup; **c** sample setup with water confinement and the protective aluminium tape

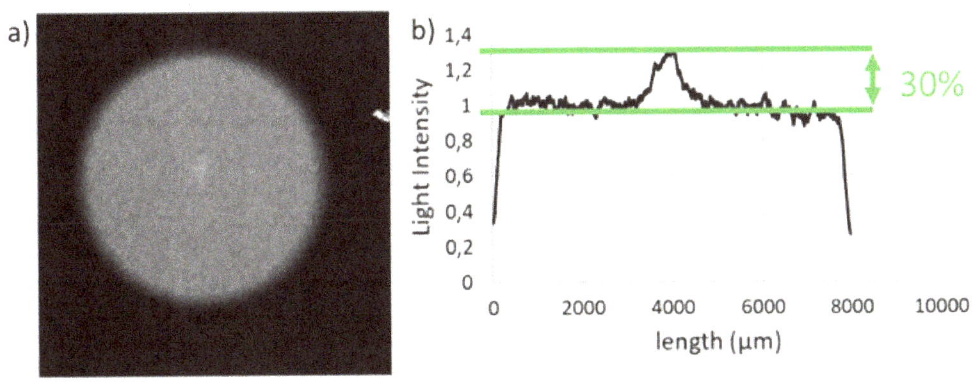

Fig. 4.15 DOE: **a** focal spot exiting a DOE and **b** light repartition over the focal spot

A series of 20 shots were realized for 11 different laser intensities, ranging from 5 to 100% of the maximum intensity. The standard deviation for each intensity level was less than 1%, meaning that each shot is reproducible and can be used for comparison purposes.

As specified earlier, ultrasound scanning is required to examine the structure after the laser shock. These are calibrated while the reference sample is being studied in situ to obtain the material data. In the case of composite layer bonding, ultrasound is needed to find the signature of the front face (blue in Fig. 4.16), the bond (yellow in Fig. 4.16), and the back echo (green in Fig. 4.16).

Before shooting the sample, a simulation must be run to optimize the laser parameters, whereby the position and width of the maximum tensile stress are analyzed and modified if necessary (Fig. 4.17). The challenge of such numerical calculations resides in the simulation of the high strain rates generated by the technology ($\sim 10^7\ s^{-1}$). Because the material used is orthotropic, it is not possible to describe the state of the matter using an equation of state, as is usually done with isotropic materials. The simulation of damage also requires more advanced laws to take into account the different damaging modes of composite materials.

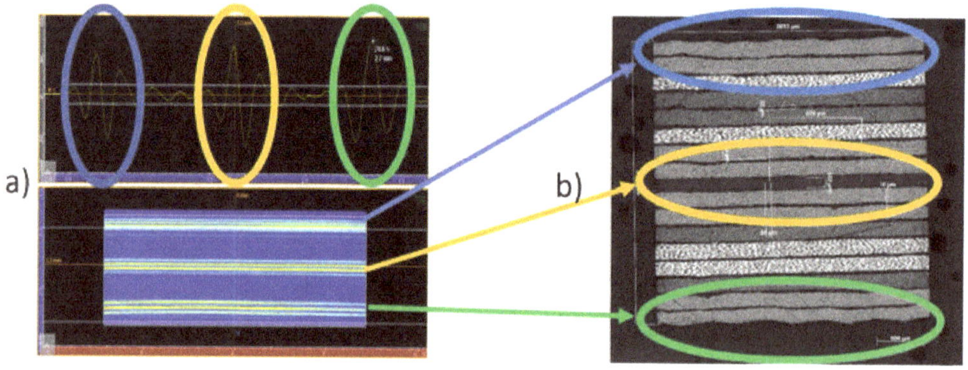

Fig. 4.16 In situ studies: **a** reference ultrasound of the benchmark structure and **b** photomicrograph of the benchmark structure

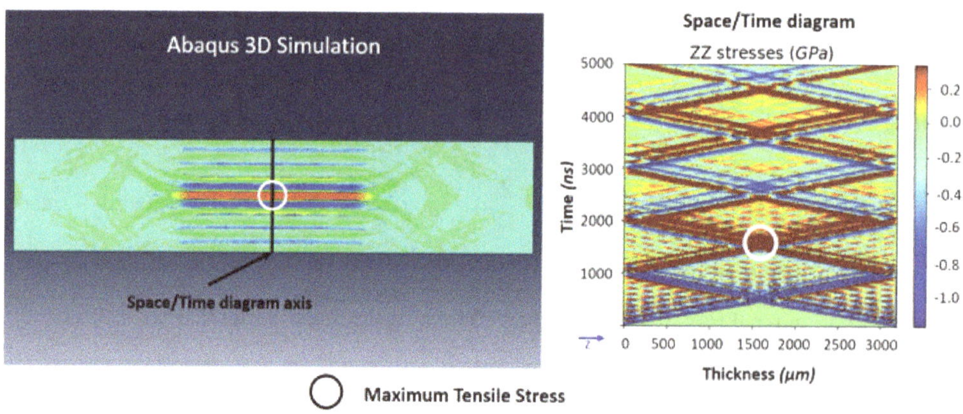

Fig. 4.17 Finite element simulation of a laser shock

4.3.2 Laser Shock Results

4.3.2.1 Laser Shock Results on Coupon Level Samples

Prior to the large-scale experiments, an initial study was performed to assess whether ultrasound scanning is efficient enough to spot a defect created with a laser shock. Each coupon was divided into 16 different areas, whereby the laser intensity (ranging from 5 to 95%) was raised from one area to the other. The first area to be opened defined the sample's bond threshold (red circle in Fig. 4.18a). This was followed by a study in which both the first opened area and the one preceding it (red circle in Fig. 4.**18**b) were cold mounted and examined using optical microscopy.

In each case, the initial default was located on the bond, and none of the composite laminate was damaged. This study also showed that a defect detected using ultrasound scanning (Fig. 4.18c) can be confirmed by a direct inspection through photomicrographs (Fig. 4.18b). These results lead to the conclusion that ultrasound scanning is a reliable technique to reveal defects generated using laser shocks.

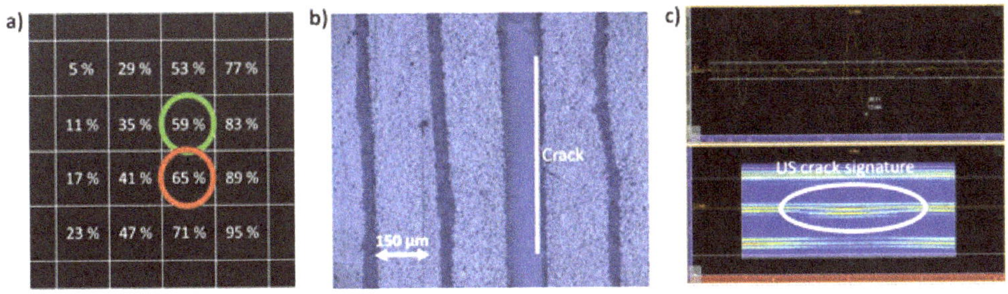

Fig. 4.18 **a** Shot layout over a coupon sample; **b** photomicrograph of an open bond; **c** related B-scan

Several procedures can be used to assess bond strength using LASAT. In this study, one specific area was subjected to incremental laser intensity. After each shot, an ultrasound scan was performed to verify whether or not the bond had failed at this given intensity. If it broke, then that energy was assumed to be the bond threshold for that sample. On the other hand, if it remained intact, the energy was increased. Samples representative of both the manufacturing and the repair scenarios were tested. In each case, the reference sample sets the standard bond threshold. When testing a contaminated sample, if the threshold found was to be lower than the standard one, the LASAT was considered successful. The results for the manufacturing samples are summarized in Fig. 4.19.

Each bar represents the energy required to open the bond. The reference sample is presented in blue, while orange is the sample from the release agent contamination, green represents the moisture contamination, and yellow is for the fingerprint contamination. The red line represents the scanning intensity. If these samples were from an actual component, then this intensity should be chosen to assess whether or not a bond is weak. Every sample that fail before this point should be considered weak. The LASAT successfully differentiated eight out of nine contaminated samples. The only

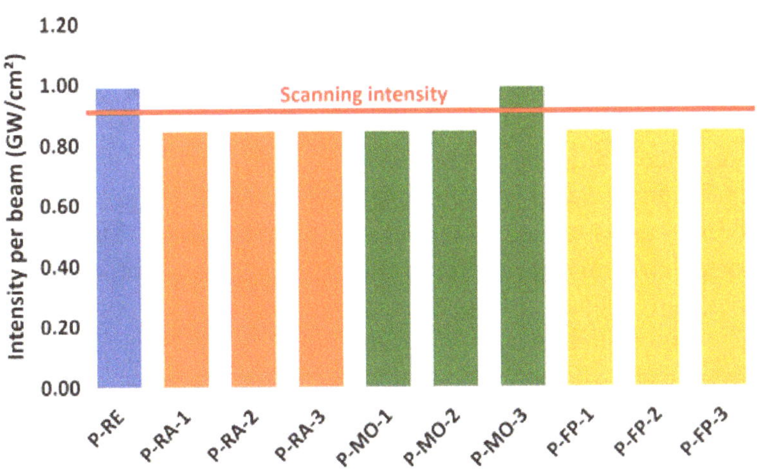

Fig. 4.19 Summary of results for the manufacturing samples

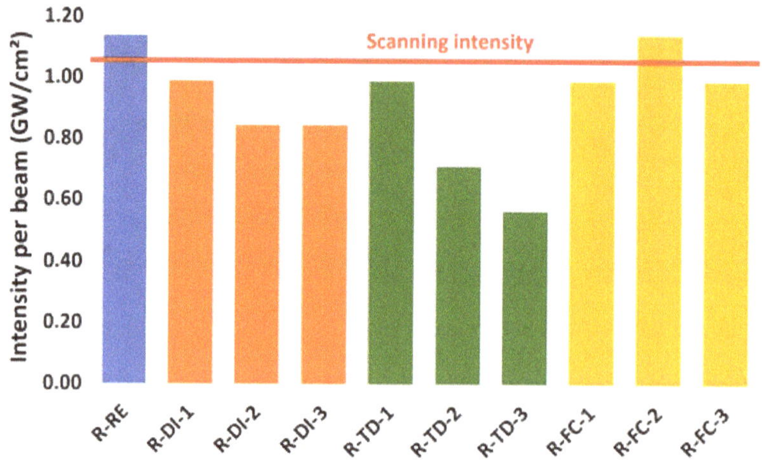

Fig. 4.20 Summary of the results for the repair samples

sample with the same intensity as the reference sample was the one with the highest moisture contamination. It has already been observed that following an increase in water intake, an epoxy-based adhesive bond can have a higher mechanical strength than a non-contaminated bond [20]. Since the bond was mechanically tested, this result is in accordance with the literature. The repair scenario samples also showed good results for the LASAT (Fig. 4.20). The reference sample is again represented in blue, while the results for the de-icer are in orange, those for the thermal degradation in green, and the samples for the false curing are in yellow. For the first two contamination types, the different levels of contamination could be detected, especially for the thermally degraded sample, whereby each level of contamination has its own threshold. The results are not as clear for the faulty curing. This type of contamination is not easy to control, and these samples had already shown delamination before any test was performed on them. Even though two out of three samples were shown to be contaminated, this is hard to state conclusively given the initial state of the tested samples.

It has also been observed that the repair reference sample and the production reference sample had different bond thresholds, namely 0.99 GW/cm^2 for the production scenario and 1.15 GW/cm^2 for the repair scenario. Photomicrographs taken from both bonds revealed a different crack pattern Fig. 4.21.

As the repair sample was ground down to the fibre before the epoxy-based adhesive was applied, the cracks tended to propagate within the bond. Meanwhile, for the production sample, a thin layer of prepreg matrix was still present, and it was noticed that the crack mainly propagated along the interface between the bond and the prepreg matrix. Studies have already shown that the interface strength can be higher than the actual mechanical strength of the glue [21]. One of the explanations for this behaviour was the difference in surface tension between the prepreg matrix and the actual composite fibre, which may have led to a different bond quality with the FM300 used for the joint.

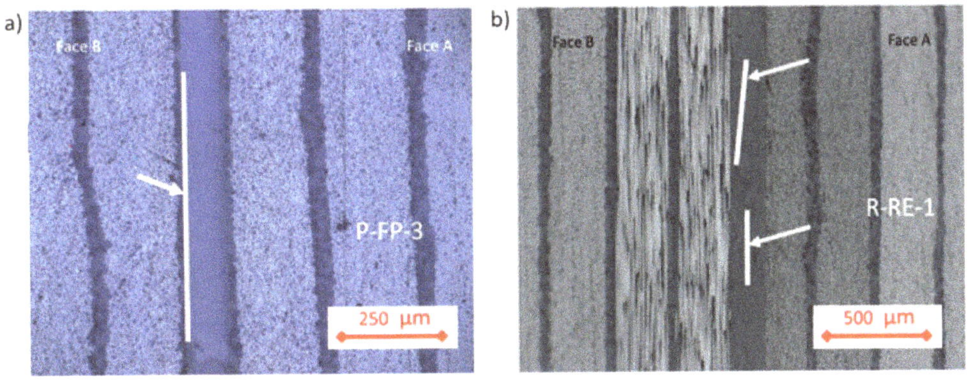

Fig. 4.21 Photomicrographs: **a** production sample and **b** repair sample

To conclude, the LASAT gave good results for the coupon samples, and these were in accordance with the results found during the G_{IC} and G_{IIC} tests [22]. In certain cases, the test was also able to identify the different levels of contamination.

4.3.2.2 Laser Shock Results for the Pilot Level Samples

Three different scenarios were tested for the pilot samples: multiple contaminations, scarf bonding, and curved samples.

The results are mixed for coupon samples with multiple contaminations (Fig. 4.22). The LASAT did not differentiate production samples with multiple contaminations from the reference sample, sample, and the intensity required to open the faulty joints was the same as for the standard. The results are also hard to explain in the repair scenario. The technology detected the least contaminated sample but was unable to clearly detect the most contaminated sample. However, the same contaminations on the scarfed samples showed much better results (Fig. 4.23). In this case, the LASAT technology was effective at differentiating a sound bond from a weak one. The difference between the results for the coupon samples and the scarfed samples is not yet fully understood; the bond geometry may have played a

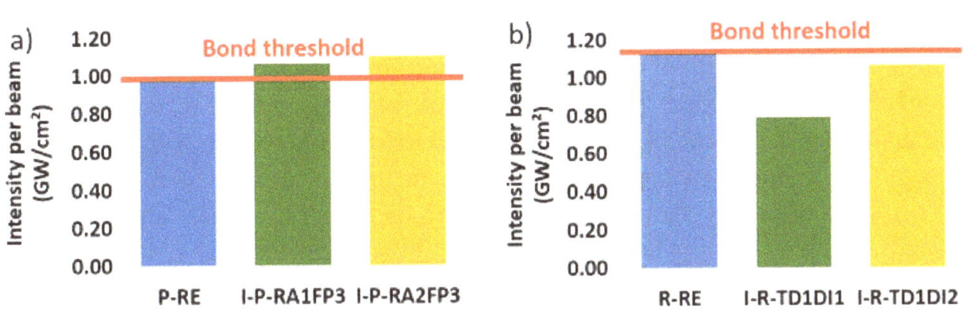

Fig. 4.22 Summary of the results of multiple contaminations on coupon samples for **a** production samples and **b** repair samples

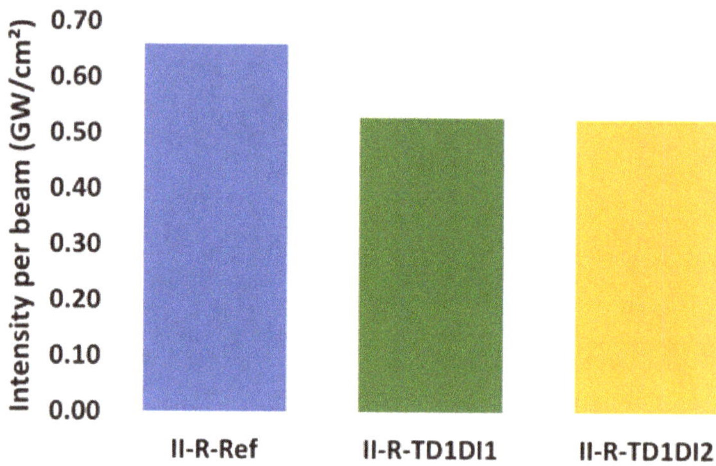

Fig. 4.23 Summary of the results for multiple contaminations on scarfed samples

part in the crack detection/propagation within the contaminated samples. This technology also heavily relies on a standard sample, and the reference used for the coupon samples may have been produced with different parameters. Furthermore, the curved samples were highly porous and did not follow the specifications. Therefore, no relevant conclusion can be derived from the results and further study is required in order to draw decisive conclusions for these samples.

4.4 Nonlinear Ultrasonic Technique

In this section, we describe the principle and instrumentation of the nonlinear ultrasonic technique (NUS) and detail the results recently obtained in the ComBoNDT project.

4.4.1 Principle and Instrumentation

Conventional ultrasonic NDT instruments used in industry and technology make use of the so-called linear elastic response of materials, which results in amplitude and phase variations of the input signal due to its interaction with any defects present. The nonlinear approach to ultrasonic testing involves the nonlinear material response related to the frequency changes of the input signal. These spectral changes are caused by the nonlinear dynamics of solids, which range in scale from the inter-atomic level for perfect materials to meso- and macro-scale nonlinearity for damaged areas [23–25]. In many cases, monitoring material nonlinearity directly reveals vulnerable areas within a material or product with sensitivity far superior to traditional ultrasonic inspection methods [26]. Numerous studies have confirmed that the increase in nonlinearity is closely related to a "softening" of the material due to, e.g., fatigue, cracking, and internal interfaces. This also correlates with the general characteristic of acoustic nonlinearity that is related to the thermal expansion of the material, which is negligible for stiff materials and increases significantly for soft solids [27].

A conventional method to evaluate nonlinearity in composite coupons is based on the second harmonic measurements for propagating plate (Lamb) modes [28, 29]. The approach proposed in this research makes use of a local nonlinear response of the adhesive bond. A contaminated bonding layer with weaker adhesion is supposed to have a higher nonlinearity that enables the difference in bonding quality to be recognized, evaluated, and visualized.

The nonlinear technique is based on the local generation of high amplitude vibrations and the detection of the higher harmonics in the excitation area. Commercial piezo-actuators (isi-sys GmbH, Germany) with a frequency response extending from the low kHz into the high kHz range (above 100 kHz) were used in the experiments. The actuators were driven by a CW voltage generated by an HP 33120A arbitrary waveform generator and were vacuum-attached to one of the sides of the sample, while nonlinear vibrations were measured on the opposite side. The generator was combined with a HVA-B100 amplifier to result in a 10–40 V input amplitude for the piezo-actuator. To measure and analyze the frequency content of the vibrations generated locally in the excitation area, a scanning laser vibrometer (SLV, Polytec 300) operating in the vibration velocity mode with a maximum frequency bandwidth of 1.5 MHz was used.

The dynamic range of the SLV measurements (100–120 dB) lies well beyond the level of nonlinear frequency components. In the experiments, the vibration amplitude was measured as being in the range of $(4-5) \times 10^{-8}$ m, thus the local strain that developed directly in the excitation area was approx. 10^{-5}. This strain is sufficient for the manifestation of noticeable local nonlinearity in composite materials.

To avoid an impact of reflections on the local vibration in the excitation area, the edges of the samples were covered with a dissipative material and a high vibration frequency was chosen (49 kHz). With these precautions, the spectra and temporal patterns of the vibrations were measured directly in the acoustic source area (a circle with approx. 5 mm radius), whereby plate wave propagation was not yet involved.

Each value for the fundamental frequency vibration velocity (v_0) and the higher harmonic components (v_n) measured in the probing area was used for the evaluation of the nonlinear ratio

$$N_i = \sum_n \frac{v_n^2}{v_0^2} \tag{4.4}$$

An average value of the nonlinear ratio in the probing area was then calculated with

$$N = \sum_{i=1}^{m} \frac{N_i}{m} \tag{4.5}$$

where m is the number of measurements; the standard deviation of the results was estimated

$$\Delta N = \sqrt{\sum_{i=1}^{m} \frac{(N_i - N)^2}{m(m-1)}} \tag{4.6}$$

The measurements were repeated at various points over the central part of the sample to provide the relative error $\Delta N/N \leq 10{-}20\%$.

The nonlinear ratio N_i is part of the vibration energy ($\sim v_0^2$) converted into the higher harmonics ($\sim v_n^2$) so that it clearly quantifies material nonlinearity. To avoid the influence of the amplitude-dependent effects in the estimation of the nonlinear ratio N, the amplitude of the input voltage of the transducer was kept constant (20 V) over the course of the measurements. As a result, in all the samples measured the fundamental vibration amplitude was virtually constant within an approx. 10% deviation.

4.4.2 Nonlinear Ultrasonic Technique Results

4.4.2.1 Nonlinear Ultrasonic Technique Results for Coupon Level Samples

For the production scenario, the following kinds of contaminants were studied: release agent (RA), fingerprint (FP), and moisture (MO). The repair scenario investigations comprised a de-icing fluid (DI) contamination, thermal degradation (TD), and faulty curing (FC). For each type of contamination in both scenarios, a set of three samples for every level of contamination was prepared to provide reliable statistics. The results of the measurements and calculations of N for all samples (30 repair scenarios [9 I-R-TD, 9 I-R-DI, 9 I-R-FC, 3 I-R-RE] and 30 production scenarios [9 I-P-MO, 9 I-P-RA, 9 I-P-FP, 3 I-P-RE]) are plotted in Figs. 4.24–4.31 to trace the impact of the contamination level (numbers 1, 2, 3) for a particular sample index (−1, −2, −3), whereby the level of contamination increases as the level number increases. For the reference samples free from any contamination, the average values of N(AV) were also calculated.

As shown in Figs. 4.24 and 4.25, the reference samples revealed the minimum values of $N \approx 3$. The contamination of the adhesive noticeably increased the nonlinearity for all samples by 1.5–2 times. The maximum nonlinear ratio N was obtained for the sample TD 3-2 ($N = 50 \pm 6$ [here and below in 10^{-3} units], outside the scale in Fig. 4.26). In this sample, the N_i values were found to depend on the position of the measurement point. This fact, along with the anomalously high value of the

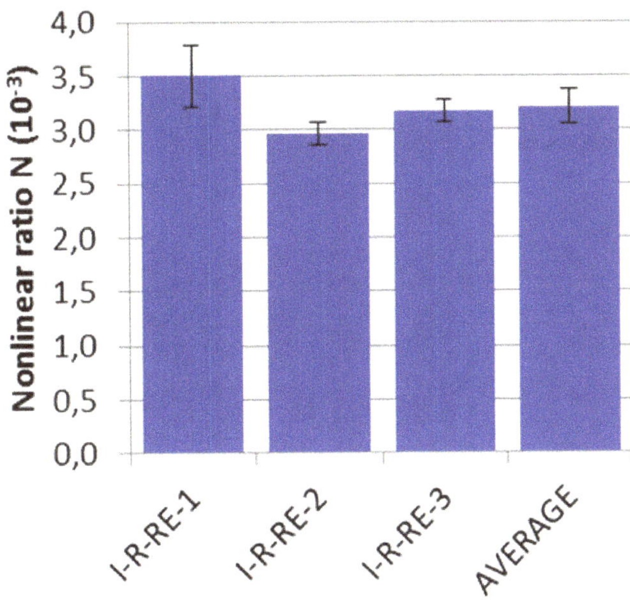

Fig. 4.24 Non-linear ratios for the reference samples of the repair scenario

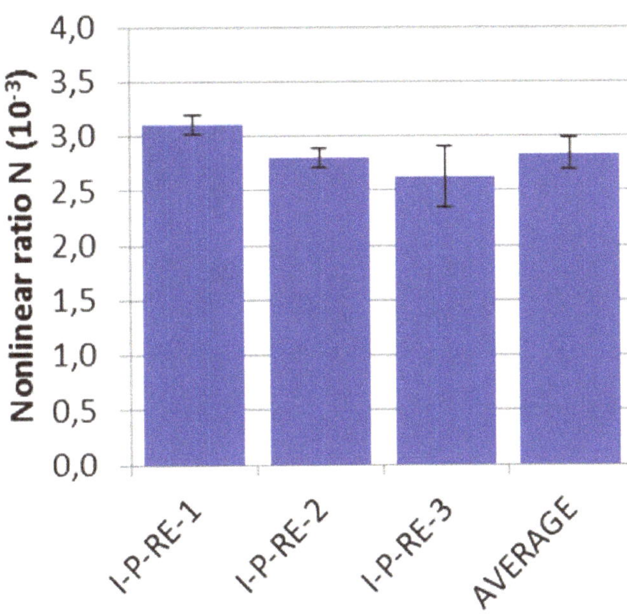

Fig. 4.25 Non-linear ratios for the reference samples of the production scenario

nonlinear ratio, indicates the presence of local delamination in the sample induced by thermal activation, which was also verified with conventional (linear) ultrasonic testing [8].

For the contamination types TD, RA, FP, DI, and FC, the material nonlinearity increased with the increase in the level of contamination (Figs. 4.26, 4.27, 4.28, 4.29, 4.30, and 4.31). For each type of contamination, the values of N changed noticeably with a variation of the contamination level, which indicates a sensitivity to the changes in the thin boundary layer between the adhesive and the adherends. According to the

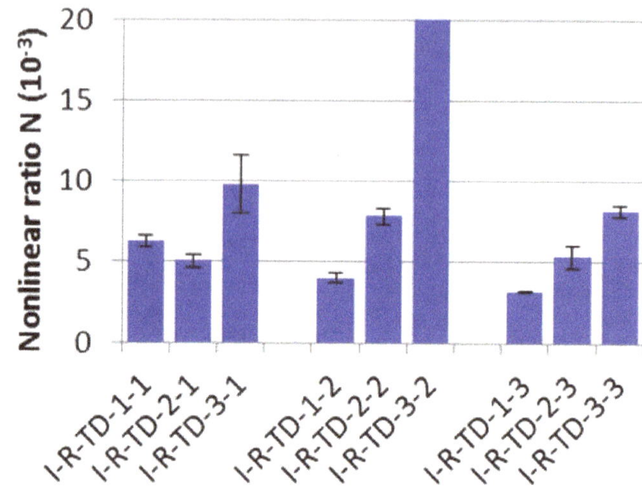

Fig. 4.26 Nonlinearity of each of the three samples (−1, −2, −3) as a function of the TD level of contamination

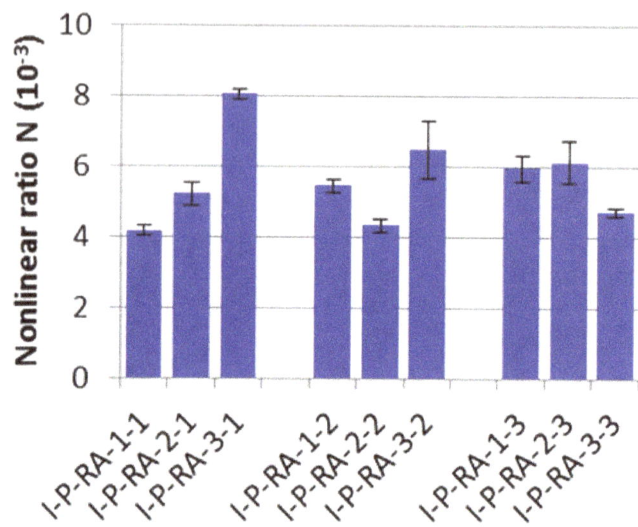

Fig. 4.27 Nonlinearity of each of the three samples (−1, −2, −3) as a function of the RA level of contamination

measurement results, the sensitivity of the nonlinear method is, therefore, sufficient to recognize the effect of contamination of the adhesive bonding.

The overall results for the nonlinearity variation for single contaminations are summarized in Figs. 4.32 and 4.33. Here, the average values of the nonlinear ratios are calculated for each group of samples with presumably the same level of contamination (groups 1, 2, 3) along with the average values for the reference samples. These data can be used to explicitly trace the nonlinearity behaviour as a function of the contamination level for each type, as well as between different types of contamination.

To correlate particular variations in N with the strength of bonding as a function of the level of contamination, we refer back to the introduction to this chapter, where it was mentioned that the increase in nonlinearity is related to a "softening" or weakening of the material. A similar conclusion can be drawn based on the calculations of

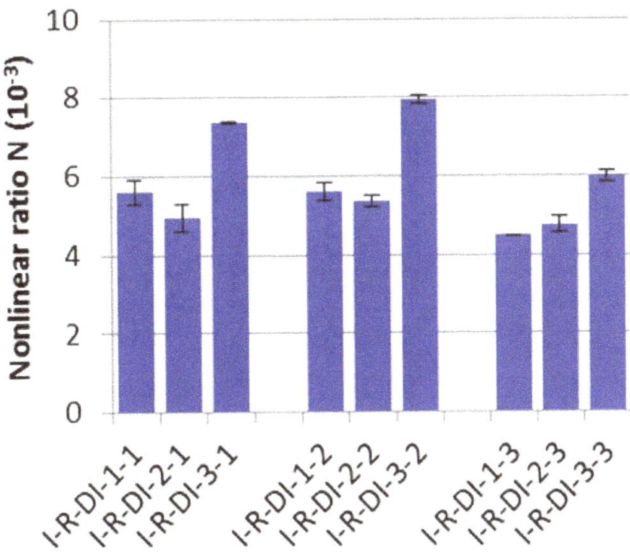

Fig. 4.28 Nonlinearity of each of the three samples (−1, −2, −3) as a function of the DI level of contamination

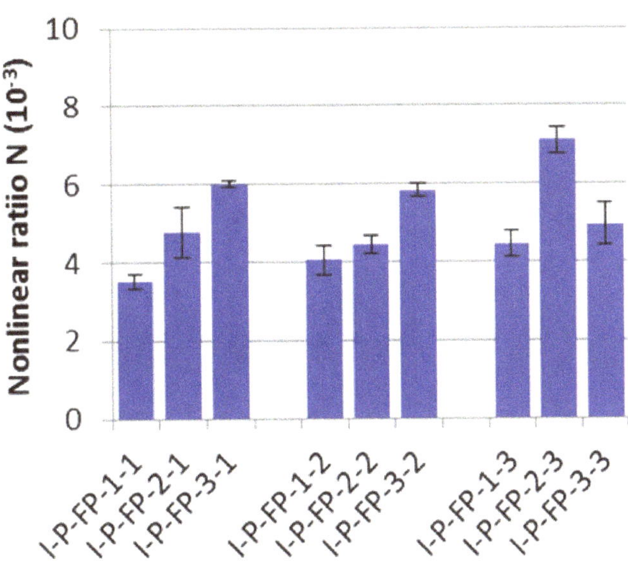

Fig. 4.29 Nonlinearity of each of the three samples (−1, −2, −3) as a function of the FP level of contamination

the stress–strain relations for the adhesive layer [30], which shows that this softening of the boundary layer increases its nonlinear response.

This fact goes some way towards explaining the lowest values ($N \approx 3$) of the nonlinear ratios in the reference samples in Figs. 4.32 and 4.33; being free from any contamination, they are supposed to manifest the highest bonding strength. From this viewpoint, the strongest decrease in bonding strength was caused by the level 3 TD contamination (average value N = 239, outside the scale in Fig. 4.10). A similar (but somewhat lower) decrease in bonding strength was also recognized for the level 3 DI (N = 7.10.6), FP (N = 5.60.3), RA (N = 6.40.9), and FC (N = 7.71.6) contaminations (Figs. 4.9 and 4.10). The MO case is not so convincing as, with

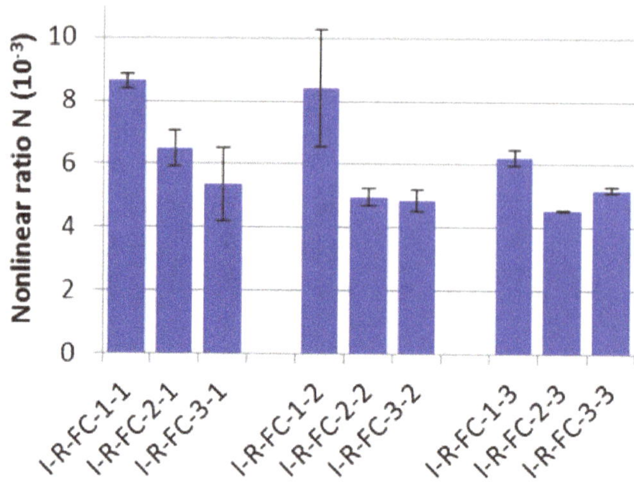

Fig. 4.30 Nonlinearity of each of the three samples (−1, −2, −3) as a function of the FC level of contamination

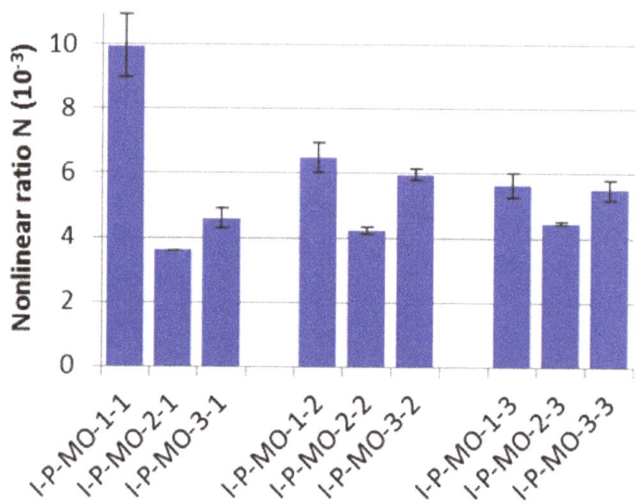

Fig. 4.31 Nonlinearity of each of the three samples (−1, −2, −3) as a function of the MO level of contamination

the exception of an abrupt kick for sample MO1, the nonlinearity did not change noticeably, so $N \approx 5$ for all levels of contamination.

The effect of the combination of contaminations was studied for 12 multi-contaminated flat CFRP samples, using two combinations of single contaminations:

- A constant low-intensity thermal degradation (TD1) with two different moderate amounts of de-icer (DI1, DI2) contamination applied to the surface prior to bonding.
- Two different amounts of release agent (RA1, RA2) and a constant high amount of fingerprint (FP3) contamination applied to the surface prior to bonding.

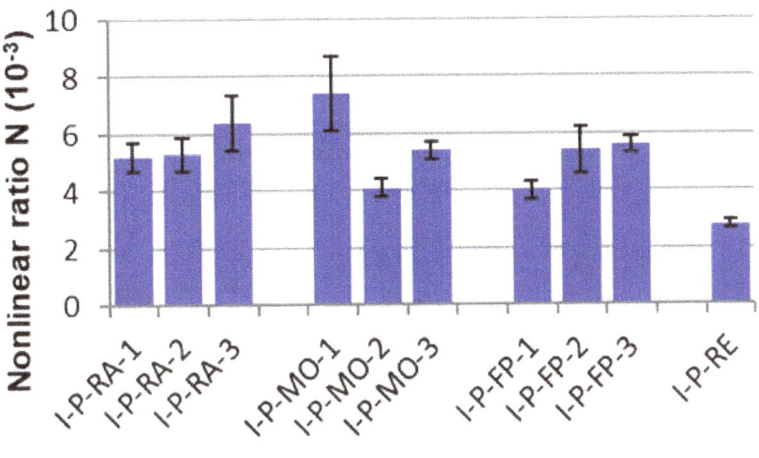

Fig. 4.32 Average nonlinear ratios as a function of the contamination levels for the samples of the production scenario

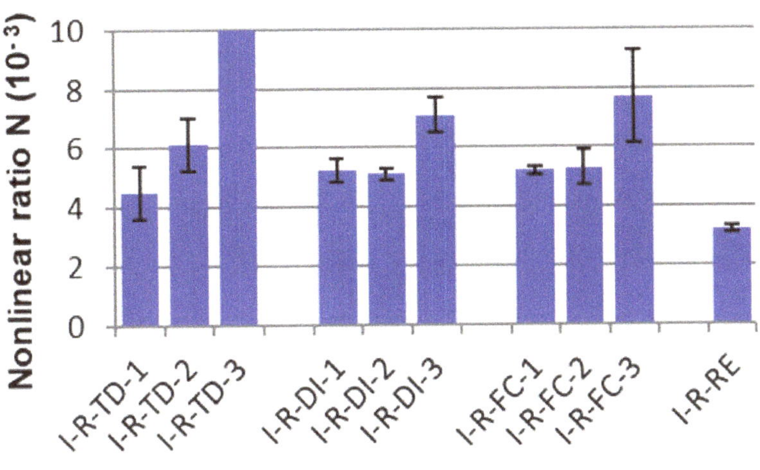

Fig. 4.33 Average nonlinear ratios as a function of the contamination levels for the samples of the repair scenario

The results of the measurements are shown in Figs. 4.34 and 4.35 along with the values of N averaged separately over the groups with the same level of contamination (RA1 and RA2 in Fig. 4.34; DI1 and DI2 in Fig. 4.35). As shown in Fig. 4.34, the addition of a low amount of RA1 to highly contaminated FP3 samples had virtually no impact on the nonlinearity of the laminate, and the average value of N for FP-3+RA-1 (5.8 ± 0.4) stayed very close to the N values that were previously measured for the single FP3 contaminated samples (5.6 ± 0.3), as shown in Fig. 4.32. However, the increase of the RA contamination level to RA2 enhanced the nonlinearity of the laminate noticeably, so that the average value N for FP-3+RA-2 in Fig. 4.34 is 6.9 ± 0.4.

A similar effect of the increase of nonlinearity was observed when increasing the level of DI contamination in combination with TD-1 (Fig. 4.35). The addition

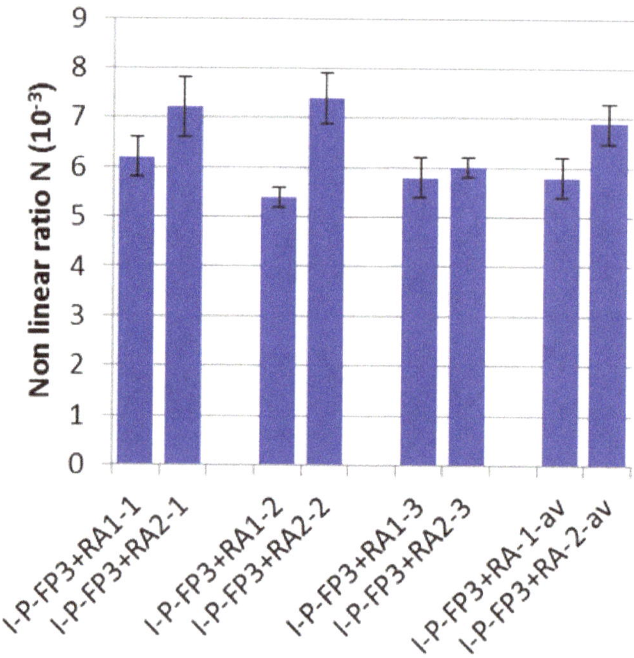

Fig. 4.34 Nonlinearity of contaminated FP3 samples (−1, −2, −3) additionally affected by two levels of RA (RA1 and RA2) contaminations

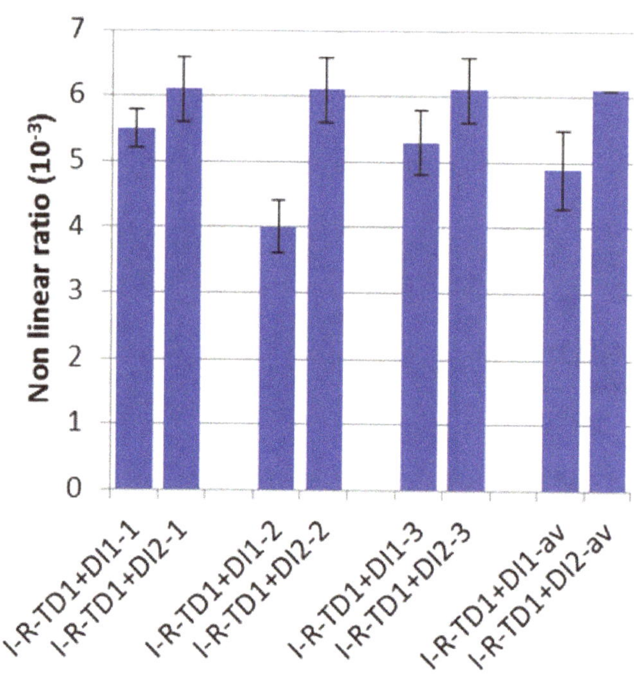

Fig. 4.35 Nonlinearity of contaminated TD1 samples (−1, −2, −3) additionally affected by two levels of DI (DI1 and DI2) contaminations

of DI-1 left the average value of N for TD-1 (4.5 ± 0.9; see Fig. 4.33) unchanged within the margin of error, namely 4.9 ± 0.6 (Fig. 4.35). A moderate level of DI-2 enhanced the nonlinearity level of the laminate substantially, whereby the N for DI-2+TD-1 increased to 6.1 (Fig. 4.35). Both combinations, therefore, demonstrated a cumulative decline in the adhesion between flat coupons caused by the increase of the DI and RA contaminations.

4.4.2.2 Nonlinear Ultrasonic Technique Results on Pilot Level Samples

The nine scarfed samples studied comprised three reference samples (without contamination) and samples with a low level of thermal degradation (TD-1) and two different levels of de-icer contamination (DI-1 and DI-2) applied to the surface of the laminates prior to bonding. These samples were glued at a slant in an attempt to mimic the repairing of composite components in practice. The results of the measurements are presented in Fig. 4.36.

For two sets out of the three samples measured, the experimental data demonstrated an increase in the nonlinearity caused by the contamination in comparison to the reference samples (sets 1 and 3; Fig. 4.36). The nonlinearity of sample set 2 did not change within the margin of error. A particularly strong effect of contamination was observed for sample set 3 (approx. 150% increase in nonlinearity). The effect of the increase of the level of DI contamination was not as noticeable, as was also the case in the adhesion of conventional flat laminates (see Fig. 4.33), whereby the average N values practically did not change between DI-1 and DI-2 (Fig. 4.36). This behaviour shows that even a moderate amount of DI leads to major deterioration of the adhesive properties, while a further increase of the contamination level does not substantially further affect the bonding strength.

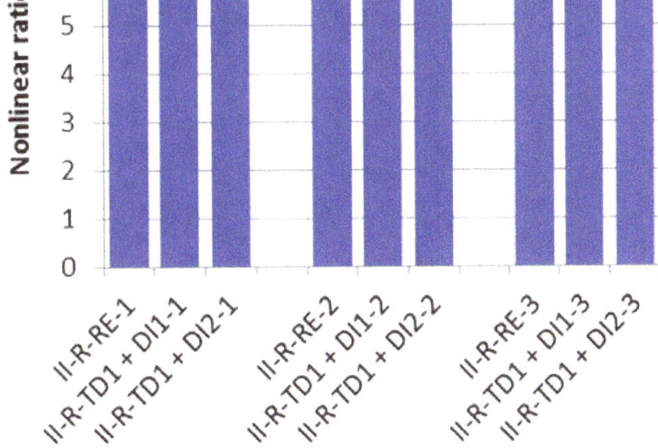

Fig. 4.36 Nonlinearity of three sets of scarfed samples: reference and TD1 (−1, −2, −3) affected by two levels of DI (DI1 and DI2) contaminations

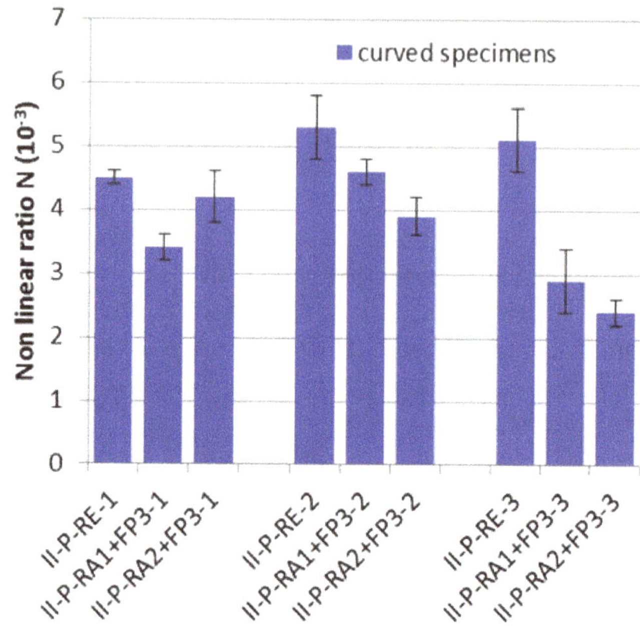

Fig. 4.37 Nonlinearity of three sets of curved samples: reference and FP3 (−1, −2, −3) affected by two levels of RA (RA1 and RA2) contaminations

The nine production samples (curved) comprised three reference samples (without contamination) and six samples with a constant high level of fingerprint contamination (FP-3) along with two different moderate levels of release agent (RA-1 and RA-2) applied to the surface of the laminates prior to bonding. The results of the measurements are shown in Fig. 4.37.

All three sets of curved samples reliably indicated a decrease in the nonlinearity (and therefore, a strengthening of the bonds) induced by the contamination, whereby the highest values of N were observed for the reference samples. For sample sets 2 and 3, the decrease caused by the contamination resulted in very low values comparable to those measured for flat non-contaminated samples. Unlike similarly contaminated flat coupons (Fig. 4.34), the nonlinearity further decreased (strengthening of the bonds) as soon as the contamination level increased from RA-1 to RA-2. An interpretation of this unconventional behaviour would need to involve more detailed information on the technology and process of adhesive bonding for the curved sample, including data on the internal stresses in the laminate.

4.5 Conclusion

In this chapter, we highlight the relevance and performance achieved when applying extended non-destructive testing (ENDT) methods for the bond line quality assessment of adhesive joints. For important application scenarios with regard to aircraft manufacture or the in-service repair of composite structures, the in-process monitoring of CFRP composites is facilitated through the use of advanced setups and approaches based on electromechanical impedance (EMI), laser shock adhesion testing (LASAT), and the nonlinear ultrasound method (NUS). Adhesive joints in the shape of flat coupons, scarfed samples or curved samples were investigated and correlations with the results from assessing design-relevant joint features were demonstrated for the obtained ENDT datasets.

The EMI method enabled differentiation of the reference samples from the contaminated samples in the production scenario. This was possible for the flat samples. In case of moisture contamination, the increasing frequency shift correlates with the contamination level. Due to the different geometry of the curved samples, another bandwidth was analyzed and a clear differentiation between the reference and contaminated samples was not possible. In the case of the repair scenario, this technique enabled the detection of almost all the flat samples with modifications. The numerical parameter for comparison was defined after inspecting the spectrum of each sample and also of the free sensors. In the case of thermal degradation and faulty curing, the highest level of modification was easily detected. This was also possible for the combined contamination case. In the case of scarfed samples, the detection of four out of five samples was possible. However, the sensitivity to the level of contamination was not obvious.

The LASAT gave good results for the coupon samples, and these are in accordance with results found during the G_{IC} and G_{IIC} tests. It was also possible in certain cases to identify the different levels of contamination. The LASAT did not differentiate the samples from the production scenario with multiple contaminations from the reference sample. The intensity required to open the faulty joints was the same as that for the reference. The results are also hard to explain for the repair scenario, where the technology detected the least contaminated sample but was not able to clearly detect the most contaminated sample. However, the same contamination on scarfed samples showed much better results, whereby the LASAT technology was effective at differentiating a sound bond from a weak one. The difference between the results of the coupon samples and those of the scarfed samples is not yet fully understood. The bond geometry may have played a part in the crack detection/propagation within the contaminated samples. Furthermore, this technology heavily relies on the standard sample, whereby the reference used for the coupon samples may have been produced with different parameters. Further studies need to be conducted in order to draw conclusions from the results for these samples. The curved samples were also highly porous and did not follow the specifications. In summary, no relevant conclusion can be drawn from these particular results.

The nonlinear measurements revealed an increase in the nonlinearity caused by contamination of the adhesive layer for most single contamination types. For each contamination type, the values of the nonlinearity ratio N changed noticeably with a variation of the contamination level, which indicates a sensitivity to the changes in the thin boundary layer between the adhesive and the adherends. According to the measurement results, the sensitivity of the nonlinear method is, therefore, sufficient to recognize the effect of contamination on adhesive bonding. For the contaminations types TD, RA, FP, DI, and FC, the material nonlinearity increased with the increase of the level of contamination. Since nonlinearity increases due to a "softening" of the material, an increase in N is an indication of a weakening of the adhesive bond.

The effect of multiple contaminations was studied for the FP3+RA and TD1+DI contaminations. Both combinations demonstrate a cumulative decline of the adhesion between flat samples caused by the increase of the DI and RA contaminations. In scarfed samples, the effects of multiple contaminations TD1+DI similarly led to an increase in nonlinearity, which indicates a decline in bond quality. However, the effect of an increase in the level of DI contamination was not as noticeable, as was the case in the bonding of conventional flat samples. This behaviour shows that even a moderate amount of DI leads to major deterioration of the adhesive properties, while a further increase of the contamination level does not affect the bonding strength substantially. The effect of multiple contaminations offers a somewhat different result for curved CFRP laminates, whereby all three sets of curved samples studied reliably indicated a decrease of nonlinearity (and therefore a strengthening of the bond) induced by the contamination. The interpretation of this unconventional behaviour should involve detailed information on the technology and the process of adhesive bonding for curved samples, including data on the internal stresses in the laminate.

In summary, our investigations reveal that in certain scenarios the ENDT methods EMI, LASAT, and NUS are sensitive to impacts affecting a bond strength reduction and can, therefore, be utilized for identifying not in order (NIO) joints during bond line assessment. In the final research chapter of this book, we will underline this perception and prognosis with findings highlighting the performance of ENDT for the monitoring of quality-relevant operand features in adhesive bonding processes involving parts of real aerospace structures with stringers.

References

1. Wadley HNG (1988) Interfaces: the next NDE challenge. In: Thompson DO, Chimenti DE (eds) Review of progress in quantitative nondestructive evaluation: volume 7B. vol 42. Springer US, Boston, MA, pp 881–892
2. ENCOMB "Extended Non–Destructive Testing of Composite Bonds" (2010–2014) European Seventh Framework Programme for research and technological development (FP7). Grant agreement ID: 266226
3. ComBoNDT "Quality assurance concepts for adhesive bonding of aircraft composite structures by advanced NDT" (2015–2018) Horizon 2020; research and innovation programme (H2020). Grant agreement ID: 636494
4. Dugnani R, Zhuang Y, Kopsaftopoulos F et al (2016) Adhesive bond-line degradation detection via a cross-correlation electromechanical impedance–based approach. Struct Health Monit 15(6):650–667. https://doi.org/10.1177/1475921716655498
5. Dugnani R, Chang F-K (2017) Analytical model of lap-joint adhesive with embedded piezoelectric transducer for weak bond detection. J Intell Mater Syst Struct 28(1):124–140. https://doi.org/10.1177/1045389X16645864
6. Malinowski P, Wandowski T, Ostachowicz W (2015) The use of electromechanical impedance conductance signatures for detection of weak adhesive bonds of carbon fibre–reinforced polymer. Struct Health Monit 14(4):332–344. https://doi.org/10.1177/1475921715586625
7. Malinowski PH, Tserpes KI, Ecault R et al (2017) Study of adhesive bonds by mechanical tests, ultrasounds and electromechanical impedance method. In: Structural health monitoring 2017: real-time material state awareness and data-driven safety assurance. Proceedings of the

eleventh international workshop on structural health monitoring, Stanford University, Stanford, CA, September 12–14, 2017. DEStech Publications, Inc, Lancaster, PA
8. Malinowski PH, Ecault R, Wandowski T et al (2017) Evaluation of adhesively bonded composites by nondestructive techniques. In: Kundu T (ed) Health monitoring of structural and biological systems 2017. SPIE, 101700B
9. Berthe L, Fabbro R, Peyre P et al (1997) Shock waves from a water-confined laser-generated plasma. J Appl Phys 82(6):2826–2832. https://doi.org/10.1063/1.366113
10. Fabbro R, Fournier J, Ballard P et al (1990) Physical study of laser-produced plasma in confined geometry. J Appl Phys 68(2):775–784. https://doi.org/10.1063/1.346783
11. Vossen JL (1978) Measurements of film-substrate bond strength by laser spallation. In: Mittal KL (ed) Adhesion measurement of thin films, thick films, and bulk coatings. ASTM International, 100 Barr Harbor Drive, PO Box C700, West Conshohocken, PA 19428-2959, 122-122-12
12. Yuan J, Gupta V, Pronin A (1993) Measurement of interface strength by the modified laser spallation technique. III. Experimental optimization of the stress pulse. J Appl Phys 74(4):2405–2410. https://doi.org/10.1063/1.354700
13. Boustie M, Auroux E, Romain J-P et al (1999) Determination of the bond strength of some microns coatings using the laser shock technique. Eur Phys J AP 5(2):149–153. https://doi.org/10.1051/epjap:1999123
14. Bolis C, Berthe L, Boustie M et al (2007) Physical approach to adhesion testing using laser-driven shock waves. J Phys D Appl Phys 40(10):3155–3163. https://doi.org/10.1088/0022-3727/40/10/019
15. Gilath I, Eliezer S, Bar-Noy T et al (1993) Material response at hypervelocity impact conditions using laser induced shock waves. Int J Impact Eng 14(1–4):279–289. https://doi.org/10.1016/0734-743X(93)90027-5
16. Bossi R, Housen K, Walters C, Works BP (2005) Laser bond inspection device for composites: has the holy grail been found. NTIAC Newslett 30
17. Ecault R, Berthe L, Touchard F et al (2015) Experimental and numerical investigations of shock and shear wave propagation induced by femtosecond laser irradiation in epoxy resins. J Phys D Appl Phys 48(9):95501. https://doi.org/10.1088/0022-3727/48/9/095501
18. Gay E, Berthe L, Boustie M et al (2014) Study of the response of CFRP composite laminates to a laser-induced shock. Compos B Eng 64:108–115. https://doi.org/10.1016/j.compositesb.2014.04.004
19. Courapied D, Berthe L, Peyre P et al (2015) Laser-delayed double shock-wave generation in water-confinement regime. J Laser Appl 27(S2):S29101. https://doi.org/10.2351/1.4906382
20. Chen J, Nakamura T, Aoki K et al (2001) Curing of epoxy resin contaminated with water. J Appl Polym Sci 79(2):214–220. https://doi.org/10.1002/1097-4628(20010110)79:2%3c214:AID-APP30%3e3.0.CO;2-S
21. Laporte D (2011) Analyse de la réponse d'assemblages collés sous des sollicitations en dynamique rapide, Chasseneuil-du-Poitou, Ecole nationale supérieure de mécanique et d'aérotechnique
22. Moutsompegka E, Tserpes KI, Polydoropoulou P et al (2017) Experimental study of the effect of pre-bond contamination with de-icing fluid and ageing on the fracture toughness of composite bonded joints. Fatigue Fract Engng Mater Struct 40(10):1581–1591. https://doi.org/10.1111/ffe.12660
23. Brenzeale MA, Philip J (1984) Determination of third-order elastic constants from ultrasonic harmonic generation measurement. Phys Acoust XVII
24. Guyer RA, Johnson PA (2009) Nonlinear mesoscopic elasticity: the complex behaviour of granular media including rocks and soil. Wiley-VCH, Weinheim
25. Solodov IY, Krohn N, Busse G (2002) CAN: an example of nonclassical acoustic nonlinearity in solids. Ultrasonics 40(1–8):621–625. https://doi.org/10.1016/S0041-624X(02)00186-5
26. Nagy PB (1998) Fatigue damage assessment by nonlinear ultrasonic materials characterization. Ultrasonics 36(1–5):375–381. https://doi.org/10.1016/S0041-624X(97)00040-1

27. Zheng Y, Maev RG, Solodov IY (2000) Review/Sythèse nonlinear acoustic applications for material characterization: a review. Can J Phys 77(12):927–967. https://doi.org/10.1139/p99-059
28. Pruell C, Kim J-Y, Qu J et al (2009) Evaluation of fatigue damage using nonlinear guided waves. J Appl Phys 18(3):35003. https://doi.org/10.1088/0964-1726/18/3/035003
29. Zhao J, Chillara VK, Ren B et al (2016) Second harmonic generation in composites: theoretical and numerical analyses. J Appl Phys 119(6):64902. https://doi.org/10.1063/1.4941390
30. Li W, Cho Y, Achenbach JD (2012) Detection of thermal fatigue in composites by second harmonic Lamb waves. Smart Mater Struct 21(8):85019. https://doi.org/10.1088/0964-1726/21/8/085019

5

Full-Scale Demonstration and Bonding Quality Assessment

Romain Ecault, Ana Reguero Simon, Célian Cherrier, Paweł H. Malinowski, Tomasz Wandowski, Mirosław Sawczak, Kai Brune, Hauke Brüning, Mareike Schlag, Johannes Derksen, Welchy Leite Cavalcanti, Laurent Berthe, Maxime Sagnard, Wiesław M. Ostachowicz, Saverio De Vito, Andreas Helwig, Rainer Stössel, Damien Segur, Apostolos Chamos, and Konstantinos Tserpes

Abstract This chapter highlights two advances towards a higher maturity of versatile extended non-destructive testing (ENDT) procedures. Full-scale demonstration tests are presented in realistic user application cases that involve typical production or repair scenarios. Subsequently, the investigations used to assess the probability

---

R. Ecault (✉)
Airbus Operations S.A.S., 316, route de Bayonne, B.P. D4101, 31060 Cedex 9 Toulouse, France
e-mail: romain.ecault@airbus.com

A. Reguero Simon
Aernova Composites Illescas Sau, Parque Industrial y Tecnológico, Avda Barajas, 3 Illescas, 45200 Toledo, Spain

C. Cherrier
Automation W+R GmbH, Messerschmittstraße 7, 80922 Munich, Germany

P. H. Malinowski · T. Wandowski · M. Sawczak · W. M. Ostachowicz
Institute of Fluid–Flow Machinery, Polish Academy of Sciences, Fiszera 14, 80-231 Gdańsk, Poland

K. Brune · H. Brüning · M. Schlag · J. Derksen · W. Leite Cavalcanti
Fraunhofer Institute for Manufacturing Technology and Advanced Materials IFAM, Wiener Str. 12, 28359 Bremen, Germany

L. Berthe · M. Sagnard
PIMM, CNRS-ENSAM ParisTech, 151 Bd de l'Hôpital, 75013 Paris, France

S. De Vito
ENEA – National Agency for New Technologies, Energy and Sustainable Economic Development, TERIN-FSD, Portici Research Center, P.le E. Fermi 1, 80055 Portici, Italy

A. Helwig · R. Stössel
Airbus Defence and Space GMBH, Willy-Messerschmitt-Straße 1, 82024 Taufkirchen, Germany

D. Segur
CEA LIST, CEA Saclay, 91191 Gif-sur-Yvette, France

of detection (POD) are detailed for the respective ENDT processes and application-relevant scenarios in a realistic environment. Although some results indicated that some additional in-depth investigations would be even more enlightening, these demonstrations still clearly showed that developments and progress described in the previous chapters have enabled some of the technologies to achieve a maturity that is sufficient to proceed towards industrial implementation. Some ENDT techniques revealed the presence of contaminants on real structural parts with unknown contaminant amounts. For the first time, POD results obtained for ENDT investigations are presented. Some ENDT procedures permitted POD results to be obtained for several scenarios, while others showed technologically relevant POD only for certain scenarios. For two ENDT techniques, determining the POD helped to enhance the respective testing and evaluation procedures. In most of the cases, it was possible to estimate a preliminary quantification of POD by giving the POD90/95. For some techniques, this value was below the lowest contamination degree.

Keywords Full-scale demonstration tests · Realistic scenarios · CFRP repair · Probability of detection (POD) · Bonding assessment

5.1 Introduction to the Full-Scale Demonstration Event

In the previous chapters, after providing an introduction to extended non-destructive testing (ENDT) and quality assessment (QA) in adhesive bonding processes relevant to the manufacture or repair of composite structures, we detailed the advances achieved so far, e.g., in the European joint research projects ENCOMB and ComBoNDT. We highlighted substantial technical progress under the framework of laboratory-scale investigations and inferred that it promises to facilitate the implementation of procedures to reach a technology readiness level (TRL) enabling their application in field-level sensing systems within industrial adhesive bonding processes used for the manufacture or repair of joints, including those based on CFRP adherends.

In this chapter, our objective was to test each EDNT technique in a real-world environment—outside the laboratory—and on realistic parts. Within the framework of the ComBoNDT project, this final test was used to assess the technological maturity with regard to aeronautical applications. It is a demonstration considered to be as close as possible to real industrial processes, and we consider the results highly representative.

To this end, as part of the ComBoNDT project [1], we organized a "Demonstration Days" week in Bremen at the Fraunhofer IFAM site, which offered a dedicated

A. Chamos
EASN Technology Innovation Services, Terweidenstraat 28, 3440 Budingen, Belgium

K. Tserpes
Laboratory of Technology & Strength of Materials, Department of Mechanical Engineering & Aeronautics, University of Patras, 26500 Patras, Greece

environment. Using a teamwork approach, the partners transported their individual mobile setups to Northern Germany, thereby moving them out of their base laboratory. The Fraunhofer IFAM site was a suitable location for the demonstration because it provided each partner with logistical and automatization capabilities. The available equipment included facilities for the preparation of industrially relevant bonded composite parts, which involved the intentional application of contaminants—an obligatory step in the bonding process of the demonstration.

The demonstration consisted of a set of process steps performed by different actors in a coordinated manner that was designed to be as linear as possible. Our basic idea was to have—in the same place and at the same time—all the steps of a technologically relevant bonding process, either in production or in repair, that closely followed industrial processes. To give our readers a rough outline of the complex procedure, we ensured that the assessed panels were first prepared for bonding using an appropriate cleaning process before being partially contaminated. Then, we applied the first batch of ENDT tools to perform the steps of the monitoring process, specifically those dedicated to adherend surface quality assessment. Afterwards, we conducted the adhesive bonding operations and performed post-bonding characterization using the second batch of ENDT techniques. We would like to highlight that these full-scale tests also acted as a blind test or round-robin test because the applied contamination degree was neither checked using laboratory-based reference analysis nor communicated to the inspection specialists. This holds true for both the production and repair panels assessed here.

In this chapter, we describe the performed full-scale tests. First, we present the parts produced or recovered for this demonstration. Then, we describe the preparation of the respective panels for testing, whereby the panels for the production user case and the selected repair user case are addressed separately. In the latter case, the performed scarfing is described in addition to the contamination preparation, while the respective bonding process is presented for both applications. The ENDT results are then given for each panel, starting with the surface quality control and following this with the bonding assessment. Thus, the chapter is organized following a real implementation scenario, as explained in the previous paragraph. Finally, we present a summary of the full-scale demonstration.

Aiming for quantitative evaluation, we also give some initial elements on the probability of detection (POD) in this chapter. This method is used systematically in aeronautics to assess the detection capability of conventional NDT approaches. In this respect, we used CIVA software for some of the analytical ENDT data gathered at the coupon specimen level (described in more detail in Chaps. 3 and 4, respectively) to model the obtained POD curves.

Finally, a few words on project dissemination are given to conclude the chapter.

5.2 Setup of the Full-Scale Demonstration: Materials, Workflow, and Operations

This section describes the planning and implementation of all the operations necessary to conduct the full-scale demonstration. The emphasis hereby is on the provision of the materials, the applied workflow, and the operational sequence. As previously highlighted, all three aspects were maintained as close as possible to a sequential industrial process. Almost all the ENDT techniques were available and employed on-site during the three-day demonstration period. In exceptional cases, further ENDT investigations were conducted elsewhere, whereby either all the panels or one specific panel were sent to the respective partner lab to perform the testing there.

5.2.1 Providing Real and Realistic Parts

In the following, we detail the geometry and composition of the realistic parts used for the production user case and the real parts employed for the repair user case within the ComBoNDT joint research project.

5.2.1.1 Manufacturing of a Flat Panel for the Production User Case

We designed a panel representing a wing application for the production user case with the intention that it should differ from the part used for the repaired fuselage application to highlight the versatility of the ENDT approaches. The manufacturing process of the panel was performed at Aernnova and comprised two different steps: the manufacture of the T-stringers and the manufacture of the skins. Hexcel M21E was used for the manufacture of all samples. Similar to the repair user case, we chose the dimensions to represent the correct order of magnitude of a tested region while providing enough space to test the application of several ENDT methods.

The stringers were manufactured following current industrial production guidelines. The web was shaped by joining two C-shape preforms that had been produced using automated tape laying (ATL). First, a flat panel was laminated and placed on a hot-forming tool to obtain the C-shape. After that, both C-shaped parts were joined. Finally, the element was trimmed to obtain the six T-stringers with the required dimensions. Four of these were dedicated to mechanical testing and two were kept for the full-scale demonstration. The stringers were inspected by ultrasonic C-scan inspections and no particular defects were revealed (Fig. 5.1).

The skin samples were manufactured with the following carbon fiber distribution Table 5.1, which was similar to the panels used for mechanical testing.

The production began with the manufacture of a 3500 mm by 1000 mm flat panel laminated using ATL. The distribution of the skin samples within this panel is depicted in Fig. 5.2.

Fig. 5.1 Image of manufactured T-stringers

Table 5.1 Carbon fiber distribution of manufactured samples

Part	Specimens	Plies	Stacking
Skin	4	20	$[\pm45/90/02/90/\pm45/0/90]S$

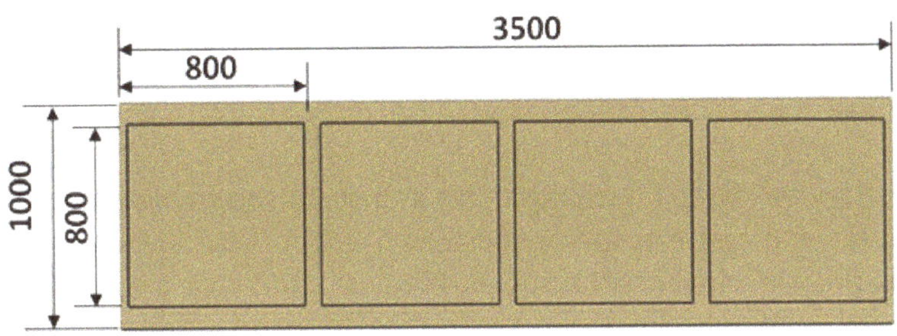

Fig. 5.2 Flat panel manufactured for four square skin samples; dimensions are given in mm

After the curing cycle, the flat panel was automatically inspected using pulse-echo equipment with water as the coupling medium to ensure the quality of the part manufactured. No defects were revealed (see Fig. 5.3). After the inspection, a trimming operation was performed to obtain the number of skin samples required for the mechanical testing (see Chap. 2) and full-scale demonstration.

The bonding operation of the stringer onto the composite skin was part of the full-scale demonstration workflow. Thus, intentional contamination needed to be done before bonding. It should be noted here that due to the dimensions of the bond line and the presence of two stringers, all the ENDT investigations could be tested using the same panel. More details are provided in the next section.

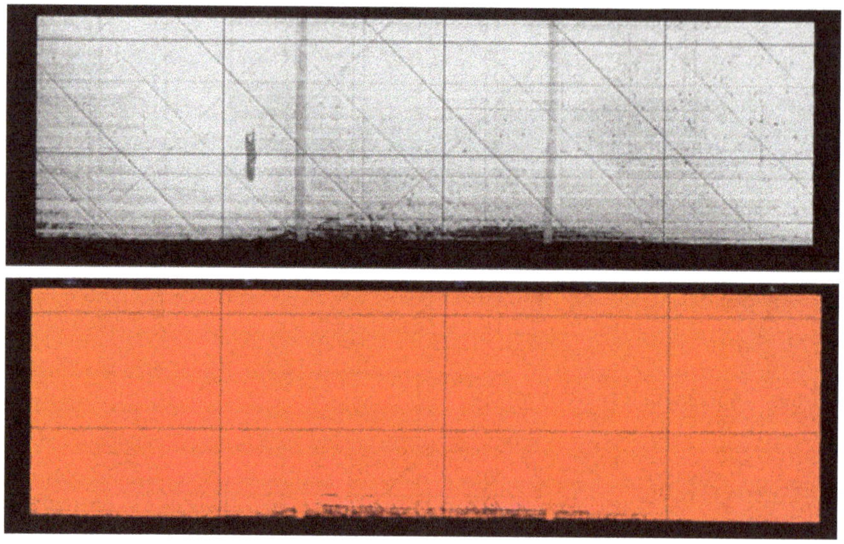

Fig. 5.3 Color-coded representation of the findings from the ultrasonic inspection of the flat panel

5.2.1.2 Tear-Down Cell for the Representative In-service Repair User Case

To demonstrate the ENDT techniques on full-scale specimens, we performed a bonded repair on a real part as we considered this to best represent a real application scenario, wherein the part being repaired has undergone a variety of loads and environmental conditions. For this purpose, fuselage panels were recovered from an A350 aircraft used by Airbus for full-scale static testing. These panels had been part of such a structural test, but had remained in good condition and could thus be employed in the demonstration (see Fig. 5.4). The size of each panel was approximately 1 m^2, which was adequate for several reasons:

1. The panels were large enough to be considered full scale, and they were in the order of magnitude of what would be tested by ENDT with a significant TRL.
2. Their size enabled a sizeable bonded repair to be made with enough space to gather data to test several ENDT tools, with the investigation performed in both the contaminated and reference areas.
3. They provided enough free edges to calibrate the measurement setup if necessary and to grasp and hold the parts in order to manipulate them manually or using a robotic arm.

As explained within the workflow and operations section, we assessed three panels. One panel was used for "intrusive" ENDT, such as monitoring based on magnetostrictive sensors and electromechanical impedance, for which it was necessary to bond sensing devices to the part. As the setups required for surface quality measurements using laser-induced fluorescence spectroscopy (LIF) and vibrometry were not moved to the demonstration event location, one panel was kept at IMP-PAN in Gdansk after scarfing and without bonded repair. Finally, one panel was

Fig. 5.4 A350 aircraft fuselage panels provided by Airbus for the demonstration of a full-scale bonded repair

used for the inspection using non-intrusive bonding assessment techniques, namely nonlinear ultrasound (NLUS) and laser shock adhesion testing (LASAT) (as outlined in the workflow and operations section).

As they were taken from the same section, these three panels were similar, but they were not identical in terms of geometry, number of stiffeners or thickness. A detailed view of two of the three panels is presented in Fig. 5.5. The skin thickness was in the

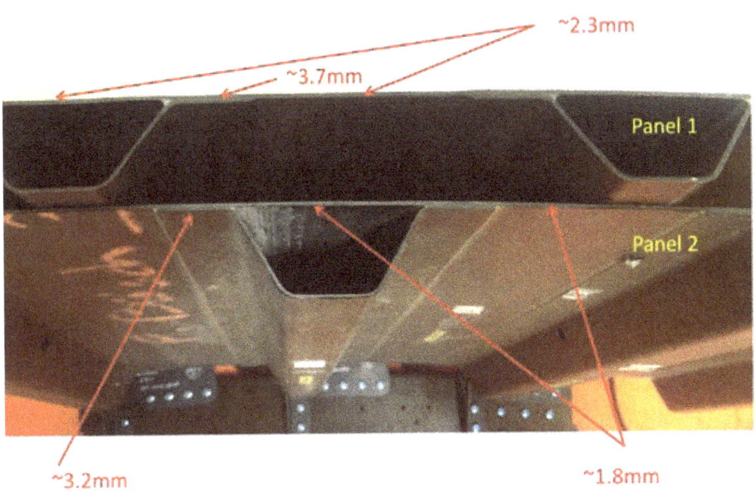

Fig. 5.5 Detailed view of the A350 fuselage panels used for the demonstration of quality assessment during bonded repair; the skin thickness is indicated at specific positions

range of 1.5–2.5 mm, which corresponds to the coupon and pilot sample geometries because—once repaired—the assembly thickness increased to about 2.5 mm. In addition, either two or three omega stringers reinforced the skin. The stringer foot was about 1.5 mm, which means that the total thickness was a maximum of 4 mm in this region. The panel material was the composite IMA/M21E (Hexcel), similar to the material used for the coupon and pilot samples.

The three panels were used to simulate a scarfed bonded repair. As the focus is on testing the adhesive bond between the scarfed skin and the repair patch, we decided to use these structures without introducing any damage to save time and costs. While it could have been subjected to, for example, delamination caused by a drop tower, this would have been removed by the subsequent scarfing operations anyway. Instead, we defined the dimensions of the bonded repair patch based on the standard for delamination that was not actually performed.

5.2.2 *Participants and Operations*

This section links the list of operations, the main steps, and the participants in the full-scale demonstration. All the information is compiled for each scenario in Table 5.2, outlining the main process steps, operations, the corresponding ComBoNDT project partners in charge, and the used techniques or procedures. It should be noted that the whole full-scale demonstration in the ComBoNDT joint research project was led by Airbus and Fraunhofer IFAM.

The following two sections detail how these operations were conducted during the full-scale demonstration event. Hereby, the focus is on the workflow, the contamination process, and the bonding operations executed within the framework of the production or repair user cases.

5.3 Production User Case

5.3.1 *Workflow Overview*

The operations within the production user case comprising distinct contamination scenarios are described in Fig. 5.6 following the ComBoNDT (with the name of the contributing partners) project. First, the skin and stringers for the full-scale demonstration were provided by Aernnova. The quality of the skin was checked at Aernnova using conventional ultrasound testing, as outlined in the previous section. In this case, one panel was sufficient for all the ENDT approaches. At the beginning of the three-day full-scale demonstration, the skin and stringer were contaminated separately with RA and FP formulation, successively. The skin was then tested using FTIR (Airbus, AGI), e-nose (AGI), and e-nose (ENEA) to conduct the

Table 5.2 List of operations for production and repair user cases comprising scenarios with the intentional contamination of an adherend surface by a release agent (RA) and fingerprint (FP, with either sweat or hydraulic oil) or de-icing fluid (DI) as performed during the three-day on-site demonstration event, indicating the corresponding ComBoNDT project partners in charge. The involved ENDT techniques comprise the aerosol wetting test (AWT), magnetostrictive sensors (MGSS), electromechanical impedance (EMI), electronic nose (e-nose), laser shock adhesion test (LASAT), Fourier transform infrared spectroscopy (FTIR), and nonlinear ultrasound (NLUS)

User case	Process steps	Operations	Partner in charge	Used procedure
Production	Supplying material	Providing skin, stringers, and bonding	Aernnova, Airbus	
	Performing contamination	Contamination applying RA	IFAM	Sweeping RA
		Contamination applying FP		Salt-based solution
	Surface assessment	Testing on separate stringers	IFAM	LIBS
		Testing on stringers	Automation	AWT
		Testing on skin	AGI	FTIR
		Testing on skin	AGI	e-nose
		Testing on skin	ENEA	e-nose
	Bonding of stringers	Curing in autoclave	IFAM	Curing cycle
	Bonding assessment	Testing on skin	IMP-PAN	EMI
		Testing on skin + loading	GMI	MGSS
		Testing on skin's rear face	CEA	NLUS
		Testing on both sides	CNRS	LASAT
Repair	Supplying material	Providing real fuselage panels and repair material	Airbus	
	Scarfing	Scarfing for repair	GMI	Stepping
	Performing contamination	Contamination applying DI	IFAM	Sweeping DI
		Contamination applying FP		Skydrol oil
	Surface assessment	Testing	Automation W+R	AWT
		Testing	AGI	FTIR
		Testing	AGI	e-nose
		Testing	ENEA	e-nose
		Testing on a separate panel	IMP-PAN	LIF
		Testing on a separate panel	IMP-PAN	Vibrometry

(continued)

Table 5.2 (continued)

User case	Process steps	Operations	Partner in charge	Used procedure
	Bonded repair	Preparing the patches	GMI	Anita setup
		Preparing the curing		
		Curing outside the autoclave		
	Bonding assessment	Testing on panel 1	IMP-PAN	EMI
		Testing on panel 1	GMI	MGSS
		Testing on panel 2	CEA	NLUS
		Testing on panel 2	CNRS	LASAT

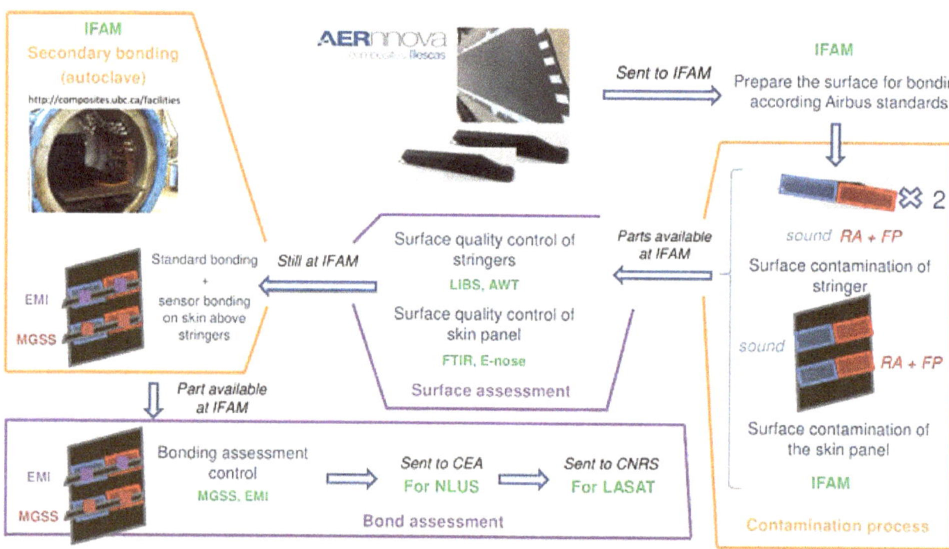

Fig. 5.6 The full-scale demonstration workflow for the panel used in the production user case

surface quality assessment. Considering that the e-nose techniques can potentially remove some contaminants during the measurements, separate areas were assigned for each e-nose technique. The stringer surfaces were also tested using AWT and (on a separate set of stringers) the LIBS technique. Indeed, as explored in the previous chapters, performing LIBS investigations can slightly modify the composite surface locally by vaporizing some of the epoxy-based matrix material. Therefore, any findings obtained with surface assessment techniques following LIBS testing are potentially influenced. After surface quality control, the stringers were bonded to the skin following a secondary bonding process in an autoclave. Hereby, the contaminated regions were placed above each other. After bonding, the obtained part was rapidly checked visually and manually to ensure that the demonstration tests did not focus on regions that were evidently not expertly bonded as this would have been trivially identified by the ENDT techniques. Subsequently, the bonding assessment was conducted. The magnetostrictive sensors were first bonded to the skin using an out-of-autoclave device (see the repair user case section) before being used for the test. Then, the EMI sensors were glued to the skin in different regions to be used for testing during the final day of the full-scale demonstration after hardening overnight. Since the NLUS and LASAT devices were not available at IFAM in Bremen, the panels were sent to CEA to be tested first by NLUS and then to CNRS to be tested by LASAT, since the latter test might be destructive for weak regions.

5.3.2 *Release Agent and Fingerprint Contamination*

In this section, we focus on the important step of applying an intentional combined contamination during the three-day full-scale demonstration event within the ComBoNDT project. The specific operation is illustrated in Fig. 5.7. First, the position of the stringers was marked on the skin, following the initial design of the stiffened panel, as illustrated in Fig. 5.7a. The decision was made to situate the stringer at a distance of one-third of the height away from each border. Then, the panel was partially contaminated with a solution containing RA; the respective positions are displayed in Fig. 5.7b. The RA contamination was deposited along two strips using a small sponge, whereby the width of each strip was about the same as the stringer foot, as shown in Fig. 5.7c. The process was somewhat different from that used for the flat coupon samples, but in this realistic user case, it was not possible to use deep-coating due to the size of the part. As a consequence, there may have been a lower homogeneity on the surface, but we expected that the amount of Si species at the surface would turn out to be within the concentration range previously investigated

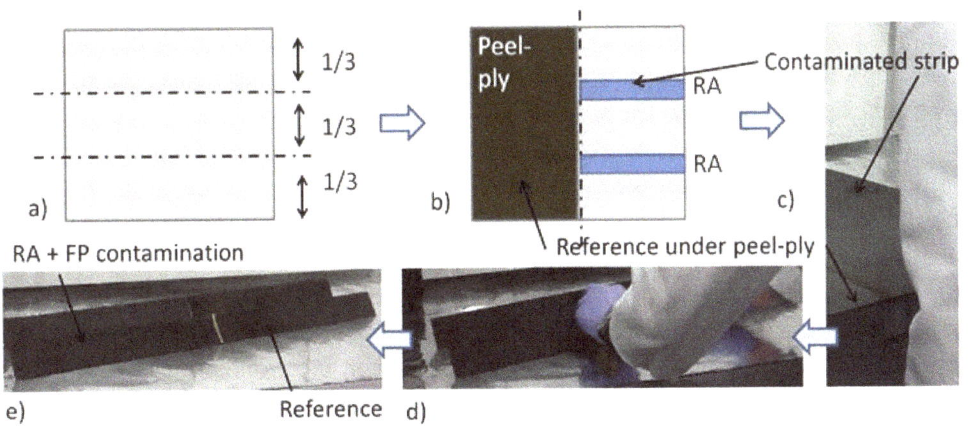

Fig. 5.7 The two-step contamination process "RA+FP" for the production user case; **a** position of the stringers, **b** regions to be contaminated with RA, **c** view of both areas contaminated with RA as well as the region where the FP contamination was randomly applied on the skin, **d** contamination of the stringer feet with RA, and **e** final contamination state of the stringer surfaces after the concluding deposition of salt-based FP

using flat specimens. Following the RA contamination, the skin was placed in an oven for 1 h at 80 °C. Subsequently, the FP contamination operation of the production user case was conducted in the same way as in the FP scenario for the coupon pilot specimens. We did not specify the distance between the applied prints in the contaminated region because we intended a random distribution, and their number should have been sufficient to be representative and unforeseeably positioned within the scanning areas of all the investigated techniques. It should be noted that during the whole contamination process, we retained the peel ply in the reference area to avoid any contamination of this region, as shown in Fig. 5.7. The stringers were then contaminated using the same process. RA was first spread on half of the finger foot surface using the same solution and a sponge (see Fig. 5.7d). They were then placed in the same oven, after which a worker marked the modified surface with FP deposits. The final state of the stringer surfaces was an "RA+FP" coated contamination area near a reference area below the peel ply, as presented in Fig. 5.7e.

After these contamination steps, the skin and stringer were ready to be tested using the various ENDT techniques. As explained, the stringers were dedicated to LIBS and AWT investigations while the skin was used for e-nose characterization. Clearly, from an ENDT point of view, this choice was arbitrary, however, it was made for operational reasons, mainly due to the geometry of the part. As we considered that the investigations with the e-nose technique might have removed some of the deposited substances from the stringer bonding surface, it was only tested by AWT (see Fig. 5.6). Thus, the stringer foot remained correctly contaminated after the first batch of ENDT and it was not necessary to repeat the contamination process.

5.3.3 Bonding Operations

After the testing of the different ENDT techniques with respect to their performance regarding surface quality assessment in a realistic production user case, the stringers were adhesively bonded to the skin.

The curing process of the stringers was quite conventional. Nevertheless, we would like to reiterate that we had decided on the techniques used in the bonding quality assessment under the consideration that one of the two adherend surfaces of each joint would be contaminated. Of course, in the curing oven, the contaminated region of each stringer was placed above the contaminated region of the skin to keep the reference bonding area away from the contaminants (as shown in Fig. 5.8a). We first applied the adhesive to the stringer foot and then positioned this onto the skin at the right place. We used adhesive tape strips to keep the partially contaminated stringer in place during loading and curing. The curing cycle in the autoclave was the same as that used for the coupon specimens.

After curing, the panel recovered from the autoclave contained both a reference area and a contaminated area, as illustrated in Fig. 5.8b. Thus, it was ready for use in the testing of the targeted set of ENDT bonding assessment techniques. In addition, to conventionally address non-expertly bonded regions of manufactured

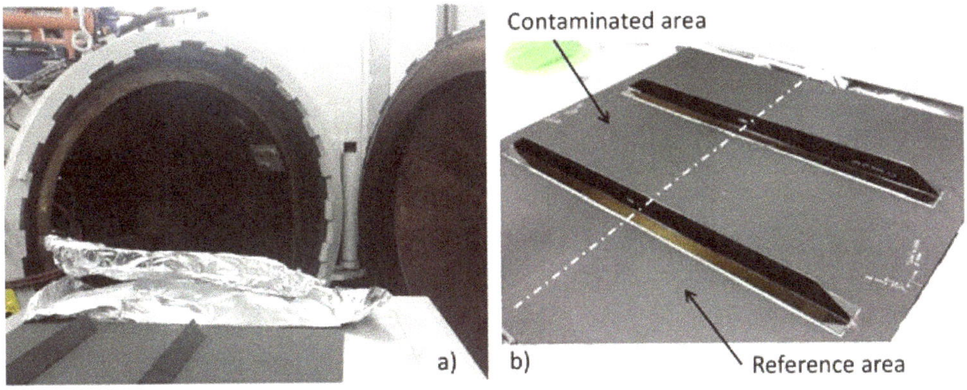

Fig. 5.8 Aspects of the bonding operation performed during the full-scale demonstration event in the ComBoNDT project: **a** panel and stringers in front of the autoclave before bonding and **b** fully cured stiffened panel

joints, we performed an ultrasonic inspection to check for unintentional defects in the bond line. This inspection revealed the presence of a region characterized by a low ultrasound amplitude, which most likely indicated some porosity. Such trivially defective regions were marked on the panel to be excluded from any subsequent in-depth ENDT investigation.

5.4 Repair User Case

5.4.1 Workflow Overview

This section describes the workflow for the repair application case in the ComBoNDT project, with the process illustrated in detail in Fig. 5.9. For several reasons, the decision was made to prepare three panels. First, the area to be tested on the repair specimens was smaller than in the case of the panel manufactured for the production user case, whereby we focused on conical repair regions with a diameter of about 20 cm. This was a good compromise because such a region was large enough to represent a real application case yet also small enough to facilitate an easy and fast repair within the three-day full-scale demonstration event of the ComBoNDT project. For example, we had decided not to involve or touch the stringers in this particular user case. Second, the magnetostrictive sensors were positioned so as to capture a significant portion of the bond line. To independently test further bonding assessment techniques, we allotted additional space because we anticipated that this more complex bond geometry, compared to the specimen in the production user case, might require different settings for each step in the ENDT testing sequence. Thus, we dedicated one repaired panel to testing the intrusive bonding assessment techniques (MGSS, EMI) and another panel to the scanning techniques. In addition to these two

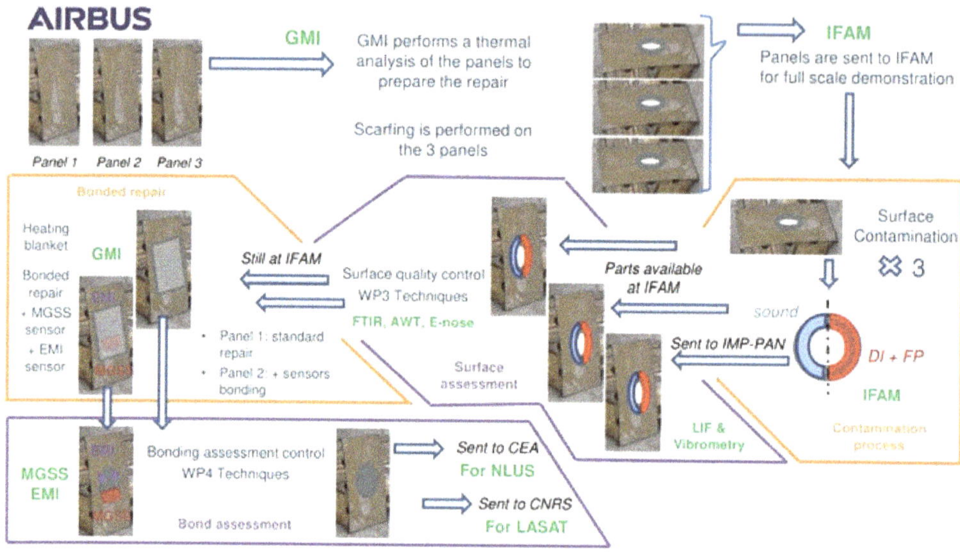

Fig. 5.9 Workflow description for the full-scale demonstration of the bonded repair user case

specimens, we agreed to provide a scarfed, but not repaired, panel for characterization using the LIF and laser vibrometry techniques, which were conducted at the IMP-PAN laboratory in Gdansk.

The three panels were first sent to GMI AERO prior to the demonstration event. There, a thermal analysis of the panels was performed in order to correctly set up the bonded repair procedures to be conducted. Then, the panels were scarfed according to the standards (as detailed in the next section) and sent to IFAM in Bremen for the full-scale demonstration.

At IFAM, the first task was cleaning the scarfed area using an Airbus cleaning procedure designed for such a geometry. Since the used product was based on isopropanol, some time was scheduled to allow the area to dry prior to the next step, i.e., the application of the contamination. The three panels were contaminated according to the same process. One panel was saved for further surface investigation using LIF and vibrometry, while the other two parts underwent the regular ENDT-assisted repair process. The surface of one contaminated adherend was tested using portable ENDT tools for surface quality control, namely AWT, FTIR, and e-nose, in that order. As a desorber device was used, some of the contaminants might have been removed during the e-nose investigation and thus the contamination process was repeated after the first of the two different e-nose techniques had been tested (by AGI-G or ENEA), as well as once more before performing the bonded repair. The panels were then successively repaired (as further detailed in the section on bonded repair). Magnetostrictive sensors were bonded onto one panel using an additional autoclave curing process. The EMI-based sensor bonding and respective investigations were performed before this final curing cycle and the corresponding MGSS testing. Finally, the last repaired panel was sent to CEA for NLUS measurement and finally to CNRS for LASAT.

The bonded repair of an aircraft CFRP panel after a more or less complex application history is a sophisticated process. Furthermore, we decided to apply a technologically relevant procedure. The subsequent sections detail the operations that were successively performed within the scope of the project.

5.4.2 *Description of the Scarfing Operation*

We first describe the repair of the coated panel. Before any adhesive bonding for repair was performed, the paint system was completely removed and the area around the repair region was prepared. Generally, the borders of the repair hole were rounded off and the hole itself was tapered to provide the best load transfer once the repair patch was bonded in. The images taken during the repair process are shown in Fig. 5.10. Scarfing, or taper sanding, is usually achieved using a high-speed grinder powered by compressed air, while stepping is an alternative method for removing material in preparation for the application of a repair patch. In stepping, the overall angle of the cone wall is achieved by removing a precise area of material per ply of the composite laminate. This is a gentle process that prepares the damaged area for the application of a repair patch. It is imperative to follow all the repair manual guidelines, and significant skill and practice on the part of the repair technician are mandatory.

We performed the stepping process using the innovative tools designed by GMI AERO. We also benefitted from a Leslie machining kit, which is a dedicated piece

Fig. 5.10 Illustration of the stepping process prior to the application of a repair patch in the bonded repair user case

of equipment used to prepare composite panels prior to the installation of repair patches. The following are the stepping characteristics chosen for this application:

- A circular repair geometry was used with a maximum diameter of approximately 200 mm
- The average thickness of one ply was 0.15 mm
- The overlap ratio was 1:60, i.e., a step width of approximately 10 mm
- Seven steps were performed to achieve a depth of 1 mm

All partners involved in the full-scale demonstration were informed of the thickness of the plies and their orientations. This was especially helpful whenever different settings had to be used per stepping step, e.g., in the case of LASAT, for which a setting per step was necessary.

5.4.3 Application of the De-icer and Fingerprint Solutions

This section provides more details about the contamination process. The images taken during the operations are presented in Fig. 5.11. Similar to the production user case, the repair user case comprised the intentional application of a combined contamination, although the respectively deposited solutions were different. Following the contamination scenario for the repair user case, the solution containing de-icing (DI) fluid was first applied. Similar to the RA contamination, a small sponge was used to spread the DI solution onto the stepped surface. Aluminum foil was then placed on the panel in such a way that it would keep half of the repair surface clean, as shown in Fig. 5.11a. As this was a manual operation and in view of the shape of the scarfing, some liquid might have moved towards the reference area. Thus, the boundary region between the reference and contaminated areas might have been

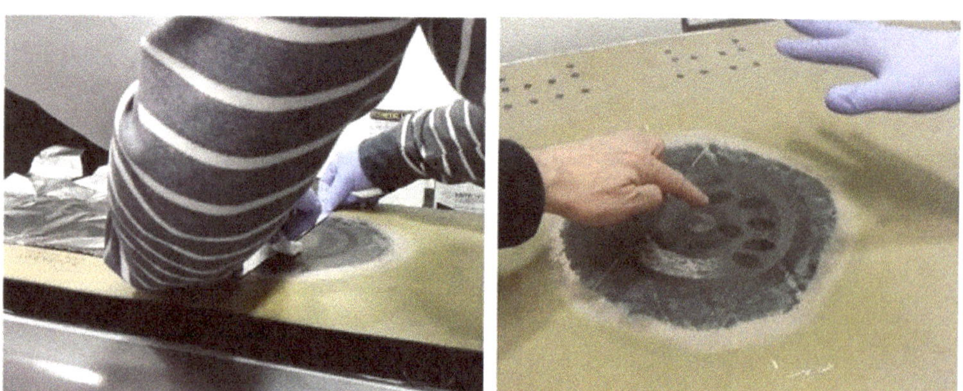

a) b)

Fig. 5.11 Contamination operation applied for the panels of the repair user case, with regionally applying contamination by **a** a liquid de-icing solution while maintaining a contaminant-free reference area, and **b** a hydraulic oil solution deposited within the fingerprinted areas larger than was considered ideal, but this was deemed to be acceptable. After the contamination operation, the panel was placed in an oven for about 10 min. Fingerprints with Skydrol hydraulic oil were then manually placed on the contaminated surface, whereby a glove was used to achieve a better repeatability of the prints (see Fig. 5.11b).

As previously outlined, these operations were repeated twice during the full-scale demonstration event to restore the contamination deposit after each e-nose measurement. We took care to employ the same contaminants and worker from the first operation until the last to ensure a minimum deviation from one contamination operation to the next. After the last contamination step, the panels were ready for repair.

5.4.4 *Description of the Bonded Repair Operations*

The repair process involved the replacement of each carbon fiber layer that had been removed during the scarfing operation by a new fiber layer exhibiting an orientation that matched that of the original layer. Additionally, a film adhesive was placed between the structure and the carbon fiber layers, as shown in Fig. 5.12a.

After each prepreg layer was stacked, the curing facilities were installed on the panel, whereby a heating blanket was placed on top and capped by a vacuum bag, as depicted in Fig. 5.12b and c, respectively. The thermal curing was then performed using the appropriate temperature. For this application, we chose the adhesive film FM300-2M (supplied by Cytec) and the carbon prepreg material M20 (supplied by Hexcel). In the ComBoNDT project, we employed ANITA EZ0901 equipment (see Fig. 5.13) to bond the repair patch. The ANITA setup provides all the necessary functions to control the heating process under vacuum for bonded repair on composite

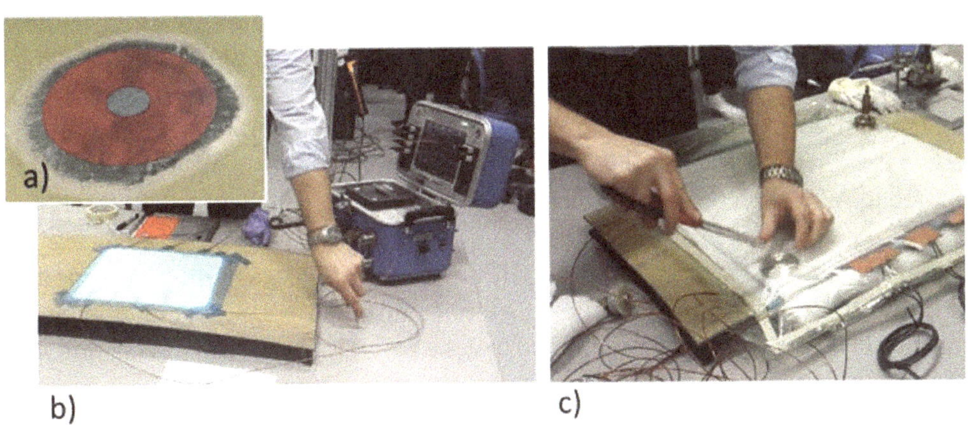

Fig. 5.12 The distinct operations conducted during the bonded repair of CFRP panels showing **a** the raw F300-2 M bonding material, **b** the placement of the heating device, and **c** the preparation of the capping vacuum bag

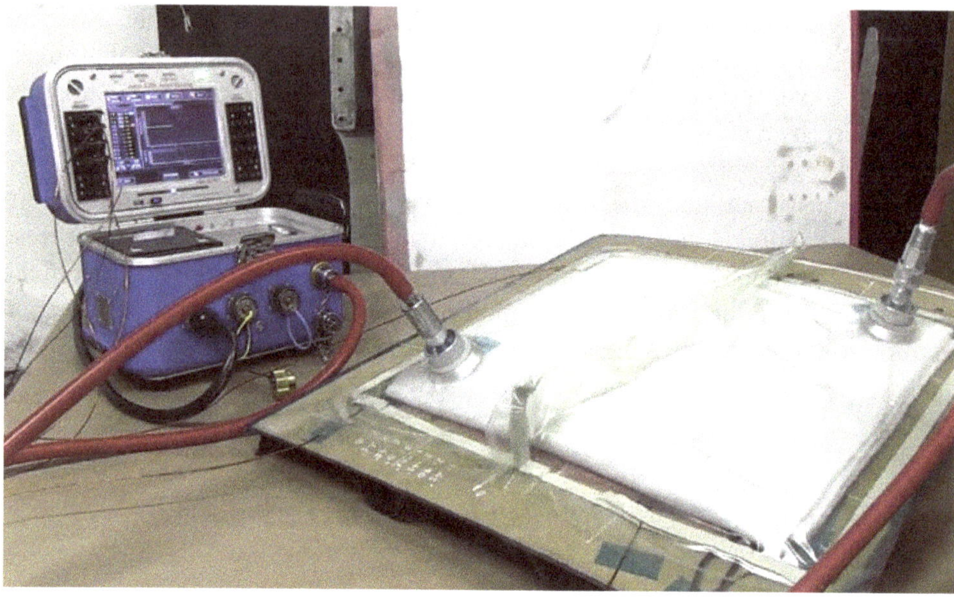

Fig. 5.13 Image taken during the full-scale demonstration of the repair process using the ANITA EZ0901 hot bonder in the ComBoNDT project

structures. The utilized device was specially conceived to achieve a maximum performance largely independent of the surface area to be repaired or the thickness of the assessed structures, and it is also suitable for structures with differentiated thermal reactions.

Appropriate curing of both the patch resin and the adhesive layer is essential to secure the integrity and sound mechanical performance of a bonded composite repair. A peculiarity of this user case, which focuses on Airbus A350 fuselage panels, results from the fact that these are carbon laminate structures reinforced with omega-shaped carbon stringers. It should be noted that, when heating a patch for repair, this complex structure makes it very difficult to achieve an even temperature distribution. This is a well-known issue and in the past has been addressed through a multi-zone heating approach. However, as this process adds complexity and is time-consuming, a compromise was identified that took advantage of the structural periodicity in the case of the A350. To this end, the AdaptHEAT adaptive heating blanket was designed, tailored to A350 structural components, which could be applied using available two-zone ANITA bonding consoles, thereby substantially improving the homogeneity of the temperature distribution. The AdaptHEAT concept adopts a multi-zone approach for the heating of a repair patch, following the heating requirements of the structure. In previous heating tests, we had identified that the part has two areas yielding different heating requirements, namely:

1. The area under which the stringers are positioned
2. The plain skin area, including the stringer foot.

5.5 Results of the Full-Scale Demonstration: The Representative Production User Case of a CFRP Stringer

In this section, we present the findings obtained for the production user case during the three-day full-scale demonstration event within the ComBoNDT joint research project. First, we report on the surface quality assessment of the adherends in manufacturing a CFRP stringer, then we detail the outcomes of the bonding quality assessment after completing the joining process.

5.5.1 Surface Quality Assessment

In the following, we present the findings based on the LIBS, AWT, and e-nose inspections used for the surface quality assessment.

5.5.1.1 Results for LIBS

Measuring setup

The complex geometry of the stringer, i.e., a realistic part, as compared to the flat coupon samples required a substantial modification of the setup and an adaptation of the measuring process to facilitate the surface monitoring based on LIBS investigations. We achieved a significant increase in the TRL by developing a robot-aided quality assessment procedure that allowed both the sensitivity and the robustness of the standard laboratory-based approach to be maintained. Essentially, we matched the control of the LIBS sensor to that of a standard industrial robot that was set to handle and position the part to be analyzed, as shown in Fig. 5.14. This arrangement offers some advantages for the LIBS measurement process since the focal point of the LIBS laser can be fixed. This fixed focal point (as shown in Fig. 5.15) was used as the origin of the coordinate system with which the robot handled the parts. Focusing

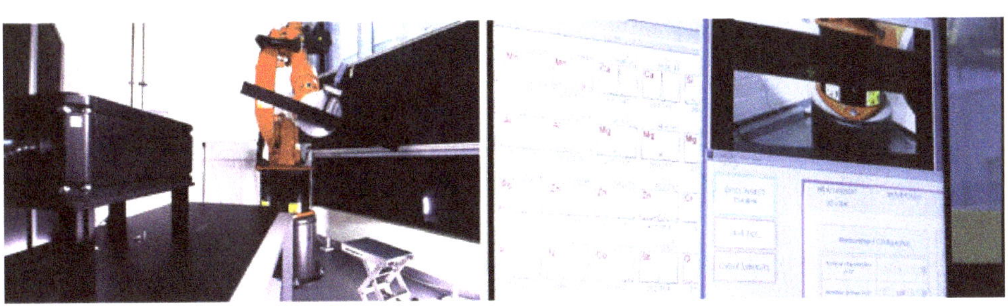

Fig. 5.14 The handling of a stringer by a fine-tuned industrial robot (left) and the well-arranged control unit for the operator (right)

Fig. 5.15 The plasma generation during the LIBS measurement on the realistic stringer specimen

on the analytical process, the LIBS measurements were performed on the CFRP surfaces after the peel ply was removed and the defined combined contamination process was performed. A laser wavelength of 1064 nm was used with a 180 mJ laser pulse energy. In order to determine the surface composition of the stringer, surface areas along three lines—each comprising 130 single measurement spots—were analyzed. An image of such a stringer is depicted in Fig. 5.16, while selected LIBS signals obtained during the investigation are presented and discussed based in Fig. 5.17.

Summary of the results

In Fig. 5.17, we present the results achieved when aiming for the detection of a contamination that arose from depositing relevant amounts of a Si-containing RA on the C-rich stringer surface. Essentially, the first 65 measurement points obtained within the intentionally contaminated half of the stringer surface indicated a clearly increased [Si] /[C] concentration ratio. This significant finding is indicative of the presence of RA contamination and is, as anticipated, in contrast to the LIBS results from the non-contaminated right half of the stringer surface. Thus, using LIBS

Fig. 5.16 An overview of the stringer surface after performing spot-by-spot LIBS measurements along three parallel lines, with the top, middle, and bottom lines composed of 130 analysis spots each; the left part of the specimen was contaminated, while the right part provided a clean reference surface

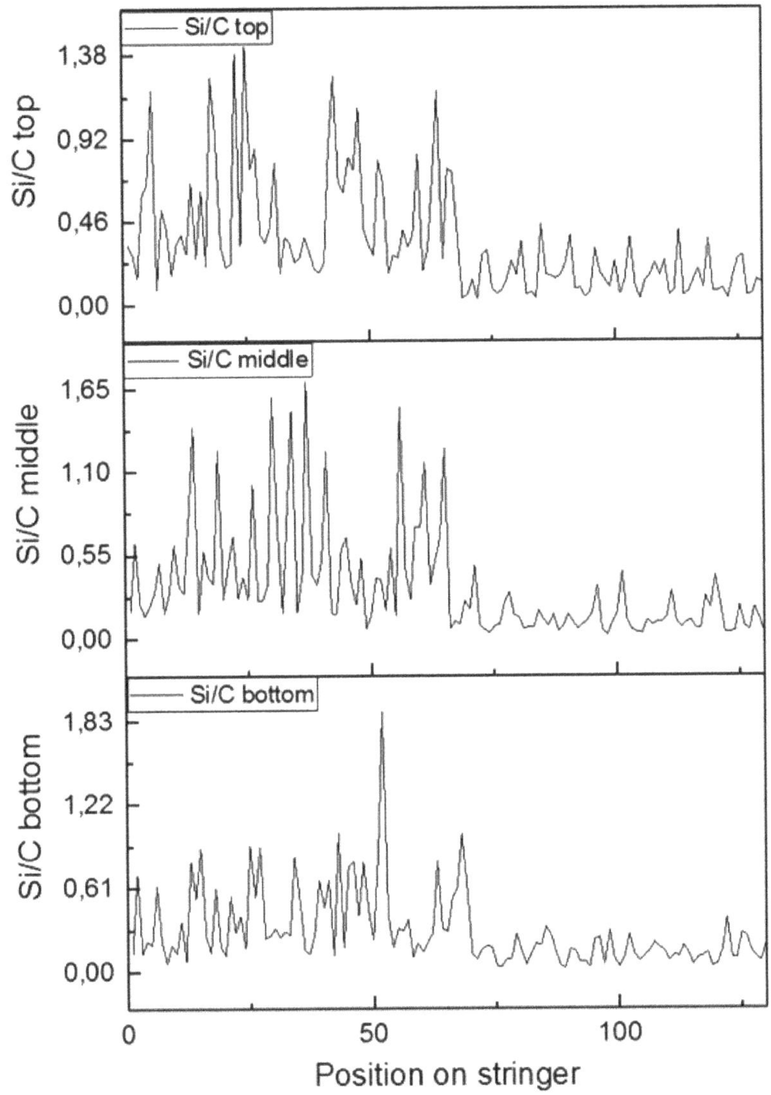

Fig. 5.17 LIBS findings from three horizontal lines composed of 130 positions on a partially contaminated CFRP, highlighting Si/C signal ratios measured on the composite surface in the contaminated area (left) and the clean reference area (right)

measurements it is clearly possible to detect and allocate the areas corresponding to either the contaminated or clean CFRP surface state. Additionally, evaluating further signals enables the type of contamination to be identified. Figure 5.17 presents the characteristics of the Na signal along with one of the lines composed of 130 measurement spots separated by 5 ± 1 mm, whereby it is possible to distinguish a locally increased intensity within the first 65 measuring points. Single peaks demonstrating an increased Na concentration indicate the presence of a Na-containing FP deposit in the contaminated area. It is more challenging to detect and locate the FP contaminations as the contaminated area was relatively limited compared to the region covered by the broader RA deposits and also because only up to three single LIBS shots were located in each FP contaminated area (Fig. 5.18).

In summary, when using the advanced robot-aided LIBS characterization process and evaluating both the [Si] /[C] signal ratio and the Na signal intensity along three parallel lines composed of measurement spots separated by 5 ± 1 mm, we achieved a clear and significant differentiation between the clean surface area and the contaminated area of the stringer, the detection of the elemental constituents of the applied contaminations, and consequently the identification of both contaminants contributing to the combined deposit.

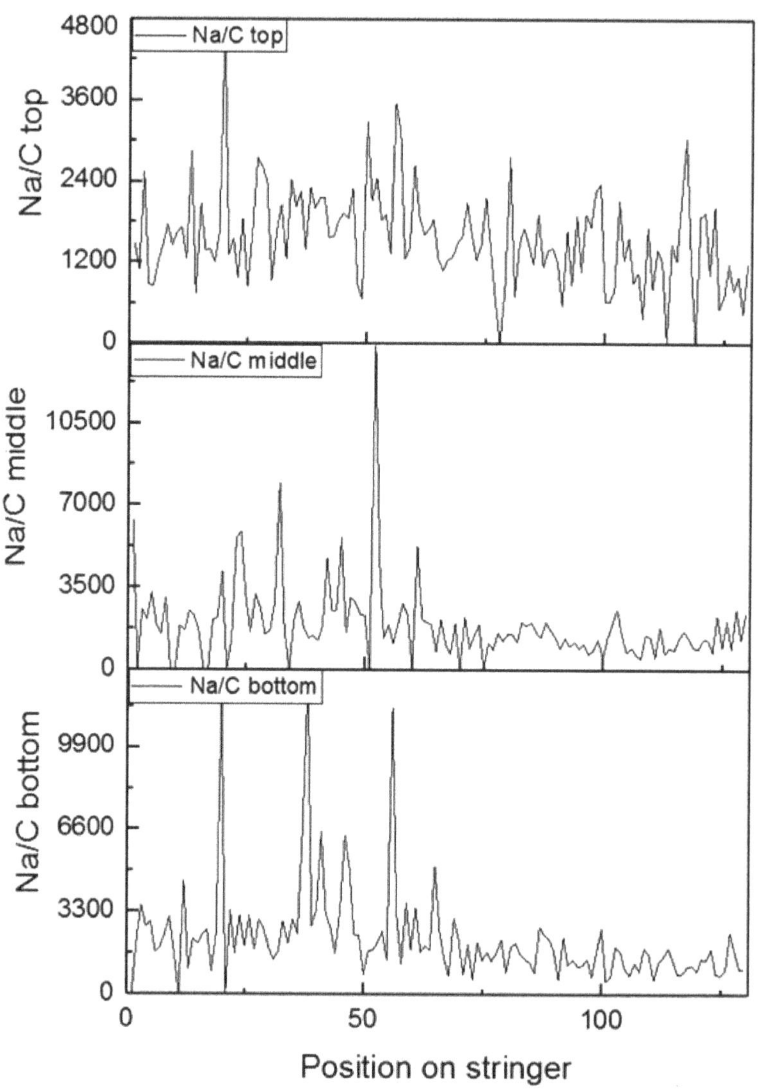

Fig. 5.18 Sodium signals for the three lines measured on the stringer surface give an idea about the contaminated areas (left) and the clean reference area (right)

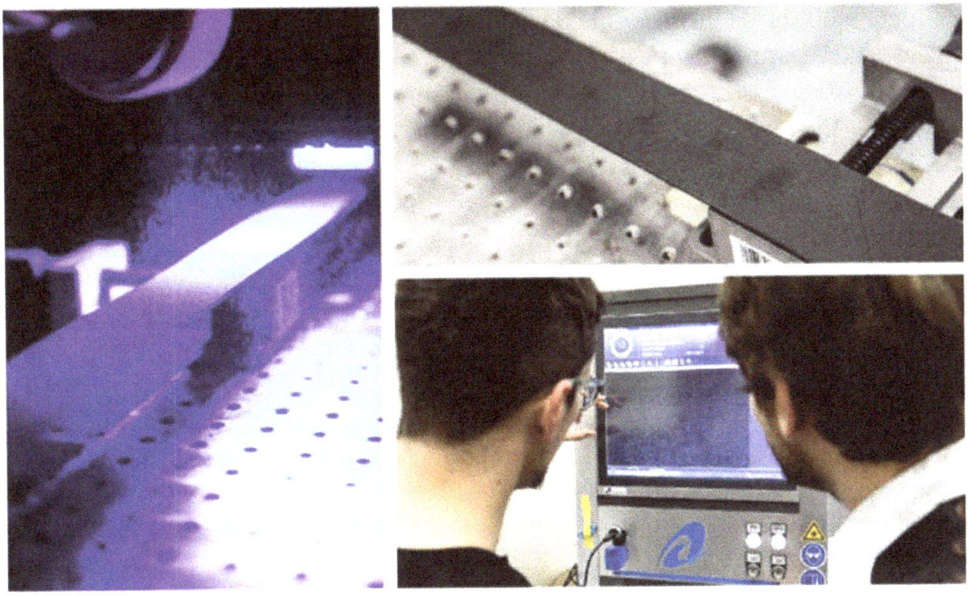

Fig. 5.19 An overview of the AWT-based investigations of a stringer surface during a full-scale demonstration

5.5.1.2 Results for AWT

The contamination on the production part was a single RA contamination on half of the part. This contaminated area was successfully detected and differentiated from the clean half. We were thus able to master the challenge of identifying and detecting laterally separated water droplets with the use of image processing based on a neural network. An overview of the measurements is given in Fig. 5.19.

5.5.1.3 Results for E-nose

Due to the high potential of e-nose techniques with respect to surface quality assessment, we tested the two distinct setups advanced in the ComBoNDT project during the three-day full-scale demonstration event.

AIRBUS E-nose setup

During the demonstration, the realistic and real parts of both main user cases were measured both with and without contamination. For the production user case, three different specimens were tested (a sample with the dimensions 20 cm by 20 cm and two production panels). Panel 1 provided a section with an intentionally contaminated area (Panel 1c) that had been prepared following the combined "RA-FP" scenario. All other samples were not contaminated.

Figure 5.20 shows the desorber device, which was developed in the ComBoNDT project by the partners from Airbus, during its application in taking measurements on the realistic parts in the production user case.

Subsequently, we discuss the evaluation of the acquired e-nose data considering two different approaches to data treatment and representation, namely the partial least squares discriminant analysis (PLS-DA) model and the principal component analysis (PCA).

When we evaluated the measured data of the production panels using the PLS-DA that had been obtained from the pilot sample test, we achieved the differentiation and correct classification of all surface states in both the "clean" and "not clean" (i.e., contaminated) classes. In Fig. 5.21, the horizontal threshold line represents the boundary signal level that can be used to determine whether a data point is indicative of a panel surface state that is attributed to the "clean" class (i.e., ready for bonding) or the opposite class (class threshold).

By performing cross-validation with the obtained, the confusion matrix shown in Table 5.3 is obtained.

Concerning the data evaluation based on a single-class identification method, a different method of signal processing was tested to identify the true class with a clean surface state of a specimen panel. In detail, six samples from the sampling areas on the clean Panels 1 and 2 were taken to determine the region in the PCA that correlates to a clean sample surface. We used these samples to build the detection model. Using a confidence level of 95%, we obtained the green domain shown in Fig. 5.22, which indicates the region in which an unknown sample would be classified as clean. Measurements that lie outside this domain would be classified as false, i.e., contaminated or not clean.

When we subsequently evaluated the remaining seven e-nose measurements of Panel 1 and Panel 2 with this built (single-class identification) PCA model, all measurements made on the uncontaminated areas of the panels were located inside the green domain. Meanwhile, all measurements made on the contaminated areas lay outside this domain, which represents clean surface areas, and were, therefore,

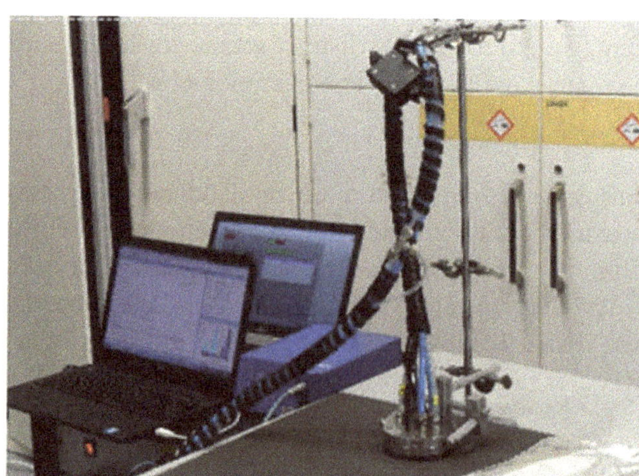

Fig. 5.20 The e-nose desorber device taking measurements on the CFRP Panel 1 within the production user case during the full-scale demonstration event

Fig. 5.21 Results of the class assignment of PLS-DA-treated e-nose data obtained on the technologically realistic parts of the production user case

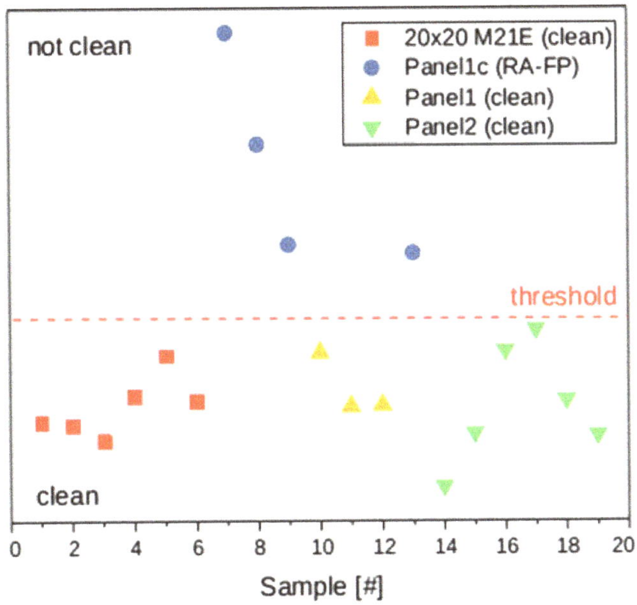

Table 5.3 Confusion matrix obtained when cross-validating the treated e-nose data shown in Fig. 5.21

Actual class	REF	RA-FP
Predicted as REF	15	0
Predicted as RA-FP	0	4

correctly identified as being indicative of contaminated samples. Figure 5.22 displays all the results for the single-class identification evaluation. We would like to highlight that all "unknown" (i.e., prepared and analyzed on-site) surfaces of Panel 1 and Panel 2 were classified correctly based on the performed intentional contamination procedure and following the e-nose measurements.

ENEA setup

The realistic part production user case was also tackled using the latest version of the ENEA e-nose, which is depicted in Fig. 5.23.

Here, we briefly describe the developed data acquisition and evaluation process. Separate models for anomaly detection were produced in order to differentiate contaminated samples from the clean reference samples. Based on nine measurements taken from a (clean) curved realistic reference specimen, we obtained a low dimensional projection using the first three principal components in the PCA rotated sample space. Hypothesizing a multivariate Gaussian distribution of the measured data's projection on the low dimensional subspace, we used a two sigma coverage to set an ellipsoidal surface acting as a detection threshold for contaminated samples. Results from the measurements of clean surface regions were uploaded to the joint Graphical User Interface (GUI) software to compute the detector surface and prepare the GUI to accept the test measurements. Then, we used findings from nine sample regions to test the developed detector: three from clean areas of the tested part and six

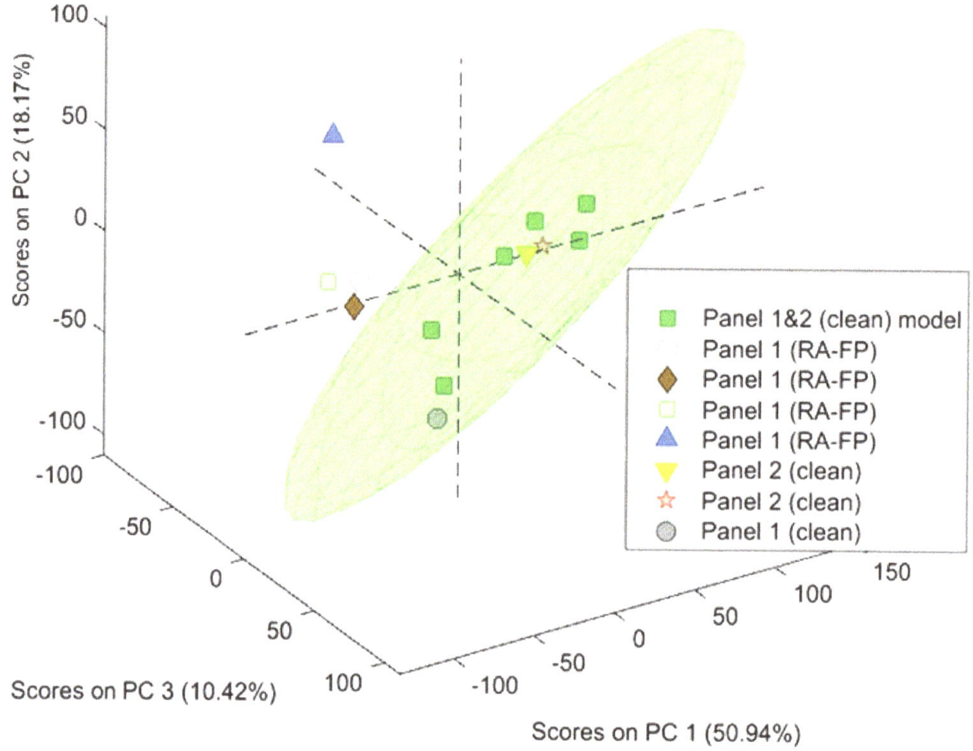

Fig. 5.22 Single-class identification of e-nose results following PCA-based data evaluation from measurements on technologically realistic parts. The green domain indicates the region of measurements on clean samples with a confidence level of 95%

Fig. 5.23 An ENEA e-nose employed for inspection of a technologically relevant realistic composite part during the full-scale demonstration event

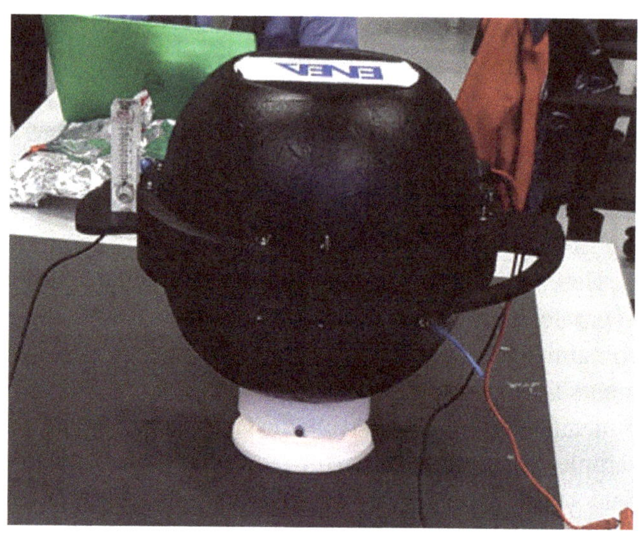

from areas that had been contaminated previously following the “RA-FP” scenario. In summary, we measured 18 samples, of which nine were set apart to build the training set of the PCA-based detector, while the other nine were used for testing purposes. The joint GUI was able to show the results in real-time immediately after the measurement was taken. A screenshot of the active GUI is displayed in Fig. 5.24.

Table 5.4 presents the confusion matrix summarizing the results obtained for the production sample scenario.

As can be seen in Table 5.4, based on the advanced process, we achieved a perfect recognition score. Furthermore, the detection threshold was overcome along a similar direction in the same sub-plane for all except one of the contaminated samples, which adds robustness and significance to the results.

Synopsis: The findings for surface quality assessment within the production user case

The LIBS measurement technique has successfully shown its potential in monitoring composite adherend surfaces, even in the case of complex shapes. During the three-day full-scale demonstration event in the ComBoNDT project, we used the technique for the automated measuring and detection of multiple contaminations with RA

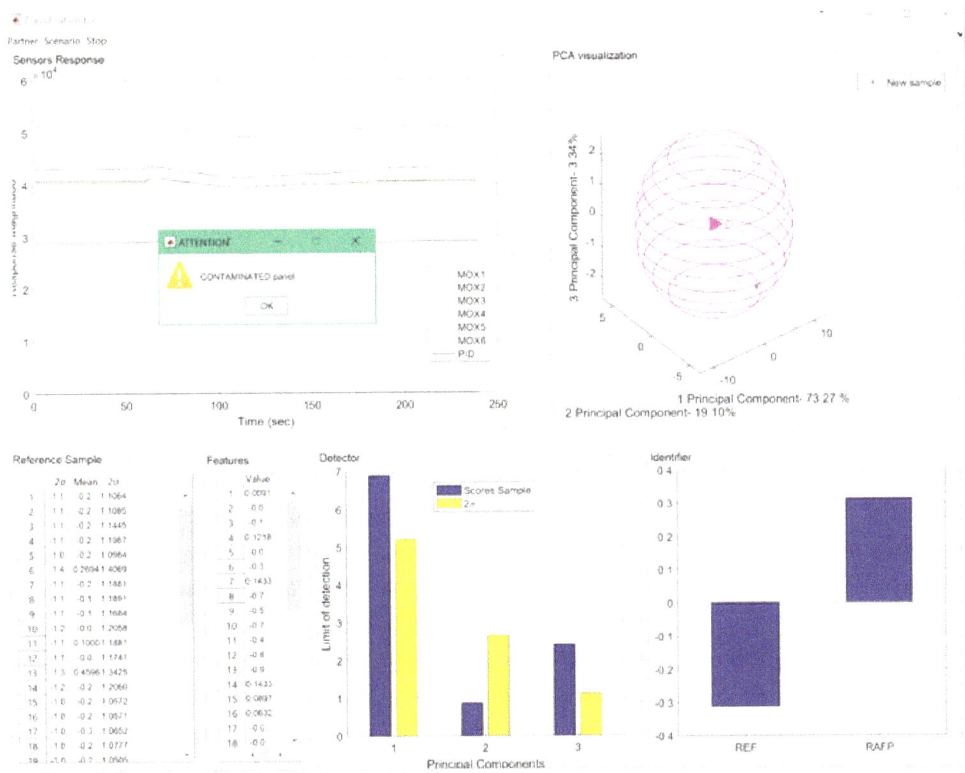

Fig. 5.24 Screenshot of the ENEA GUI employed for the detection and identification tasks based on the e-nose investigations of contaminated realistic CFRP parts within the production user scenario during the full-scale demonstration event

Table 5.4 Confusion matrix obtained when employing the ENEA e-nose for contamination detection on realistic composite parts from the production user case

Actual class	REF	RA-FP
Predicted as REF	3	0
Predicted as contaminated	0	6

and FP, whereby both contaminants were detected and revealed. If a more clearly localized detection of single contaminant-based FP were aspired in the future, then a higher statistical basis, i.e., more measuring points than the 390 measured here for each stringer, would be required.

Using AWT to inspect a realistic production CFRP part, a single RA contamination was detected by means of advanced data evaluation based on a neural network. Finally, a robust and significant detection of a complex combined contamination was achieved using both distinct enhanced e-nose techniques employed for surface quality assessment during the full-scale demonstration event, applying PCA for data evaluation and benefitting from a user-friendly GUI.

5.5.2 *Bonding Quality Assessment*

In the following, we present the findings based on the EMI, NLUS, and LASAT inspections used for bonding quality assessment in the production user case assessed during the three-day full-scale demonstration event within the ComBoNDT project.

5.5.2.1 Results for EMI

The realistic part prepared for the production user case was a CFRP panel with an adhesively bonded stringer, as depicted in Fig. 5.25. One half of the bonding area for the stiffener was intentionally contaminated before bonding. The rear side of this region is shown in the left part of Fig. 5.25a. The combined contamination scenario comprised the application of RA and FP. The second half of the stringer surface was maintained in a clean state and its rear side is shown in the right part of Fig. 5.25a. Technically, within the EMI-based inspection process, this area is used for reference purposes.

Specifically, we bonded two piezoelectric sensors with cyanoacrylate glue to the adherend surface of the CFRP panel (Fig. 5.25a) in order to assess the bonding quality of its joint with the stringer. We placed the sensors on the rear side of the panel in the area where the stringer was bonded. The DD1 sensor was placed on the area of the clean bond, while the DD2 sensor was bonded to the contaminated (following the "RA-FP" scenario) region. The electrical characteristics of both sensors were measured three times. First, the free sensors were measured for referential purposes.

a)

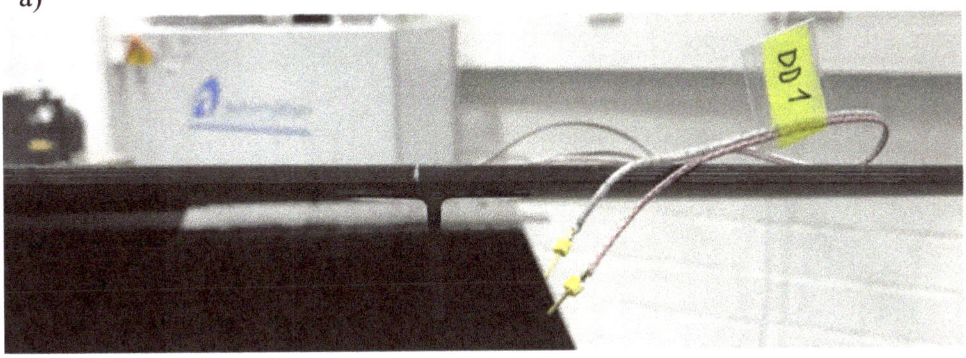

b)

Fig. 5.25 The realistic CFRP panel used as one partially contaminated adherend within the production user case, with **a** the piezoelectric sensors for EMI measurements and **b** the stiffener bonded to the front side. In (**a**), the sensor on the left-hand side (DD2) is bonded to a contaminated surface area and the sensor on the right-hand side (DD1) is bonded to a clean adherend area

Second, the bonded sensors were measured while the panel was lying on the laboratory table (as depicted in Fig. 5.25b). Third, a measurement was made while the panel was being subjected to the external loading situation used for strain measurements. The loading was applied by attaching the panel to the laboratory table with adhesive tape, as demonstrated in Fig. 5.26.

With respect to the sensor signal output and the subsequent data evaluation, Fig. 5.27a presents the obtained root mean square (RMS) values for the conductance obtained in the 3–5 MHz frequency range. After bonding the sensors, a drop in the RMS value was obtained as compared to the free sensors. We observed higher RMS values for the sensor bonded to the contaminated area, which may indicate a weaker adhesive bond. Additionally, when applying an additional load to the panel, the two sensors showed significantly different signal changes. We observed an RMS increase for the clean region, while a decrease was found for the contaminated bond region. As a second characteristic, the conductance maximum peak shift was also tracked, as displayed in Fig. 5.27b. Taking as a reference the frequency at the conductance maximum of the freely oscillating sensor to be bonded close to the bonding region with the clean adherend, the difference to the frequency at the conductance maximum of the second freely oscillating sensor was below 0.5%. After bonding both sensors, a significant increase of this characteristic frequency was observed,

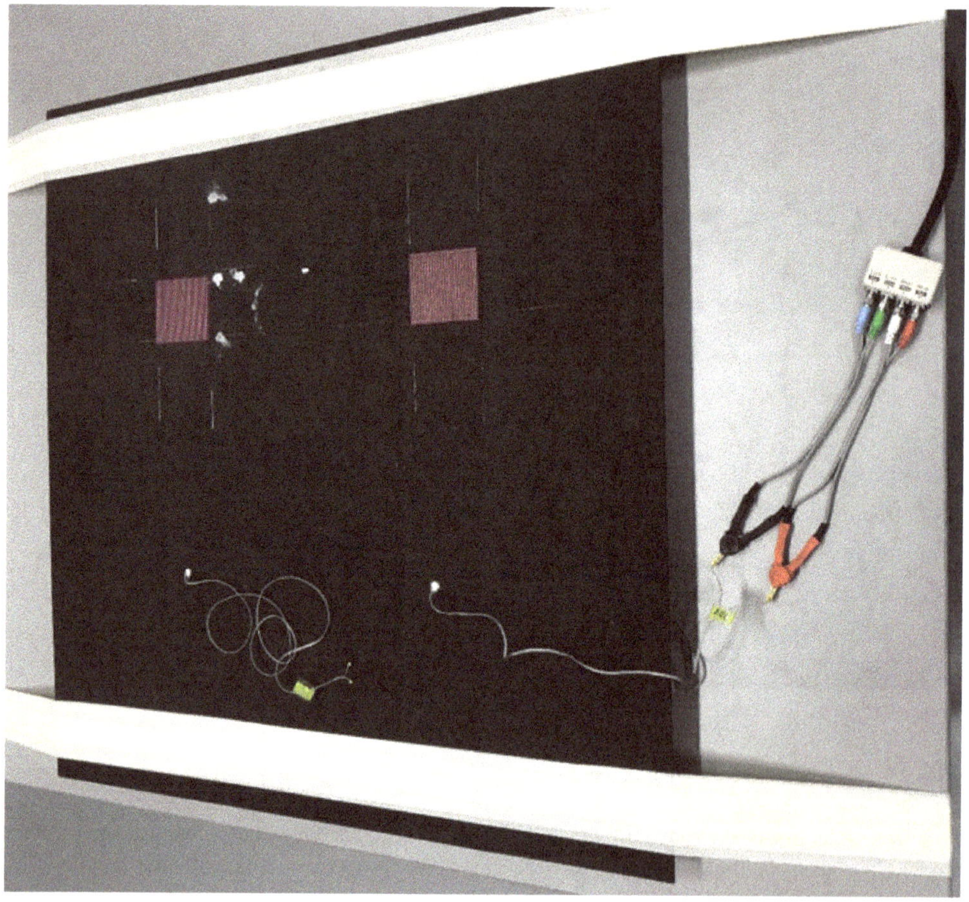

Fig. 5.26 The curved realistic part for the production user case with a loading situation introduced by fixing the CFRP panel in a flattened adjustment using adhesive tape. The square patches with pink color are strain sensors introduced by GMI AERO

which was more pronounced for the bond comprising the contaminated adherend. After mechanically loading the panel, the peak for the clean bonding region shifted leftward towards a lower frequency, while the one for the contaminated region shifted rightward, reaching a shift of almost 2%.

After inspecting the measured conductance curves in more detail, local resonances in the range of 3.75–3.90 MHz were observed. These affected the RMS values calculated for this frequency region as compared to the RMS values for the free sensors (see Fig. 5.28a). Similar behavior was observed for the resistance curves but in a slightly different frequency range, namely from 3.6 to 3.9 MHz, as shown in Fig. 5.28b. As compared to the regions close to the bonds with clean adherends, we observed lower RMS values around the contaminated area, which may indicate a weaker adhesive bond there.

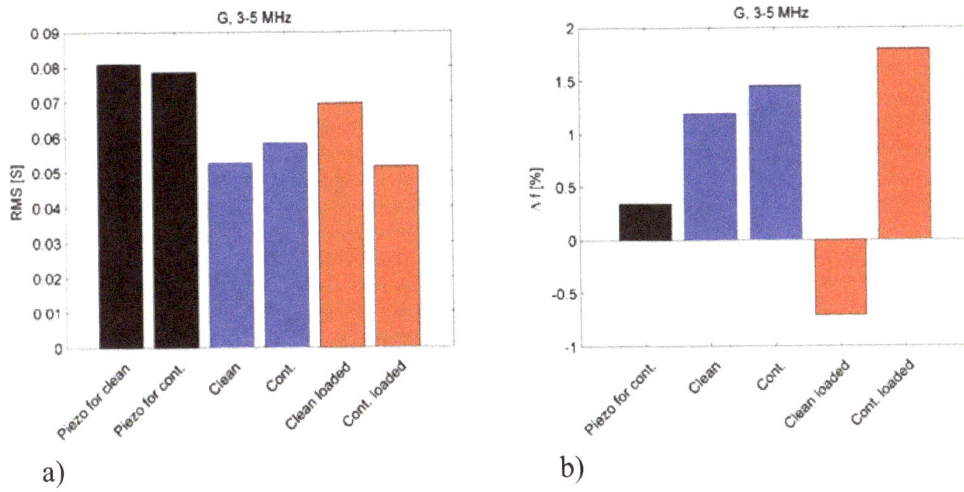

Fig. 5.27 The EMI results for the six investigated bonding states for realistic parts from the production user case obtained in the 3–5 MHz frequency region; **a** RMS values of conductance and **b** frequency shift of the conductance peak

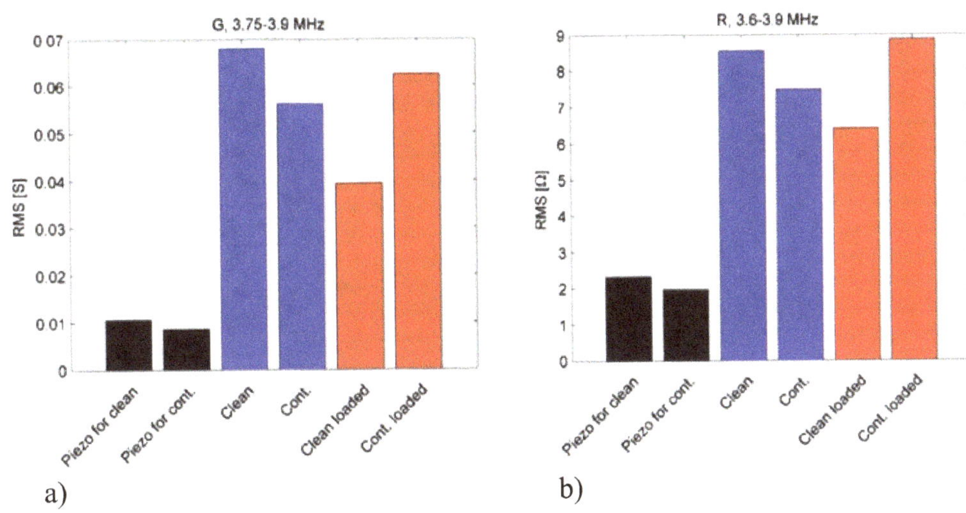

Fig. 5.28 Further details of the EMI findings revealing local resonances in the RMS values of the **a** observed conductance signals and **b** resistance curves

5.5.2.2 Results for NLUS

For the investigation of the bonded CFRP specimen from the production user case, the ultrasound excitation transducer was positioned symmetrically in relation to the longitudinal extension of the stringer to be tested, as shown in Fig. 5.29. Such a position provided a symmetrical insonation field for this stringer, and the findings for the fundamental frequency 20 kHz are displayed in Fig. 5.29.

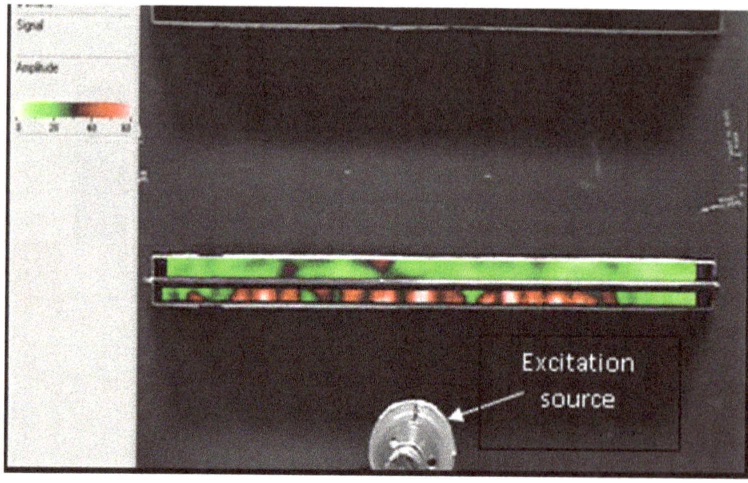

Fig. 5.29 Distribution of the fundamental ultrasonic frequency (20 kHz) amplitude along the stringer in the realistic bonded CFRP specimen from the production user case, as observed during the NLUS inspection

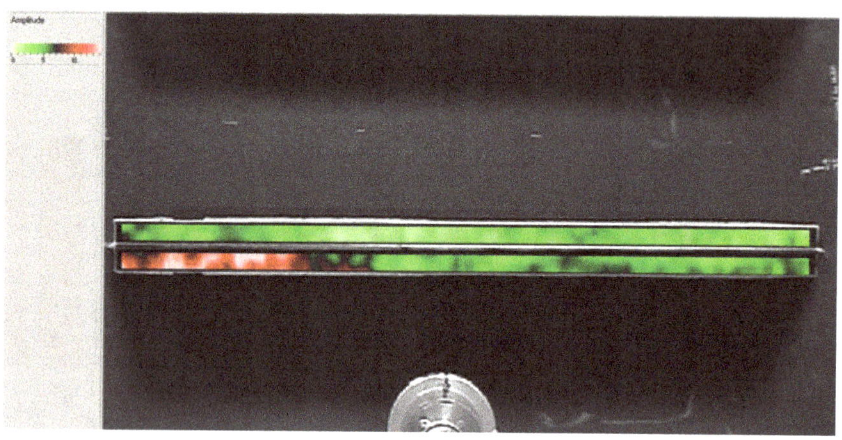

Fig. 5.30 Distribution of the second-highest harmonic (40 kHz) along the stringer length in the bonded production CFRP component, as observed during the NLUS inspection

Additionally, the second-highest harmonics were measured in the production specimen. The findings shown in Fig. 5.30 demonstrate quite different signal distributions, whereby the higher efficiency of the local nonlinear generation in the left-hand-side of the stringer can be clearly perceived and identifies a bonding region based on a contaminated adherend surface area with a lower bonding quality.

5.5.2.3 Results for LASAT

Before detailing the results obtained when applying the laser shock adhesion test (LASAT) for investigating the realistic bonded CFRP specimen from the production user case, we provide a short and demonstrative introduction to the full-scale test performed during the demonstration event of the ComBoNDT project.

Introduction to the full-scale test

In order to facilitate the LASAT investigations on large-area realistic or real parts with complex, e.g., curved, shapes, we advanced the Hephaïstos setup by implementing profound modifications. With the aim of handling specimens of the required size, the optics had to be substantially rearranged and a robot had to be installed to manipulate the sample. The resulting setup is displayed in the on-site environment in Fig. 5.31.

The second challenge was the confinement. So far, the plasma generating the shock wave within the sample had been confined using a thin water layer. Even though this ENDT technique offers an easy setup on a smaller scale, the new upscaled configuration would have introduced substantial operating costs, e.g., the installation of special pipes to supply the investigated sample region with water as well as to ensure water drainage. To overcome these challenges, solid confinement was used. Numerous materials were tested, and for some of these, the results for the flux-dependent plasma pressure are displayed in Fig. 5.32. This graph describes the generated pressure as a function of the laser intensity for confinements based on three different materials. The LASAT operating regime is around $1-3\,\mathrm{GW/cm^2}$, and in this flux interval, water, epoxy, and a thick transparent PPE adhesive tape turned out to have the same confinement capability. With respect to the aspired application, the latter was eventually chosen for its ease of use, and the setup is shown in Fig. 5.33. The aluminum sacrificial layer was not modified.

Briefly, the workflow for testing was as follows. For each sample, an initial threshold identification was performed in different areas. On each calibration area, a series of laser shots were performed. If no damage was inflicted, the laser intensity over the area was increased. The opening energy of a sound part was defined as the "reference threshold". Once this threshold had been identified and confirmed for

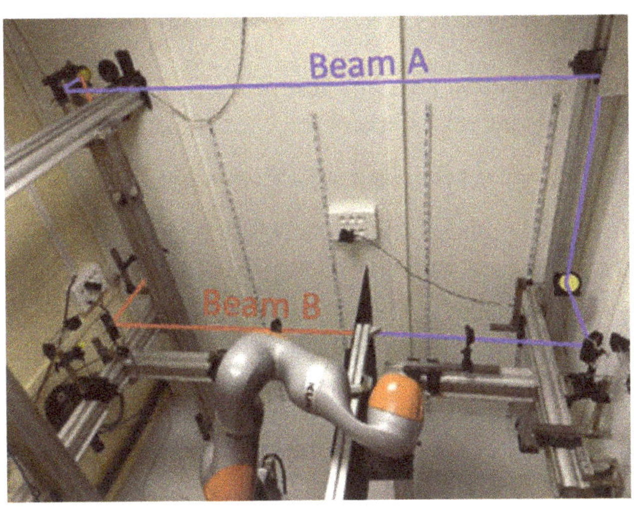

Fig. 5.31 The advanced Hephaïstos symmetrical LASAT setup with a robot to manipulate the bonded CFRP specimen in the production user case

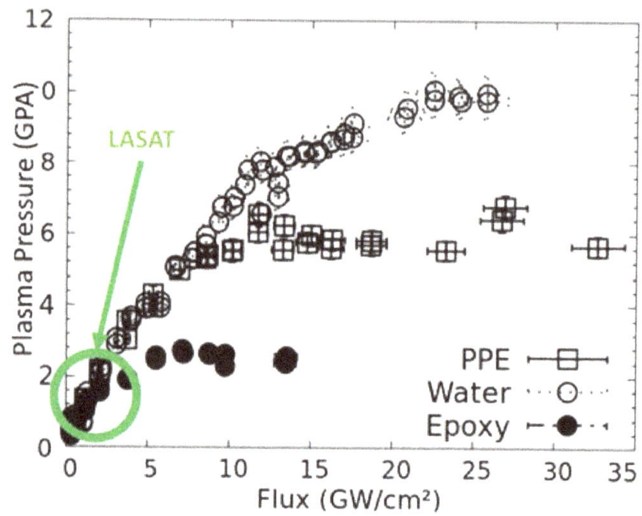

Fig. 5.32 Generated plasma pressure as a function of flux with epoxy, water, and a thick transparent PPE adhesive tape used for confinement during the LASAT investigations

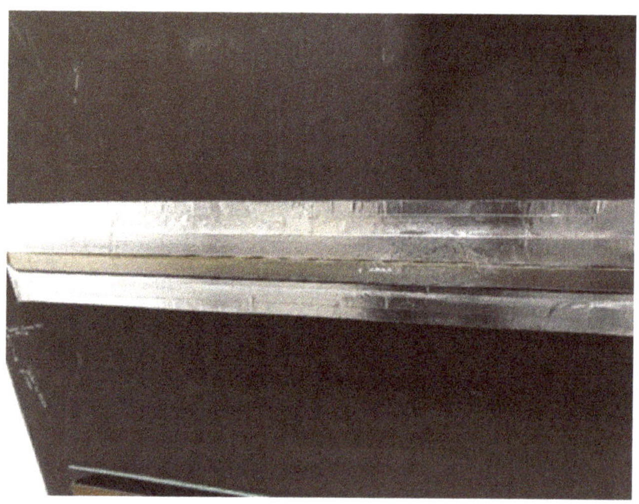

Fig. 5.33 Aluminum sacrificial layer and the adhesive tape-based confinement applied to a realistic production CFRP panel for the LASAT investigation

each calibration area, the rest of the sample was characterized by shooting at 80% of this value, regardless of the contamination state.

LASAT results for the production panel

To test the bonded CFRP panel of the production user case, the whole stiffener was spot-wise shot with an offset of 20 mm between consecutive shots, as presented in Fig. 5.34. This spacing was selected to avoid compromising an area with a crack that could have appeared in an adjacent tested zone.

The threshold of the bonded regions based on either the contaminated or the clean adherend surface area was isolated and a total of 53 shots were realized on one of the stiffeners. The achieved results are summarized in Table 5.5.

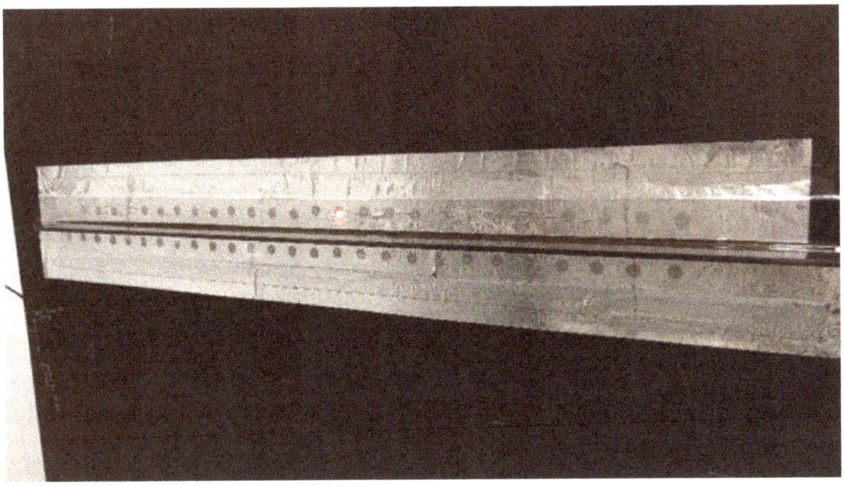

Fig. 5.34 A prepared production CFRP panel after S-LASAT investigation

Table 5.5 Summary of the LASAT results obtained in bonded regions based on either the clean (sound) or contaminated adherend surface area of the CFRP panel of the production user case

	Threshold (GW/cm^2)	Total test shots sym 80%	Opened bond sym 80%
Sound area	0.85	32	0
Contaminated area	0.74	21	21

Briefly, two distinctive opening fluxes were found. Furthermore, during the performed 53 laser shots, none of the tested areas on the sound part of the stiffener were opened. Finally, all contaminated areas failed in the test, meaning that the symmetrical LASAT (S-LASAT) results were conclusive for the production panel.

Synopsis: The findings for bonding quality assessment within the production user case

The advanced setups and workflows for the EMI, NLUS, and S-LASAT inspections used for assessing the bonding quality of the realistic bonded CFRP parts investigated in the full-scale production user case within the ComBoNDT project enabled a differentiation of bonded regions based on clean or contaminated adherend surface areas. The outcomes of the three-day demonstration event in the ComBoNDT project demonstrated the enhanced technology readiness levels of these techniques for production user cases involving the adhesive bonding of composite adherends.

5.6 Results of the Full-Scale Demonstration: The Representative Bonded Repair User Case of an Airbus A350 Panel

In this section, we report on the findings achieved for the repair user case during the three-day full-scale demonstration within the ComBoNDT joint research project. First, we show the outcomes of the surface quality assessment of an Airbus A350 panel prepared for bonded repair, after which we detail the bonding quality assessment. Recalling the workflow of the repair user case, the bonding process was performed after scarfing the CFRP part to be repaired. Moreover, the intentionally deposited contamination refers to the application of DI and FP from gloved fingers coated with hydraulic oil.

5.6.1 Surface Quality Assessment After Scarfing

Here, we detail the results achieved with the AWT, FTIR, e-nose, LIF, and laser vibrometry inspections used for the surface quality assessment of scarfed CFRP surfaces before performing the bonded repair. In more detail, the scarfed area of a convex fuselage part was split into two semi-circles. One side was contaminated with the combined "DI-FP" contamination, while the other side was left blank.

5.6.1.1 Results for AWT

For the repair user case, we inspected a circular shaft with a defined contamination consisting of DI and FP and approximately 200 mm in diameter. The measurement and analysis of the repair part were conducted as follows and as depicted in Fig. 5.35.

First, the calibration of the robot for manipulating and positioning the head of the AWT measuring device was performed. The acquisition of original image data was achieved by conducting measurements of different sections in the area of interest. All the obtained images were then processed using the Convolutional Neural Network (CNN) for droplet detection. Such processing by the neural network results in the generation of heatmaps that allow an assessment of the droplet count (and the surface density of the droplets), the droplet diameter, the certainty, and the wetting ratio. At this point, the evaluated measurement results were available. The next step was classifying the local state of the surface. We achieved this material surface and process-related interpretation with a support vector machine (SVM) based on the local values of the various heatmaps. A class training was performed using a by-hand delimitation of the defined surface states (namely clean and contaminated by DI or FP). The training of the SVM with the hand classification of the assessed surface states was based on 40% of the data for each class. Finally, the classification results for the entire surface were obtained through the trained support vector machine.

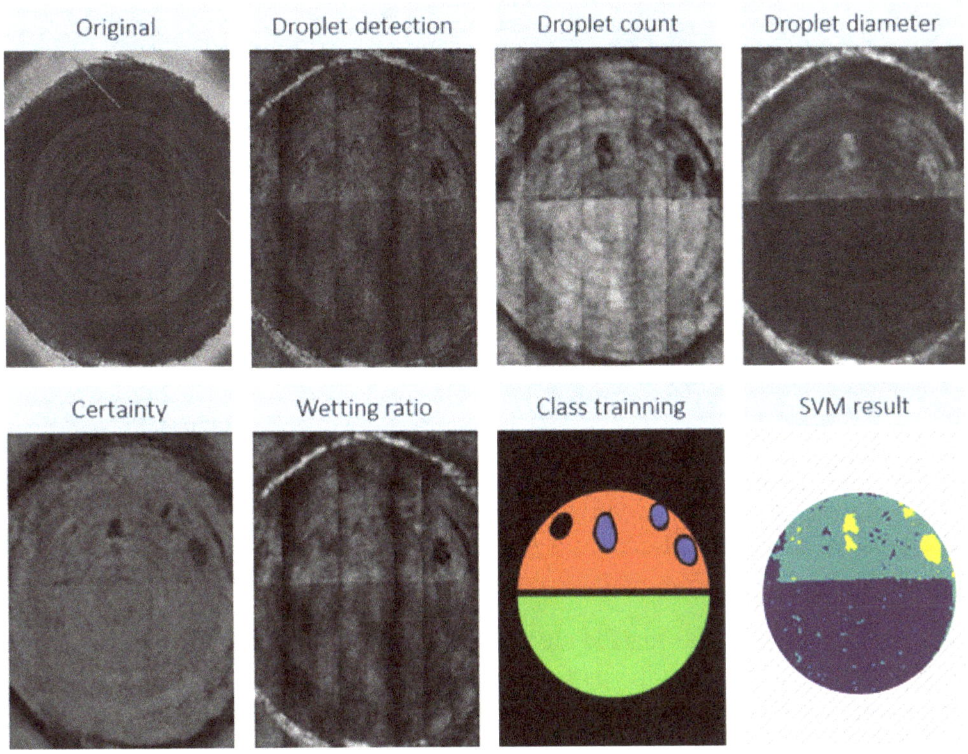

Fig. 5.35 Image-based sketch showing various steps of the AWT-related image treatment, starting from the captured image and going towards the identification of the contamination class. The off-color final result presented in the chart labeled "SVM result" indicates clean (purple) surface regions and regions contaminated by DI (green) or FP (yellow)

As a final result, we achieved a local classification of the contamination state indicated by an off-color map. This is the output of the SVM trained with the standard local output parameters of the CNN, namely droplet diameter, wetting ratio (percentage), certainty of the neural network output, and droplet count for determining the areal density of the droplets. The findings presented in Fig. 5.36 clearly indicate that the surface regions with either type of intentionally deposited contamination can be differentiated from the clean surface regions.

5.6.1.2 Results for FTIR

To characterize the surface of the technically realistic CFRP part, FTIR spectroscopy was applied. An impression of the measuring process using the portable spectrometer is presented in Fig. 5.36. Based on the spectral findings and the application of a partial least squares (PLS) algorithm for the surface region that had been intentionally contaminated with DI, a potassium concentration in the range of 9 at.% was indicated, which is considered plausible. However, a fairly high amount of DI was detected unexpectedly on the reference side of the CFRP surface, which is some centimeters

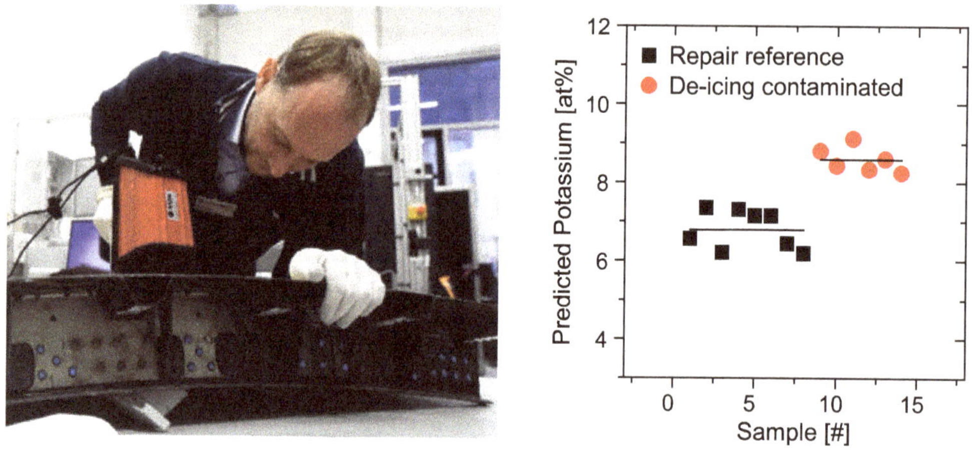

Fig. 5.36 (Left) image showing the performance of the FTIR investigations and (right) plot indicating the amount of potassium on 14 sampling regions as obtained with a PLS algorithm from FTIR spectra recorded on a technologically realistic part

away from the half of the scarfed area that had been contaminated with DI during the sample preparation. As the part was subsequently stored in an oven at elevated temperatures, we suggest that this treatment may have caused some of the DI to be transported to originally clean surface regions, e.g., following migration or after partial evaporation and subsequent condensation on the specimen surface.

5.6.1.3 Results for E-nose

During the three-day full-scale demonstration event, we tested the two distinct e-nose setups advanced in the ComBoNDT project.

Airbus Group Innovations' desorber device

Figure 5.37 illustrates how we applied the Airbus desorber device to conduct measurements on the technologically realistic parts.

Figure 5.38 presents an overview resulting from evaluating the data obtained on technologically realistic or real parts either in the production or the repair user case, benefitting from the PLS-DA model established during the testing of the pilot samples. We infer that the findings of all measurements can be separated into classes corresponding to clean or contaminated CFRP surface states, even for the complex geometries being tested. Once again, the threshold in the graph indicates the applicable parameter level differentiating surface regions that may be assigned to either class.

Performing cross-validation with the obtained, we obtain the confusion matrix of cross-validation, which is shown in Table 5.6.

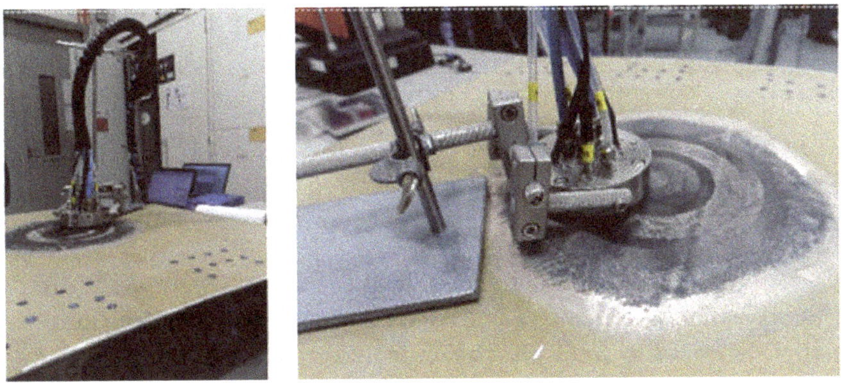

Fig. 5.37 (Left) the e-nose desorber device in use while conducting measurements on a CFRP fuselage part and (right) a close-up view of a scarfed area of a fuselage part

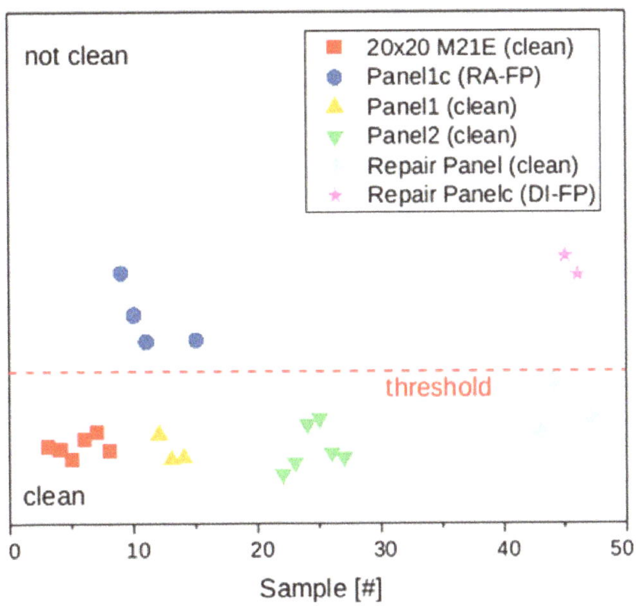

Fig. 5.38 Results for the class prediction of the PLSDA of data obtained from desorber device e-nose measurements on technologically realistic or real parts of the production or repair user case

Table 5.6 Confusion matrix obtained when cross-validating the treated e-nose data shown in Fig. 5.38

Actual class	REF	RA-FP	DI-FP
Predicted as REF	18	0	0
Predicted as RA-FP	0	4	0
Predicted as DI-FP	0	0	2

ENEA setup

Using the ENEA e-nose setup, the same investigation process was performed for the repair user case as for the production user case. Specifically, six measurements were taken from areas in the clean CFRP panel region. These were used to build the ellipsoidal 2-sigma coverage detector scheme for samples contaminated following

Table 5.7 Confusion matrix obtained when employing the ENEA e-nose for contamination detection on real composite parts from the repair user case

Actual class	REF	DI-FP
Predicted as REF	3	0
Predicted as contaminated	0	3

the "DI-FP" scenario. Six measurements in the scarfed region were recorded for testing purposes: three from clean areas and three from contaminated areas.

The elicited response from the contaminated samples was significant and the contaminated samples were revealed without any false assignments. The obtained confusion matrix is presented in Table 5.7.

5.6.1.4 Results for LIF

In the final step of the research, the LIF technique was used to study the surface composition of a real aircraft part, namely a CFRP panel with a stepped area for bonded repair, as displayed in Fig. 5.39.

In Fig. 5.39, we can distinguish eight numbered circular rings that represent separate areas of the inspection. The material layer numbered 1 incorporates the outer

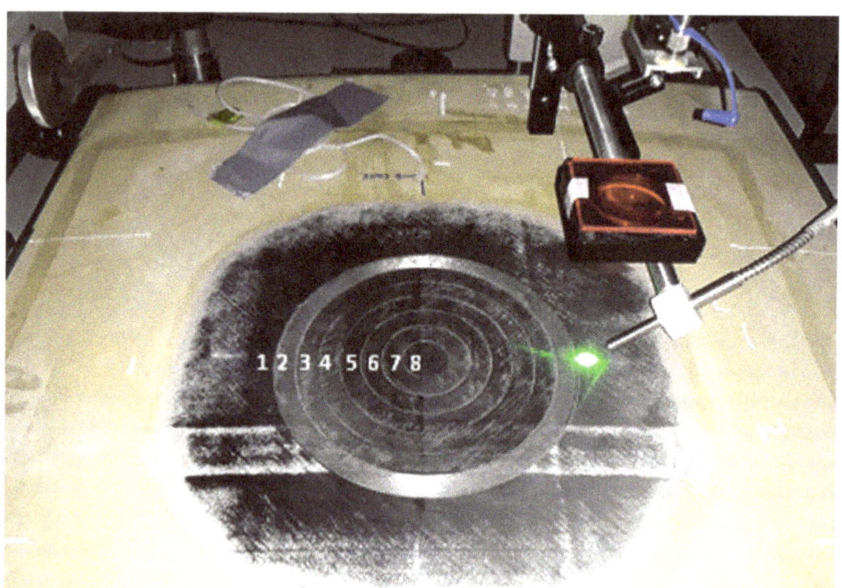

Fig. 5.39 Inspection of a real aircraft part using the LIF technique. The left semi-circular scarfed subarea appears darker and is contaminated, while the brighter right subarea is clean; the circular rings of the stepped surface prepared for bonded repair are numbered, with number 1 indicating the topmost material layer

layer of the CFRP part without primer and copper mesh. The subsequent lower-lying seven rings (numbers 2 through 8) constitute the steps used (as adherends) for the bonded repair area. The left half of the specimen's surface area is contaminated. The right side is clean and provides the reference area for the LIF measurements, which were made at three points at each of the eight areas. Separate measurements were made for the clean region and the contaminated part. The obtained fluorescence intensities at each of the three points were averaged and the standard deviation was calculated. The thus obtained results are presented in the plot in Fig. 5.40. In this presentation, the blue bars correspond to clean areas, while the red bars correspond to contaminated areas. It should be noted that clean areas are characterized by higher LIF intensities, except for areas 2 and 3. In the case of areas 1 and 5, the intensity obtained in the clean subarea is almost double that of the contaminated part. For the remaining cases, the differences are less significant. In conclusion, based on these investigations, the LIF intensity did not allow the surface states to be distinguished across all of the two distinctly prepared areas of the realistic aircraft part. This could be associated with the type of contamination used ("DI-FP"). However, based on the comprehensive set of LIF inspection data acquired here, there is evidence of a significantly inhomogeneous distribution of fluorescence intensities across the scarfed surface regions.

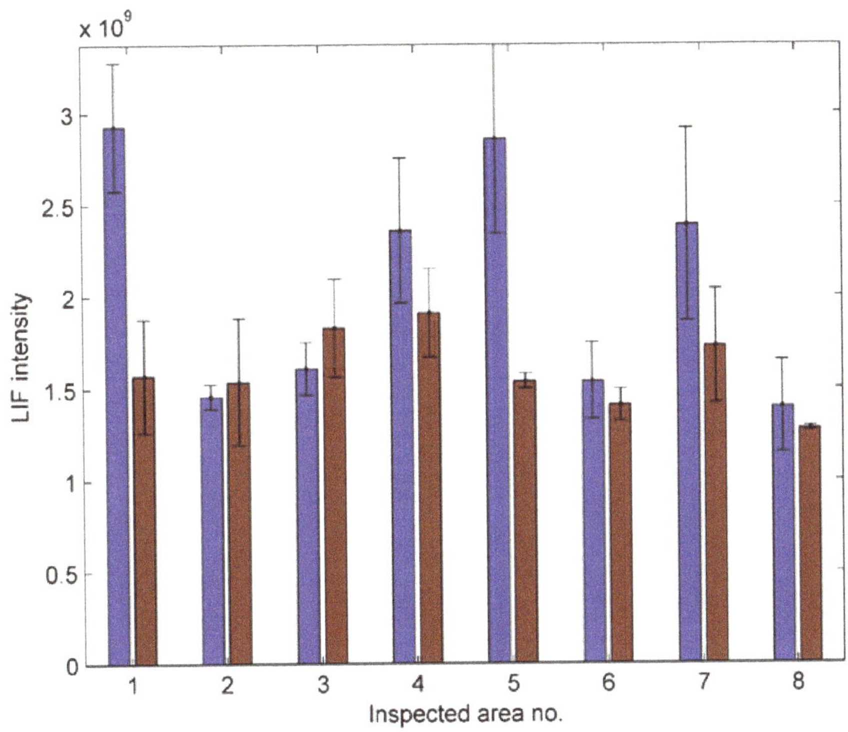

Fig. 5.40 LIF results from the inspection of a real CFRP aircraft part; the blue bars correspond to clean areas; the red bars correspond to contaminated areas; the numbers of inspected areas denote the ring zones defined in Fig. 5.39

5.6.1.5 Results for Laser Vibrometry

In the final step of the surface quality assessment of an Airbus A350 panel prepared for bonded repair, the vibrometry technique was used to study a realistic aircraft CFRP part.

As displayed in Fig. 5.41, the specimen to be inspected had been scarfed for bonded repair, and thus, a stepped region was locally exposed. In order to excite elastic waves in the part, a piezoelectric sensor was mounted close to the surface region of interest. As can be seen in Fig. 5.41, the disc-shaped piezo element was located on the panel surface area covered with primer, meaning it was some centimeters away from the grinded area (where the black CFRP is contrasted in Fig. 5.41) that surrounds the mechanically treated stepped region prepared for bonded repair. Again, we can distinguish eight (height) levels within the inspection area. The first is the CFRP without primer and copper mesh and the next seven levels correspond to the surfaces of the steps introduced by scarfing. The right half of the blackish area was intentionally contaminated. The left side was left clean and the surface areas in this region served as references for the inspection. As shown in Fig. 5.41, the piezoelectric disc was placed alongside the borderline between the two areas in a somewhat symmetrical position so that the elastic wave could propagate in both areas (the clean and contaminated halves of the surface) in the same way. The spatially resolved plot for the root mean square (RMS) of the vibration energy as obtained from the vibrometer software is depicted in Fig. 5.42. This is overlaid on the top view photograph of the inspected region of the CFRP part. As in Fig. 5.41, the contaminated area is on the right side. The total area prepared for bonding is located between two omega stringers, as may be perceived in the foreground of Fig. 5.41. The excitation of an elastic wave in this area results in concentrations of the wave energy between the stringers, which manifest as vertical lines in Fig. 5.42. Moreover, the central image section of the displayed RMS plot shows a circular pattern corresponding to the stepped area on the specimen's surface. The data processing used for this part followed the whole area scan employed for the moisture-enriched samples and pilot

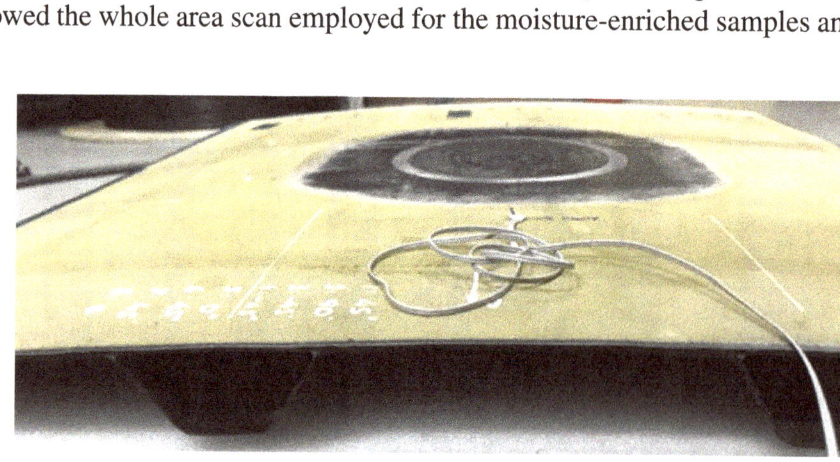

Fig. 5.41 The setup for the laser vibrometry inspection of a real CFRP panel with the scarfed and stepped areas for bonded repair

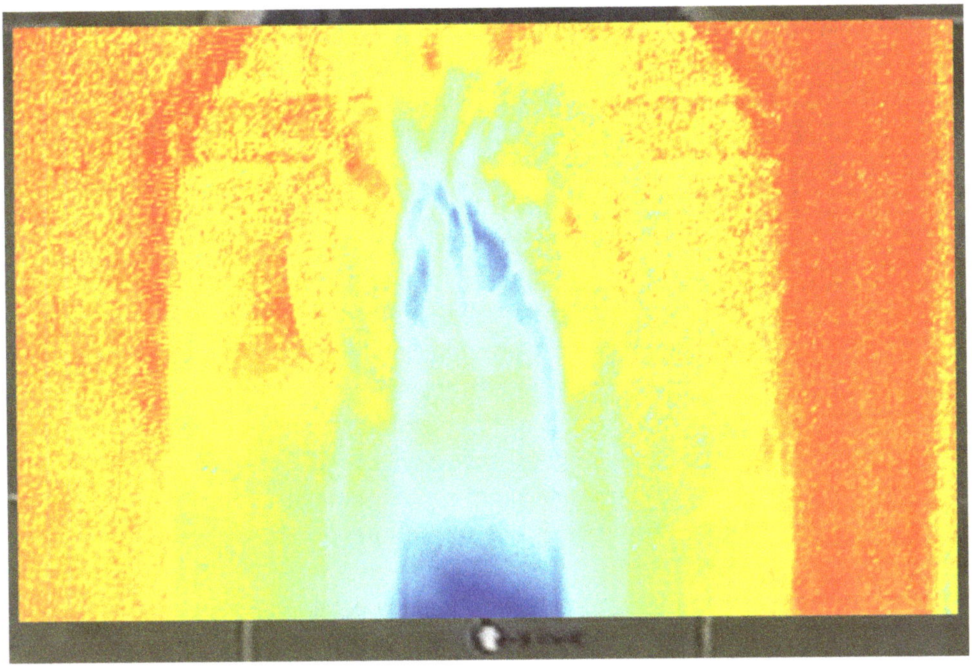

Fig. 5.42 Space-resolved RMS plot obtained from the vibrometer software. The image section is roughly half of that shown in this figure. As illustrated by the lateral view, the roundish delimitation of the area prepared for bonding is located between two omega stringers. Excitation of an elastic wave in this area results in concentrations of the wave energy between the stringers, as indicated by greenish and blueish colors

level samples. Specifically, only the data points corresponding to the stepped area were considered here. Next, this set of data points was divided into two halves. The data points obtained in the surface region shown on the left half represented the referential case, and by processing them, the E (proportional to vibration energy) and 3σ values were calculated. The same procedure was conducted for data points from the right stepped half of the panel surface. The clean area is characterized by the values $\mathrm{E} = 3.58 * 10^{-7} V^2$, $3\sigma = 9.96 * 10^{-7} V^2$, while the contaminated area showed $\mathrm{E} = 3.57 * 10^{-7} V^2$, $3\sigma = 9.68 * 10^{-7} V^2$. We discern that the E values are practically identical and that there is a slight difference in the 3σ values. We conclude that these parameters do not allow the two areas of the realistic aircraft part to be distinguished, although the state of half the panel's surface could be associated with the application of the contamination used in the "DI-FP" contamination scenario.

Synopsis: Findings for the surface quality assessment within the repair user case

The challenge of the ENDT-based surface quality assessment consisted of identifying intentionally deposited combined DI and FP contamination on a scarfed real Airbus A350 CFRP panel. The surface area in contact with DI was more than one order of magnitude larger than in the case of FP. Moreover, indications were obtained that the surface area affected by potassium-containing substances from the DI expanded after the panel was oven-heated at elevated temperatures. Depending on the assessable area for each inspection method, the level of detail of the gained insights was different.

The findings achieved with the AWT clearly indicate that surface regions with either type of intentionally deposited contamination can be differentiated from clean surface regions.

The results obtained when we applied the enhanced e-nose techniques make us confident of the accuracy achieved with these devices for detecting "RA-FP" or "DI-FP" mixed contaminations at the presented levels of contamination, both in the production and repair user cases and with realistic and real parts.

Comprehensive LIF inspection data provided evidence of a significantly inhomogeneous distribution of fluorescence intensities across the scarfed surface regions, but the evaluation of the data subsets did not enable a reliable local differentiation between clean and contaminated regions. Based on the FTIR spectroscopic outcomes obtained with a portable spectrometer and the subsequent application of a partial least squares (PLS) algorithm, indications for an intentional deposition of DI were successfully substantiated.

Given the ample specimen geometry and the expectable substantial expenditure of time, the stepped surface area of the scarfed and stepped CFRP specimen was not investigated by laboratory-based reference analysis for reviewing the actually applied contamination degree.

As indications displayed by AWT and other inspection methods like FTIR differ with respect to the extent of DI-related contaminations, we recommend to utilize the presented enhanced ENDT approaches in future projects in order to perform additional tests for validating the data post-processing, e.g. by applying a PLS algorithm, when inspecting realistic parts.

5.6.2 Bonded Repair Assessment

Subsequently, we report on the outcomes obtained using the EMI, NLUS, and LASAT inspections conducted for the bonding quality assessment after performing the bonded repairs.

5.6.2.1 Results for EMI

In Fig. 5.43, we present a photograph of a real aircraft CFRP panel with a tailored and adhesively bonded circular repair patch. The two piezoelectric sensors attached to the top of the patched area are also visible. On the left, the sensor labeled DD5 is positioned over the clean bonding area, while the second sensor (DD8) shown on the right is placed over the contaminated region. Both sensors were attached in a gluing process using a cyanoacrylate adhesive.

The selected electromechanical impedance characteristics of the sensors where measured two times. First, the free sensors were characterized before gluing for

Fig. 5.43 Photograph showing a real aircraft CFRP panel with an adhesively bonded circular repair patch

referential purposes. Second, the signals from the bonded sensors were acquired on-site while the panel was lying freely on the laboratory table and in an arrangement similar to the one displayed in Fig. 5.43.

When we investigated the realistic part after the bonded repair for the repair user case, we did not observe any additional local resonances using the sensors. This is in contrast to the observations made for the production panel that was structured in a significantly different way. The results of the conductance (G) analysis are depicted in Fig. 5.44a. It can be seen that gluing the sensors causes the RMS values to drop.

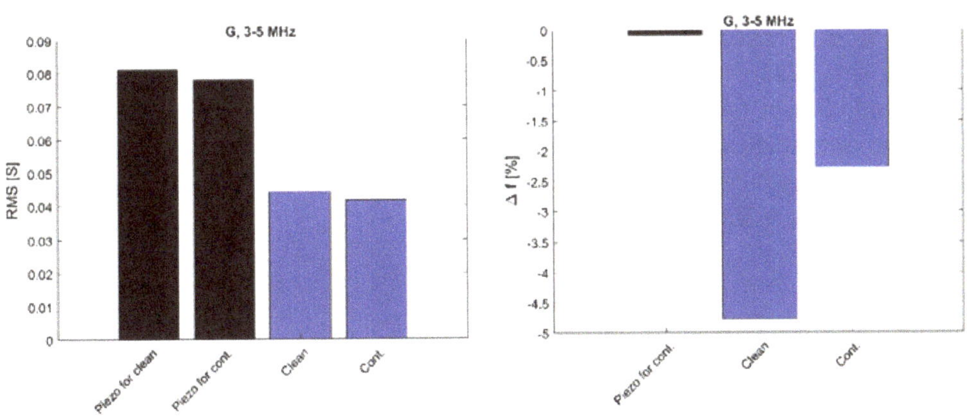

Fig. 5.44 EMI results in the 3–5 MHz frequency region for the six joint states of a realistic CFRP aircraft part that was repaired and investigated within the repair user case; **a** RMS values of conductance and **b** frequency shift of the conductance peak

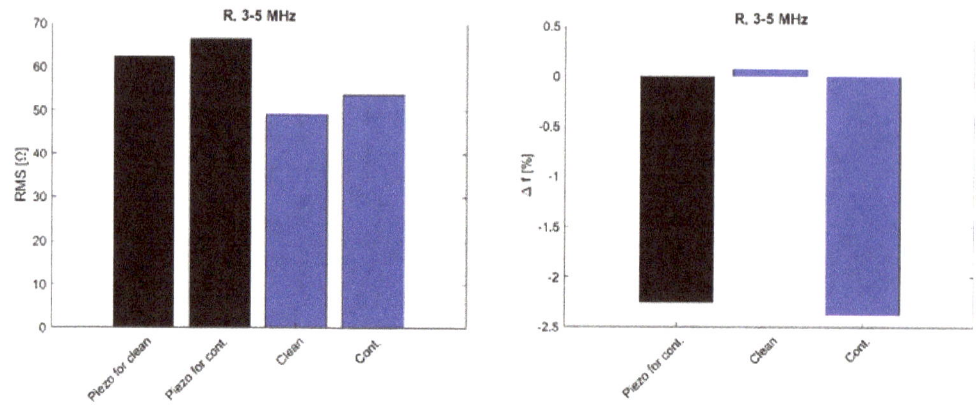

Fig. 5.45 EMI results for 3–5 MHz frequency region for six investigated states of a realistic part from the repair user case; **a** RMS values of resistance presented and **b** frequency shift of the resistance peak

This finding indicates that the sensor vibrations have a lower amplitude, which is attributed to damping effects. The findings for the sensor over the clean region were quite similar to those for the one over the contaminated region. In Fig. 5.44b, we present the measured frequency shift of the conductance maximum. The respective shift for the sensor positioned over the clean bonding region was more than double for the sensor glued over the contaminated region. So, the frequency shift indicates a significant change in comparison to the behavior of the RMS values.

Next, we investigated the resistance curves in the same frequency range (3–5 MHz). The gluing of the sensor resulted in a drop in the respective RMS values (Fig. 5.45a). But the drop was not so significant as for the conductance. Interestingly, we observe that the differences between the RMS values of the two sensors are almost the same for the two investigated sensor states, i.e., the free sensor state and the glued sensor state. The values for the respective differences between the two sensors were found to be 4.25 and 4.53 Ω, respectively. Also, the frequency shift did not differ significantly between these two states (Fig. 5.45b), since almost the same value of shift is observed between the two free sensors and between the free sensors and sensor bonded to the contaminated bond area. The results for the resistance are interpreted as a low sensitivity of the high-frequency region of resistance to the supposed structural differences between the clean and intentionally contaminated bonding regions in the inspected part.

5.6.2.2 Results for NLUS

The fundamental frequency field obtained through the NLUS inspection of the real aircraft CFRP panel of the repair user case with the bonded circular patch is visualized in Fig. 5.46, and it shows a rather conventional standing wave pattern over the whole area of the repaired region. In contrast, the higher harmonic fields measured in the patch region (shown in Fig. 5.47) demonstrate quite different nonlinear field

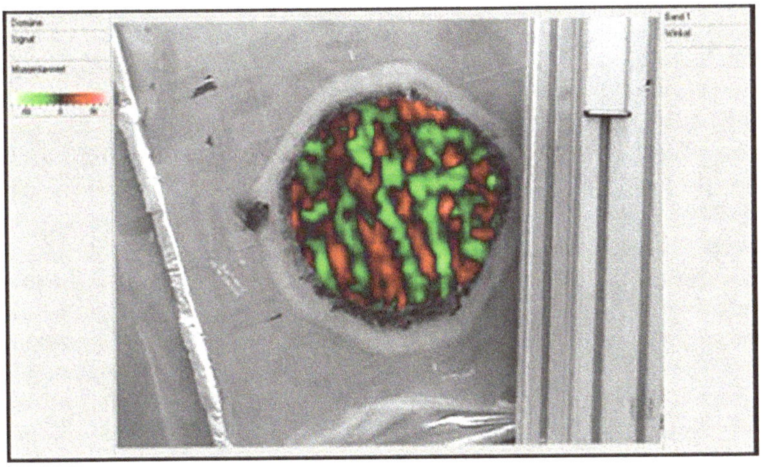

Fig. 5.46 NLUS fundamental frequency field excited in the bonded repair patch region of the CFRP specimen from the repair user case

distributions. The second and third harmonics presented at the top and bottom of Fig. 5.47 reveal a strong enhancement of the nonlinearity and, therefore, indicate a lower bonding quality in regions of the patch that are discernible at the right-hand side and lower part of the patch.

To reiterate here, we also observed an enhancement of nonlinearity in joint regions with weaker bonds when we were mapping and imaging bonded joints based on locally contaminated adherends in realistic aircraft components. In both user cases, the accuracy of the accordance between the positions of the intentionally prepared areas with weakened bonds and the regions identified and visualized with the NLUS approach was confirmed by the respective manufacturers of the contaminated specimens.

5.6.2.3 Results for LASAT

In view of the challenges of a LASAT inspection, the repaired CFRP panel of the repair user case comprised an adhesive bond at different depths, depending on the positioning of the tested subzone with respect to the region of the repair patch. In consideration of this fact, we defined a specific time delay for each inspected subzone of the bond to set the maximum tensile stress induced during a LASAT shot on top of the respective bond location. Moreover, the CFRP panel featured stiffeners on one side. Therefore, we modified and adapted the testing procedure since access to both sides of the panel was not viable for every shot.

Figure 5.48 presents a sketch diagram showing the selected shot pattern as well as the general geometry of the panel.

The panel is partitioned into three distinct main areas, namely the stiffener area (highlighted in dark brown in Fig. 5.48 where the panel can only be accessed from one side, the stiffener flanges (lighter brown), where the stiffener is bonded to the

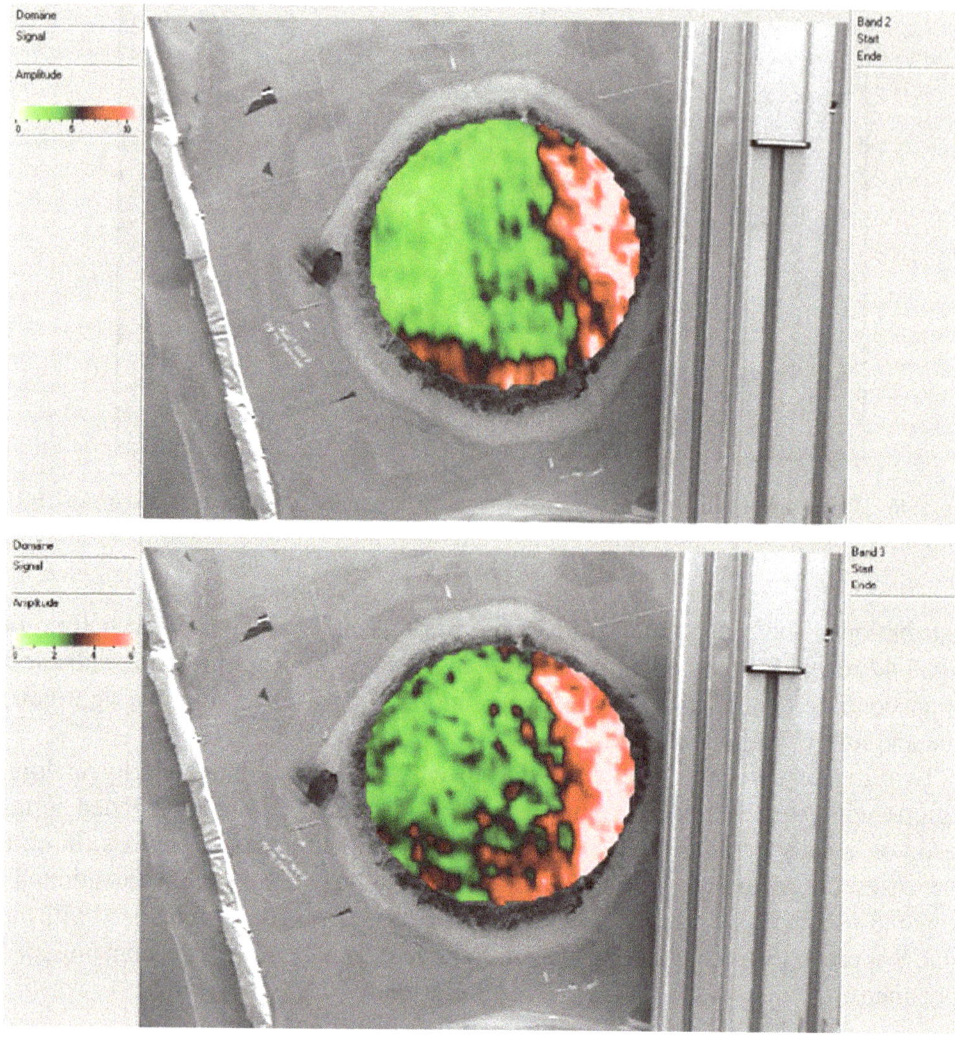

Fig. 5.47 The second harmonic (40 kHz, top) and the third harmonic (60 kHz, bottom) NLUS fields in the repair patch of the fuselage panel from the repair user case

panel, and the actual panel (light brown). As the whole panel was tested, S-LASAT was complemented by standard mono-shot LASAT in the stiffener area. The obtained results are summarized in Table 5.8.

Without applying the enhanced LASAT technique, we were unable to achieve a clear differentiation between a contaminated and a sound region of the patch on top of a stiffener. Both thresholds were alike, and from the ten LASAT shots performed on the contaminated area, only two showed indicators characteristic of an opening. The comparatively low energy required to open the bond might be explained by the actual state of the repaired panel. Numerous porosities were found by the ultrasonic test performed after the repair. Following the line of action for a quality assurance procedure of a repair process, this finding would be considered an indication that the

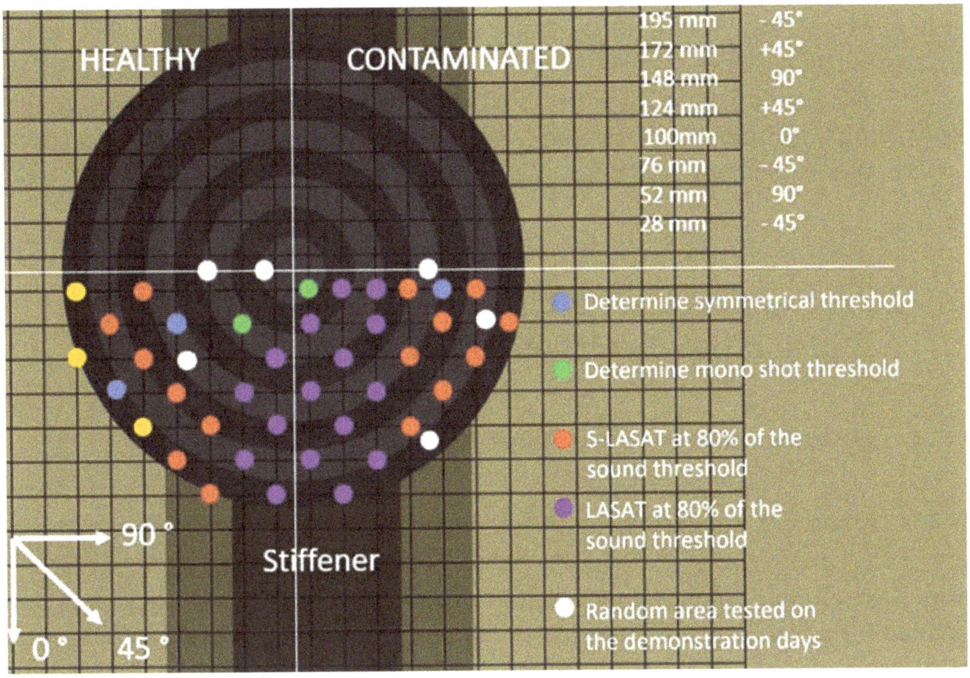

Fig. 5.48 Diagram of the LASAT shot partitioning on the adhesively bonded repair patch within the CFRP panel of the repair user case. Contaminated joint regions (in contrast to clean ones) are based on intentionally contaminated adherend surface regions

Table 5.8 Summary of the LASAT results obtained for the repaired CFRP panel of the repair user case for sound and contaminated regions of the repair patch

	Threshold S-LASAT (GW/cm^2)	Total test shots mono 80%	Opened bond mono 80%	Threshold S-LASAT (GW/cm^2)	Total test shots sym 80%	Opened bond sym 80%
Sound area	0.63	7	0	0.72	8	0
Contaminated area	0.63	10	2	0.61	8	8

workers did not perform a high-grade repair task. During the full-scale demonstration, we performed further inspection. Although the state of the panel did not meet the demands, we achieved proper discrimination between the repair region on the contaminated adherend surface area and the one on the sound adherend surface by applying S-LASAT. We observed that only the area tested on the contaminated part failed the LASAT test, and thus it was identified.

Advancing the S-LASAT technique enabled us to differentiate, for both inspected panels, the contaminated area from the sound one. We faced some limitations of the basic LASAT technique in assessing inspection regions with complex geometries: If one side of the CFRP panel was not accessible, the technique could not be applied and a non-optimized testing approach, such as the mono-shot setup, performed inadequately.

We consider the LASAT double-shock setup highly promising for overcoming the existing limitations in the case of complex panel shapes and we will be ready to perform further challenging studies in this vein.

5.7 First Evaluation of ENDT Procedures Introducing a Probability of Detection Approach

5.7.1 Introduction and Motivation

In this section, we conduct in-depth evaluations based on a probability of detection (POD) approach to the findings achieved with the distinct enhanced ENDT procedures. To this effect, we introduce the POD methodology and demonstrate its aptitude for selected applications in several of the user cases introduced above. In more detail, we propose cutting-edge evaluations highlighting initiatory POD for the ENDT procedures advanced in the ComBoNDT project [1] and we base these on selected findings from the previously described intentionally contaminated CFRP adherend surfaces from distinct user cases. To scale the expectations, we consider such investigations a prerequisite to any certification step. Indeed, performing a POD evaluation is mandatory for any new non-destructive testing technique deployed in service, including our applications. The subsequently described examinations were performed as an important forward-looking contribution of the ComBoNDT project to initiate prospective in-depth approaches for selecting technically appropriate ENDT methodologies and procedures for specific quality assurance tasks following a POD evaluation.

We are aware that we are unable to present a complete POD evaluation that complies with technical standards. For example, a first aspect that we consider is that none of the presented ENDT procedures is mature enough to have a fixed testing workflow with a well-defined threshold value that allows surface states to be robustly identified as "ready to bond" based on the extensive (accelerated) testing of the respectively manufactured adhesive joints. Therefore, we are as yet unable to define an exact POD procedure for evaluating ENDT procedures. For this reason, we use POD as applied for conventional NDT as a guideline, which allows us to outline the background and basic principles of such an evaluation. We repeatedly highlight in this report that there are some constraints on the formal resilience of the used data. A second argument is that real POD is very time-consuming and expensive—with good cause as it requires a large amount of reproducibly prepared specimens with representative defect types and sizes that are intentionally introduced within a representative range. However, we here merely include the three contamination scenarios that were available for our demonstration. Third, it would be necessary to involve various operators in order to account for the effects of individual human factors. In addition, to ensure a formal evaluation approach, the operators should not be scientists with expertise in the evaluated ENDT technique but rather adequately trained operators, and to enable blind testing, they should not have prior knowledge of the defect introduced into the samples. Clearly, for the data gathered in the ComBoNDT research project, these conditions were not fulfilled.

In spite of these differences, we consider the promising quality of the acquired data to be interesting enough to start a simplified investigation. Therefore, we propose a first POD evaluation here, albeit not for all the ENDT techniques and procedures applied and not for every contamination scenario. Rather, we advantageously select already available procedures and datasets that appear conveniently close enough to typical POD procedures.

Subsequently, in the next section, we first provide general information about POD. Then, we describe the selection and inquiry processes applied for gathering suitable datasets from the ComBoNDT project partners for POD calculations. In this context, we also explain some choices, especially regarding the values for the requested contamination levels within a distinct scenario. We present the thus obtained results in the main part of this section, whereby they are grouped by ENDT technique with no specific prioritization. In the last part of this section, we draw some preliminary conclusions and propose a synopsis table on the POD quantification to summarize the main results obtained in this work.

5.7.1.1 General Information on POD Procedures and Tools

Following our layout, a POD procedure is applicable for a given list of quantifiable features related to instantiated (sub)sets based on:

- a given material/structure to be inspected
- a given defect type
- a given NDT technique
- a given testing procedure

Regarding the user cases and scenarios from the ComBoNDT project, this means that each POD calculation is valid for one specimen type (i.e., contaminated CFRP for surface quality assessment, and assemblies comprising adhesive joints based on contaminated adherends for bonding quality assessment), one contamination scenario, and one ENDT technique and its associated diagnostic procedure. If one of these instantiations or elements changes, then the actual POD approach might change.

We designed the POD approach following the targeted purpose of accounting for the uncertainties in the capability of an ENDT technique to detect a given defect. Regarding the POD procedure for non-destructive examination (NDE), "*The POD(a) function is defined as the proportion of all cracks of size that will be detected in a particular application of an NDE system*" [2]. The NDT result is made non-deterministic because of the potentially numerous uncertainties. For example, the actual defect (e.g., fatigue cracks), procedure, material, mechanical repeatability, and human factor will influence the signal response of an NDT inspection. Therefore, a POD evaluation is a statistical estimation of the performance of an NDT with a consideration of the uncertainties, accounted for via the dataset. The POD estimation uses a sample set to gather data and is thus rather more "sample-dependent" than inherently representative of the basic population.

In practice, there are two main approaches to estimating a POD, namely the hit/miss model and the signal/response model. Some key elements are provided below, and more details can be found in the literature, such as Berens [2, 3], and other publications related to non-destructive inspection (NDI) procedures, e.g., ultrasonic inspection [4] and scheduling NDI [5].

Hit/miss approach

The hit/miss approach refers to a binary diagnosis providing as its output a logical variable with the feasible values "detected" or "not detected". For each flaw size a of the dataset, the "observed POD" $POD_{obs}(a)$ is calculated as shown in Eq. (5.1). The POD function for all a is then obtained through a parametric regression on the POD_{obs} values. The Berens parametric regression model uses cumulative normal or log-normal functions.

$$POD_{obs}(a) = \frac{\text{Number of Hits } (a)}{\text{Number of trials } (a)} \tag{5.1}$$

It is generally recommended to have a minimum of 60 flawed inspection sites with at least three operators.

Signal/response approach

In this approach, the signal response $S(a)$ from the defect with the flaw size a is used to give a quantified diagnostic with regard to a decision threshold value sth ($S(a) > sth$ or $S(a) < sth$). For a given defect size a, $S(a)$ is considered a normally distributed random variable. The probability density function of the signal value, g_a, has a mean value μ_a and a standard deviation ε independent from a (see Eq. 5.2). Then, the POD for flaw size a is defined as in Eq. (5.3)

$$S(a) = \varphi(a) + \varepsilon a(M) = \mu_a + \varepsilon \tag{5.2}$$

$$POD(a) = \int_{sth}^{\infty} g_a(s)ds \tag{5.3}$$

Some hypotheses must hereby be verified:

$\ln(S)$ and $\ln(a)$ (or S and a) are linearly linked: $\ln(S) = \beta_0 + \beta_1 \ln(a) + \delta$, with β_0 and β_1 representing regression parameters and δ being a normally distributed variable with zero mean and variance σ_δ; σ_δ is independent of a.

Then, the POD function is a cumulative log-normal(μ, σ^2) distribution function, with parameters shown in Eq. (5.4). The POD curve is then evaluated to estimate the β_0, β_1, and σ_δ that best fit the signal data.

$$\begin{cases} \mu = (\ln(s\text{th}) - \beta_0)/\beta_1 \\ \sigma = \sigma_\delta/\beta_1 \end{cases} \tag{5.4}$$

It is recommended to have a minimum of 40 flawed inspection sites with at least three operators.

Regardless of the approach, the estimation of POD-based on several datasets for the same measuring problem leads to several POD curves and facilitates the evaluation of a confidence interval that is the dispersion in the POD results for a given amount of data obtained for the sample:

- For a given flaw size a, the probability of detection with 95% confidence, labeled "PODa/95", is to be understood such that the estimated probability of detection "PODa" has a 95% chance of being higher than PODa/95;
- For a given POD, the flaw size detectable with 95% confidence, labeled "aPOD/95", is such that the estimated flaw size "aPOD" has a 95% chance of being lower than aPOD/95;
- Typically, the most common value is POD90/95, labeled a90/95 in this report. For a given flaw size or contamination degree, this means a probability of detection of 90% with a 95% confidence. As this value is the one generally communicated to the authorities for aeronautics applications, we selected this one in this work.

To separate and classify the measured analysis data, one important parameter is the detection threshold, and its determination can strongly influence the POD. We recommend defining it in a way that ensures that the defect signature is identified and differentiated from the noise and the saturation signal. In the case of the subsequently evaluated data obtained in the ComBoNDT research project, a saturation of the signal detection systems was not reached, and we consider the noise to be according to the signal in the case of the reference noise level was defined considering the signal on reference specimen. Therefore, we adjust the detection threshold so that it is, on the one hand, superior to the reference values obtained using cleaned adherend surfaces and, on the other hand, inferior to the lowest contamination values that could be differentiated from the reference values.

POD model implementation in NDT software

The model described above uses CIVA software, which is one of the main simulation software packages for NDT modeling.

In the upcoming section, which reports the results for selected example ENDT investigations, we always display the POD findings using the same layout, which is composed of three windows, as shown in Fig. 5.49. First, the top-left window represents the material-related data points as obtained with the respective techniques in the ComBoNDT project. The underlying system of coordinates shows the contamination

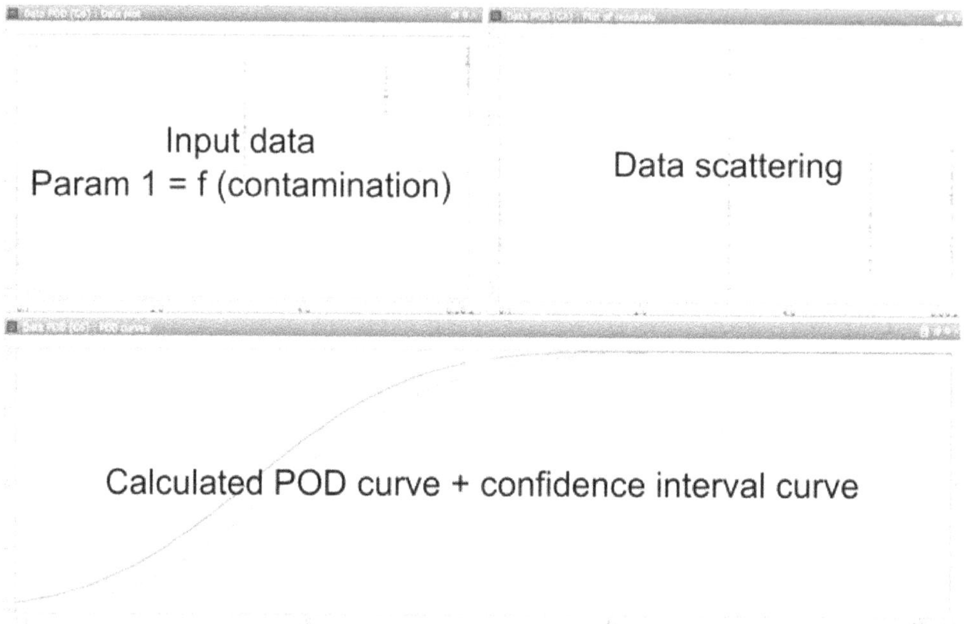

Fig. 5.49 The graphical user interface provided by the CIVA simulation software package for NDT modeling. Most importantly, the upper-left window shows the measured material-related input data and the lower window displays the calculated POD curves and the probability of detection *a*90/95, as introduced in the text

values along the abscissa, as determined using a laboratory-based reference analysis. Meanwhile, the values for the feature being quantified based on the ENDT inspection are displayed along the ordinate. The dimensions and characteristics are individual for each technique and they might be, for example, amplitude values. We used the graphical user interface of this chart to adjust the detection threshold. Second, the top-right window displays the data scattering and a control parameter for the numerical model hypothesis, which is not detailed. Finally, the bottom window depicts the calculated POD curves, upon which the introduced probability of detection *a*90/95 is generally displayed.

5.7.2 Input Data for POD Calculation and Compiled Hypotheses

For the POD calculations, we gathered example datasets acquired with advanced ENDT techniques and procedures for distinct and intentionally applied contamination scenarios within different user cases. As indicated above, these datasets were originally not measured following established POD approaches and instead may be considered instances of a challenging supplementary blind testing. In detail, we considered raw data because we wanted to maintain the statistical distribution within the data, and as many points as possible were required for the selected ENDT tasks. As at least 30 points per contamination degree within a selected contamination scenario were required according to the model hypothesis, for the aspired POD demonstration, we primarily relied on the ENDT data obtained for the smooth coupon samples. Unlike with the more complex realistic and real CFRP specimens, the fact that numerous coupon samples were available allowed the inspection specialists to complement the ENDT datasets for which a POD calculation was performed, with a wide score of datasets obtained using LIBS, AWT, E-nose, EMI, and NLUS.

As this section aims to provide a vivid and demonstrative presentation of our POD approach, we focused on only one characteristic feature per ENDT technique. This technique-specific feature was selected to be "differentiating" for the respective contamination scenario. By this, we mean that the contemplated ENDT testing procedure was selected based on this characteristic feature so that not only could the contamination be detected compared to a clean reference specimen, but also different contamination levels could be distinguished. For this purpose, a feature value could be a measured amplitude or velocity or a value deduced from measured quantities using a model. Moreover, the correlation between this assessed feature and the contamination should be linearizable to respect the model hypothesis (see Fig. 5.50). The data input was requested as shown in Fig. 5.50 as an example contamination scenario. Hereby, we tested various scenarios and respective ENDT-based features.

For one contamination scenario (e.g RA)

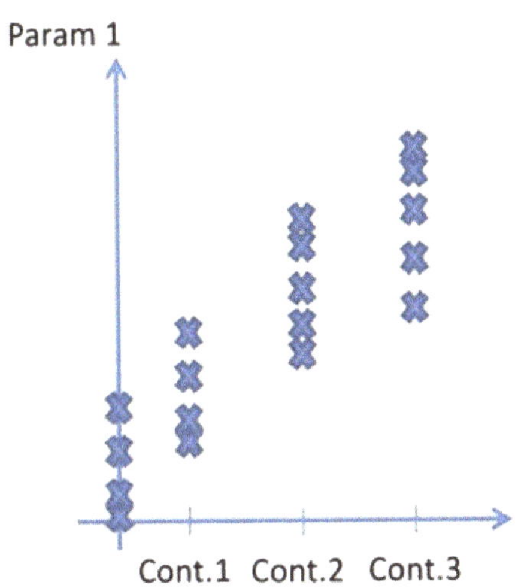

Contamination	Param 1 ()
C0	0,2
....	
C1	4,1
C1	3,8
...	
C2	7,2
C2	8,1
...	
C3	15,6
C3	14,7
....	

Fig. 5.50 Schematic description of the input data required for the exemplified POD approach. The contamination degrees Cont. 1, Cont. 2, and Cont. 3 were adjusted according to an accurate intentional contamination procedure and were characterized by laboratory-based analysis, while the ordinate values for Param 1 were obtained from the ENDT inspection

5.7.2.1 Choice of Contamination Values

Here, we explain our cutting-edge approach and highlight the caution that guided how we proceeded during our proof of concept.

As, to our knowledge, the POD approach has never been used for ENDT techniques and the associated material-features, and as we aimed to introduce a quantitative probability of detection for contaminants, we adapted the contamination values included here in a way that the POD-based model converges. Indeed, the model was established for defects that are conventionally monitored in aeronautical applications, such as cracks or delaminations, which are generally quantified referring to a length in millimeters, with relatively small values. Pragmatically and to avoid any numerical issues, we thus decided to adapt and scale the contamination-related values from the laboratory analysis and ENDT to remain in an expected range for the used numerical tool. We list the thus obtained values in Table 5.9, and we briefly outline our value implementation for each contamination scenario. Essentially, we used a consistent procedure for all the contamination scenarios which had been realized centrally and characterized by surface-sensitive XPS laboratory investigations in a standardized way. For the local implementations of contamination scenarios, we relied on the feature values that were individually measured in laboratories. Furthermore, in the case of double-digit feature values, we scaled them as described subsequently.

- The contamination deposited on the CFRP specimens following the P-FP fingerprint scenario within the coupon sample production user case was quantified using the atomic surface concentrations [Na] and [Cl] given in at.% as measured by XPS. For our POD calculation, we arbitrarily selected the [Na] values since they were in the expected range for the input parameters of the POD software.
- For the RA contamination of the respective P-RA contamination scenario, the [Si] as obtained by XPS was also directly used in the POD calculation.
- For the R-DI contamination scenario, which was based on depositing de-icing fluid on CFRP specimens within the repair user case, we used the [K] measured

Table 5.9 Adapted and scaled values from laboratory analyses for distinct contamination levels in the different contamination scenarios intentionally applied to CFRP specimens. Further details are given in the text

Contamination	Reference	Level 1	Level 2	Level 3	Unit
P-FP	0.01	0.2	0.5	0.7	at.% of Na
P-RA	0.1	3.2	5.1	6.2	at.% of Si
P-MO (surface assessment)	Measured locally and individually, values were not merged				
P-MO (bonding assessment)	0.01	0.3	0.75	0.98	Rel. humidity
R-DI	0.1	6.4	10.9	12.0	at.% of K
R-FC	0.1	1	2	3	dimensionless
R-TD	0.01	0.4	0.8	1.0	°C/10
R-FP	0.01	0.2	0.5	1	% of Skydrol oil

by XPS since the obtained values were in a range that complied with a conventional defect.

- In the case of the P-MO scenario regarding moisture uptake, we proceeded in a slightly different way because the exposure of the CFRP specimens to humid environments was performed and characterized locally. We did not merge these values but maintained their individual qualities for all assessments of surface quality. In contrast, as the adhesively bonded specimens were prepared locally and then passed on for bonding quality assessment, we used the respective communicated mean values.
- In the case of the false curing R-FC scenario, we proceeded in a slightly different way. As outlined above, the applied procedures for producing a weak bond did not rely on any contamination quantification. The only quantified characteristic was the curing temperature of the device used to bond the composite skin. However, this temperature could not plausibly be directly used as a contamination value. Therefore, we decided to use arbitrary values (see Table 5.9) in a way that a higher value represents a stronger deviation from the qualified bonding process.
- In contrast to the FC scenario, we used the overheating temperature as a value related to the incident level in the case of the thermal degradation R-TD scenario because the excess temperature was applied directly, similar to a thermal "contamination" on the skin. As the increments +40 °C, + 80 °C, and +100 °C constituted too great dimensionless values for the model input parameters, we divided them by 10 to remain in a smaller number range.
- In the case of the R-FP scenario, which was based on applying a fingerprint with Skydrol oil, the set value for the percentage concentration of Skydrol in the fingerprinted liquid was used as a contamination value.

Finally, we would like to note that for all the contamination scenarios, the respective reference values were deliberately set as hypothetical values, and in short, are supposed to represent features of cleanly prepared CFRP specimens. We chose each value to be at least ten times inferior to the value of the lowest applied and investigated contamination level. To be specific, it cannot be equal to zero, neither in practice nor in the model that is supposed to correctly converge. We estimated this specimen state and the respectively measured abscissa values to be a good material-related approach to provide the noise level for each ENDT technique.

5.7.3 Examples of the First POD Evaluations

In this section, based on the abovementioned considerations, we present the first POD calculations performed for the selected ENDT techniques and contamination scenarios, for which a promising data basis was provided within the production and repair user cases detailed in Chaps. 3 and 4. Since this was a preliminary investigation, we did not anticipate that any robust conclusions could be drawn for any ENDT technique since the certified POD requirements were not accounted for during the data acquisition. Rather, and from the point of view of the POD methodology, the aim was to establish whether our model hypotheses stood the first test. The two most prominent and restrictive boundary conditions under question were, first, the monotone evolution of the chosen and evaluated feature (specific to each ENDT method) with an increase of contamination, and second, the presumption of a comparable scattering within the data for each contamination level. We may anticipate here that the subsequently presented results appear encouraging, offer interesting trends, and provide a compass for the upcoming establishment of procedures for estimating ENDT-related POD—indeed just what we aimed to elucidate.

5.7.3.1 Evaluating the AWT Surface Quality Assessment

As a first example, we present our findings of the POD calculation for a feature obtained from the aerosol wetting test (AWT) data obtained by using the bonNDTinspect device for the RA contamination scenario within the production user case.

The first feature that we investigated was the average droplet diameter observed after temporarily depositing primary droplets from a water spray onto intentionally contaminated CFRP surfaces. The respective AWT results are presented in Table 5.9. In this case, the POD model hypotheses are respected because the data are linearly correlated with the various contamination levels and the scattering reduces with the increasing contamination. In the next step, we positioned the detection threshold to differentiate the findings of the minor contamination level "1" from those for the cleaned reference surface. We determined that the set of measured feature values obtained for level 1 partially overlap with the reference dataset. Specifically, one point exceeds this tentatively set threshold. This finding affects the POD for the AWT technique for this scenario, although the detection probability remains good and achieves **$a90/95$ = 3.0 at.% for Si** atoms (Fig. 5.51).

From the technological point of view, we would like to highlight that we will always be able to prepare CFRP surfaces with RA deposits that are so low that their AWT droplet pattern can barely be differentiated from that of a reference sample (or a set of reference samples). Thus, the technologically relevant aspect here is to highlight the importance that those contamination levels that significantly affect the fracture toughness, as an example of design-relevant criterion, can be reliably detected. As shown in Chap. 2 for the RA-1 contamination scenario, depositing an amount of Si-containing RA that corresponds to 3.0 at.% Si (as measured by XPS) is related to a loss of the average fracture toughness. In detail, G_{IC} and G_{IIC} values are reduced by almost 18% and 37% of the value respectively observed for joints prepared with CFRP adherends with surfaces treated according to a qualification process. The second AWT-based feature that we investigated was the wettability instead of the droplet diameter. Based on essentially the same raw dataset, the thus obtained results were slightly better, as can be inferred from Fig. 5.52. Indeed, the evaluated data points for the lowest investigated contamination level (level 1) were not in the range of the data for the reference surface state. When we used this feature, the detection threshold thus completely separated the reference data from the level 1

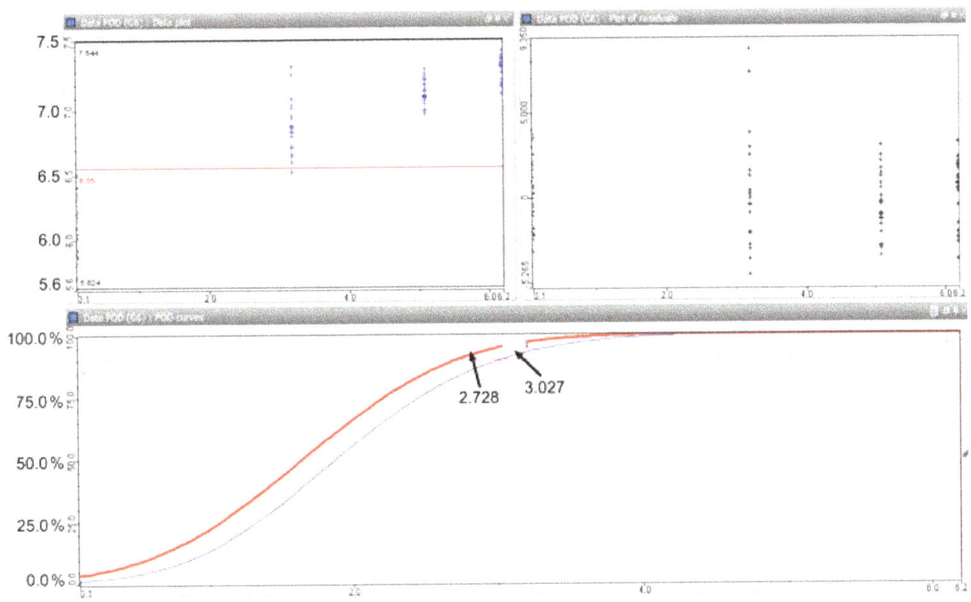

Fig. 5.51 POD curve for the AWT data obtained for the RA contamination scenario within the production user case when using the droplet diameter on the CFRP surface as the feature for evaluation

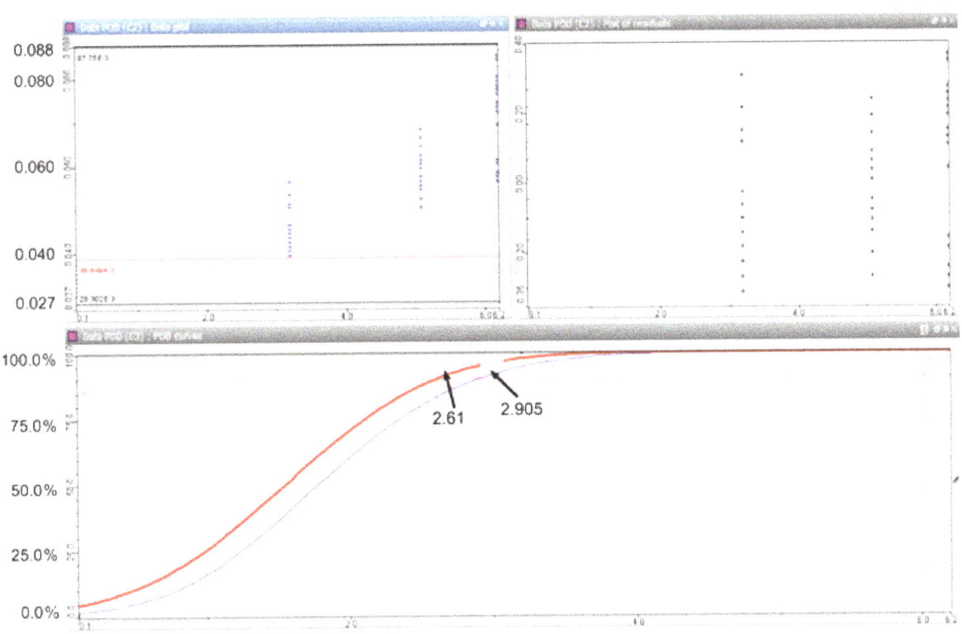

Fig. 5.52 POD curve for the AWT data applied for the RA contamination scenario within the production user case when using the wettability feature for evaluation

data. As a consequence, the POD curve was more significant. The calculated ***a90/95* value corresponded to 2.9 at.% of Si, which is very similar to the outcome of applying the droplet diameter feature**.

As a conclusion of these data post-processing options for the AWT findings, our POD investigations have shown that in the case of the RA contamination scenario, the performance of the testing procedure depends on the selection of the feature to be evaluated. We found a better detection performance when using the wettability feature compared to the droplet diameter feature. As a next step, from the data post-processing point of view, we suggest performing a POD calculation by coupling the two approaches to determine whether the performance can be extended even further. From the test setup point of view, the area density of the applied aerosol droplets could be further adapted to be optimal for this scenario. We would like to highlight that at the end of the day, the multi-dimensional optimization of several parameters would be desirable, which would require extended modeling resources.

5.7.3.2 Evaluating the FTIR Assessment in the Contamination Scenarios

We performed POD calculations for several infrared spectroscopic (FTIR technique) datasets that had been performed for various contamination scenarios in both the production and repair user cases, namely moisture uptake in CFRP adherends, the deposition of de-icing fluid, and the thermal degradation of specimens. The respectively obtained results are presented as plots shown in Figs. 5.53, 5.54, and 5.55. For

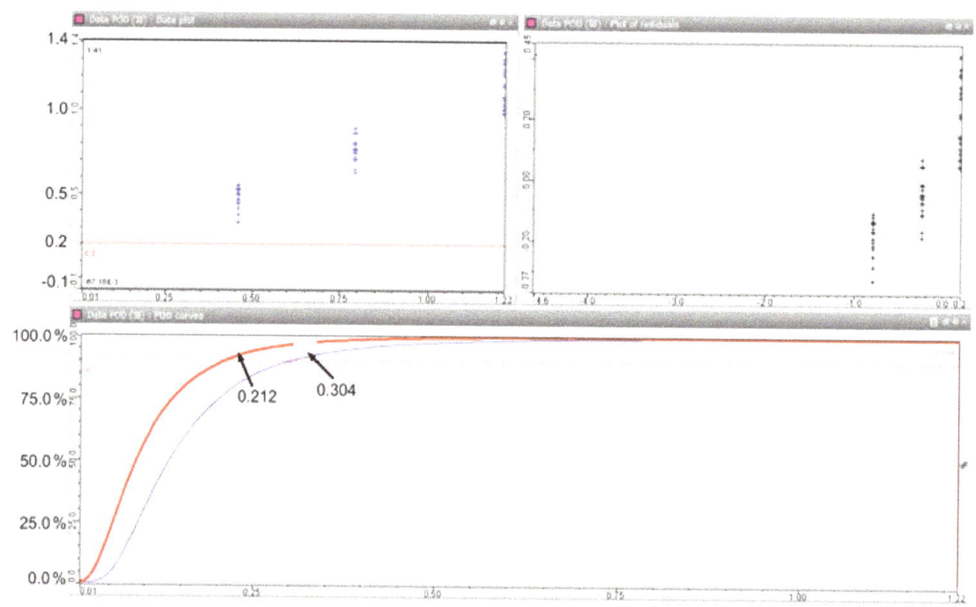

Fig. 5.53 POD curve for the FTIR datasets in the case of the MO contamination scenario within the production user case

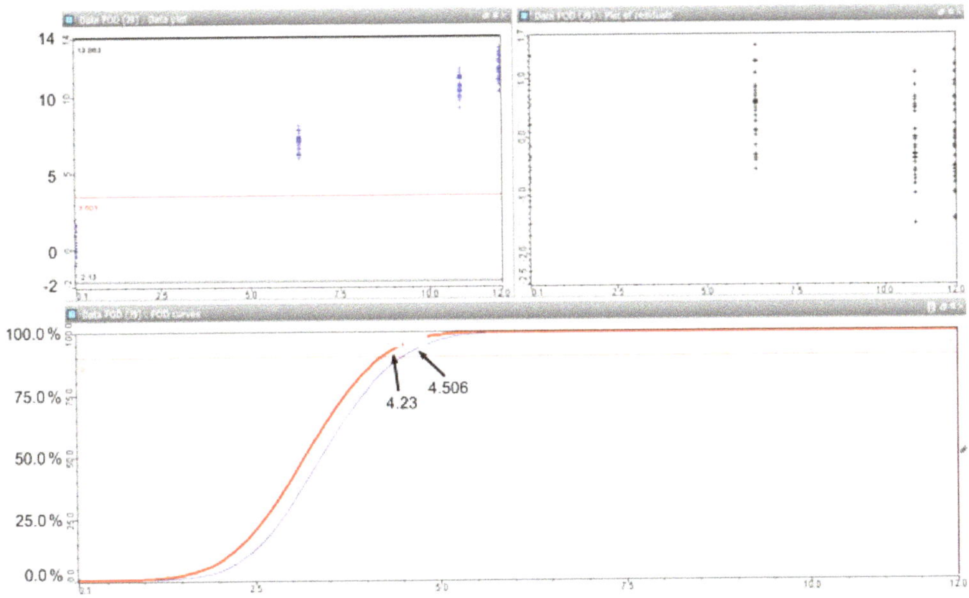

Fig. 5.54 POD curve for the FTIR datasets in the case of the DI fluid contamination scenario within the repair user case

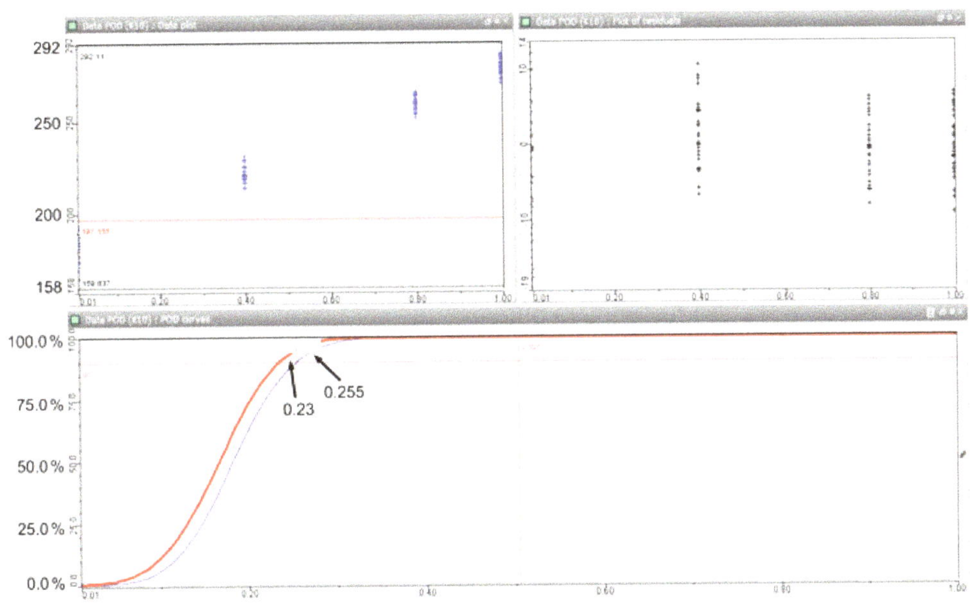

Fig. 5.55 POD curve for the FTIR datasets in the case of the TD scenario within the repair user case

all the scenarios, the data are well suited for POD calculation; only the scattering in the moisture scenario is slightly diverging. Otherwise, the detection threshold can be easily placed to separate the reference values from the contamination values. With no overlap between the reference and the contamination data, the POD performances are good. The values are presented in the following:

- In the case of the MO scenario, $a90/95 = 30.4\%$ of moisture uptake.
- In the case of the DI scenario, $a90/95 = 4.5$ at.% of K. This is good because the value is below the first contamination degree (6.4 at.% of K).
- In the case of the TD scenario, $a90/95 = 25.5$ °C of overheating. This performance is good as the value is below the lowest contamination degree, which is equal to 40 °C of overheating.

5.7.3.3 Evaluating LIBS Procedures for Surface Quality Assessment

For the LIBS technique, we performed in-depth POD investigations and here present the findings related to the aspect of assessing and increasing the robustness and reliability of the technological procedure. We highlight the LIBS findings here because numerous datasets, each with a considerable number of data points, were available or were made available upon request. This was supported firstly by the fact that LIBS is a spectroscopic, and thus multimodal, technique that provides several quantifiable features and secondly by the sub-millimeter width of the spot-like CFRP surface area assessed with a single LIBS shot. We performed these investigations for the RA contamination scenario of the production user case for CFRP surfaces. For LIBS, the actual testing procedure during data acquisition includes a high number of measurements across surface regions that are small compared to, e.g., the lateral dimension of the intentionally applied contaminations. This means that during the data evaluation, an average calculation within a (grouped) set of data points may be performed to compensate for the variability among the locally acquired data.

Based on these considerations, we evaluated in more detail how different approaches to data processing affect the quality of the procedure in assessing quality-relevant material surface features. In other words, our QA approach can be outlined as follows: First, the user operates the LIBS device to perform several measurements at various spots in a given area. During the processing of the thus achieved spectroscopic datasets, a spectroscopic feature is chosen—most often an intensity ratio of two spectral signals with one representing the RA deposit and one comprising the CFRP substrate. Then, an average is calculated for a group of feature values and is considered as a quality-relevant data processing result for the respective area. In the LIBS investigations described in Chap. 3, we experimentally established the number of involved LIBS shot measurements based on a confidence interval. Subsequently, we demonstrate how the procedure for refining this crucial criterion in the testing procedure may be enhanced by applying the POD tool. Specifically, for calculating the average values, we vary the size of the groups formed by the data points obtained from the raw data of the LIBS measurement that were acquired at neighboring surface positions.

Our starting point was a group size of one data point, meaning that we did not actually calculate the average and instead considered each single measurement point to be the result for the respective area that serves as the input data for the POD tool.

The result of the POD calculation obtained for this boundary condition is presented in Fig. 5.55. In this case, 60 data points were used per contamination level (also including the reference surface state, which was also attributed to a contamination degree of 0.1). Clearly, these are raw data and no averages were calculated. The displayed data show that along with the increase of the contamination degree, there exists a considerable overlap between the reference data range and the data range for the contaminated samples. This finding hindered us from correctly positioning a detection threshold since either many points from contaminated samples would be below this threshold or some data points from non-contaminated surface positions would be above it. This uncertainty surrounding the POD result means that it is not satisfactory. The *a*90/95 does not even appear on the graph (as presented in Fig. 5.55). These outcomes indicate, as already outlined in Chap. 3, that it is mandatory to perform average calculations during the data processing to increase the reliability of QA approaches based on the LIBS technique.

Therefore, we moved on to develop testing procedures based on calculating the average values from the raw data obtained from several LIBS shots. First, we calculated the averages over groups of ten points, thereby obtaining the outcomes of the POD calculation, as presented in Fig. 5.56. We worked with 120 points per contamination degree, including the reference. To some extent, we used some of the same data points as in the abovementioned first calculation. This means that we used 12 (averaged) values per contamination degree as input data for the POD calculation. Initially, this dataset appears somewhat inadequate, but it was worth further inspection, and indeed, the POD calculation offered a better curve than previously. It was

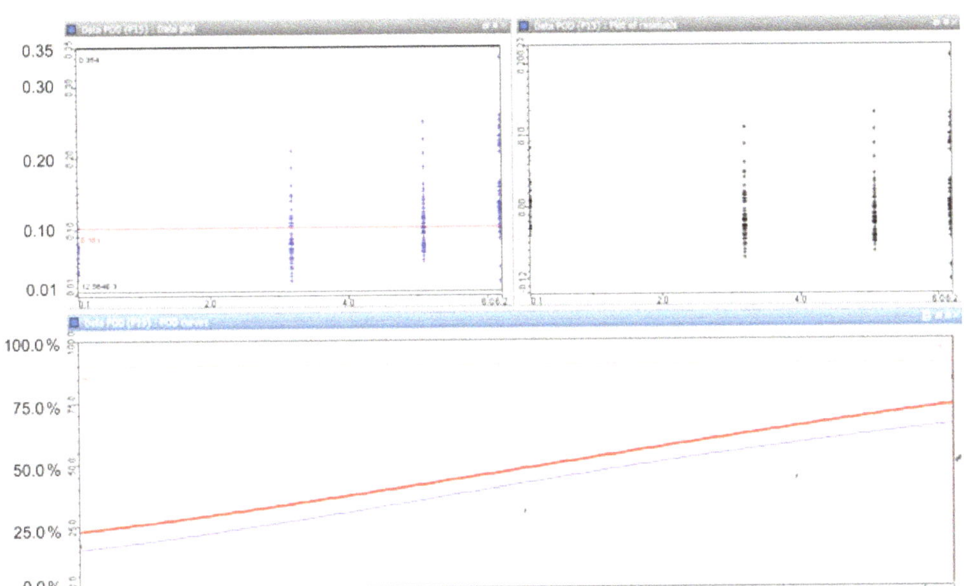

Fig. 5.56 POD curve for the LIBS datasets in the case of the RA contamination scenario within the production user case; the POD findings shown here were obtained by considering an evaluation procedure without averaging

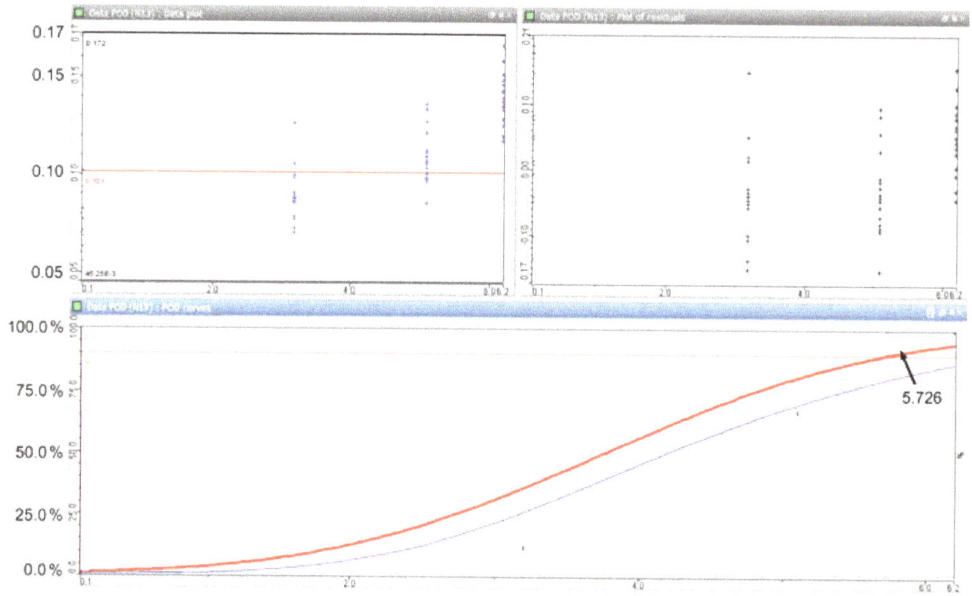

Fig. 5.57 POD curve for the LIBS datasets in the case of the RA contamination scenario within the production user case. The POD findings shown here were obtained considering an evaluation procedure involving an average measurement per group of ten data points

still not possible to perfectly differentiate data points for the reference specimen from the points for the first two contamination levels, but the third level was correctly determined based on the detection threshold. This enhanced approach thus led to a better probability of detection, even if the *a*90/95 value was not yet obtained.

Hence, we had gained sufficient evidence to suggest another series of measurements that were performed specifically for our POD investigations. Within this advanced procedure, 36 measurement points were included from each surface region of the CFRP specimen, and a total of 20 regions were inspected for each contamination degree (including the reference). The POD calculation results presented in Fig. 5.57 are thus based on an input dataset comprising 20 data points per degree, with each point being an average of 36 measurements. The achieved results show a much better probability of detection. We did not find any overlap between the data ranges of the different contamination degrees, and we could identify and even reduce the detection threshold. As all data points from the contaminated surface regions were above this detection threshold, the final POD result was better. **The *a*90/95 was evaluated to be about 3.2 at.% of Si.** As shown in Chap. 2 for the RA-1 contamination scenario, depositing an amount of Si-containing RA that corresponds to 3.2 at.% Si (as measured by XPS) is related to a loss of the G_{IC} and G_{IIC} values by almost 18% and 37%, respectively.

Briefly, in addition to this POD quantification, this in-depth study of surface QA processes based on the LIBS technique for the RA scenario has provided some interesting implications for establishing this testing procedure. For this contamination scenario, it emerged that doing averages over ten measurements for a total of

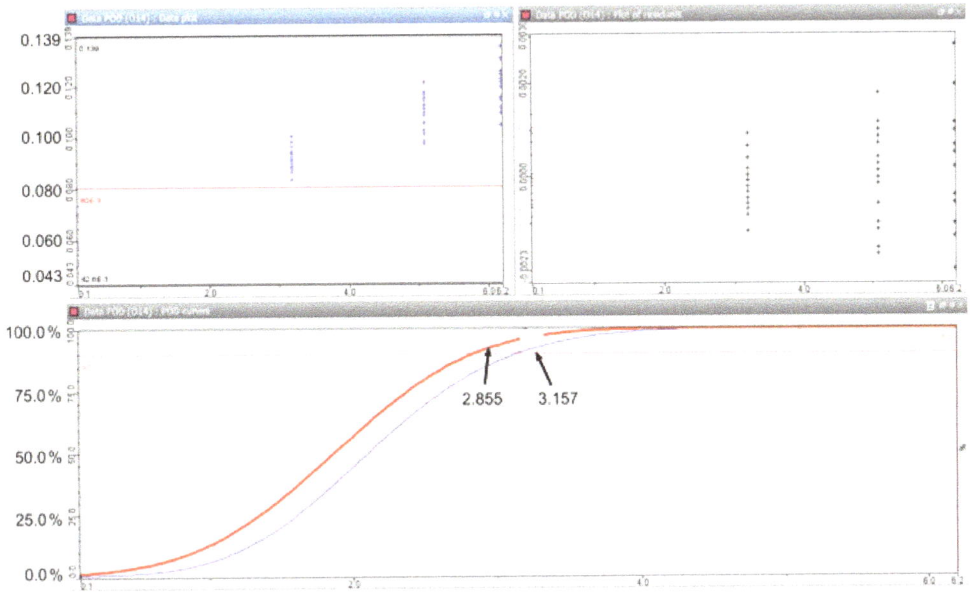

Fig. 5.58 POD curve for the LIBS datasets in the case of the RA contamination scenario within the production user case. The POD findings shown here were obtained considering an evaluation procedure involving an average measurement per group of 36 data points

120 LIBS shots is not sufficient to obtain a good POD. By averaging 36 measurements and performing 720 analytical shots, the POD improves to the point that it can even be quantified. This overall data acquisition and evaluation process can be maintained if the technique meets the detection requirements set for testing the design-relevant mechanical performance of adhesive joints manufactured based on the assessed CFRP adherends. We expect that the detection performance could be further improved by evaluating more and/or larger average groups, as might be revealed by another POD investigation. Clearly, establishing the optimum lateral distance between neighboring LIBS spots should also consider the total measurement time required per CFRP surface area. In this respect, many of the presently applied aeronautical user cases that require a significant number of manually performed operations offer a sufficient time "buffer" for LIBS-based surface quality monitoring (Fig. 5.58).

5.7.3.4 Evaluating NLUS for Bonding Quality Assessment

As an example of ENDT techniques for bonding quality assessment, we performed POD investigations based on datasets obtained using nonlinear ultrasound (NLUS) for five different contamination scenarios from the repair and production user cases. The respectively obtained results for the P-FP, P-MO, P-RA, R-DI, and R-FC scenarios are presented in this section in, Fig. 5.59, Fig. 5.60, Fig. 5.61, Fig. 5.62, and Fig. 5.63, respectively. For this ENDT technique, the hit/miss approach was used.

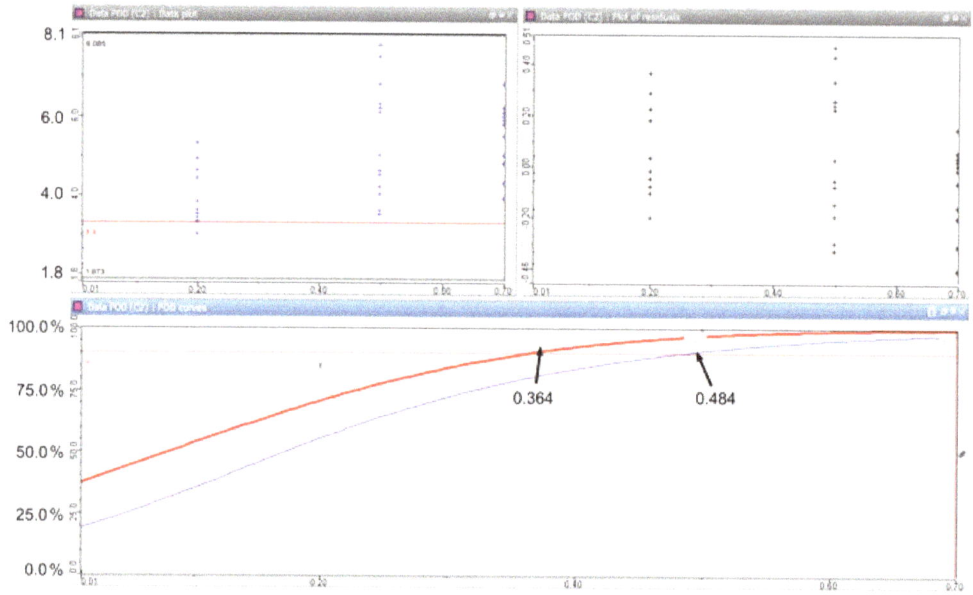

Fig. 5.59 POD curve for the NLUS datasets from the FP contamination scenario, applying a hit/miss model approach

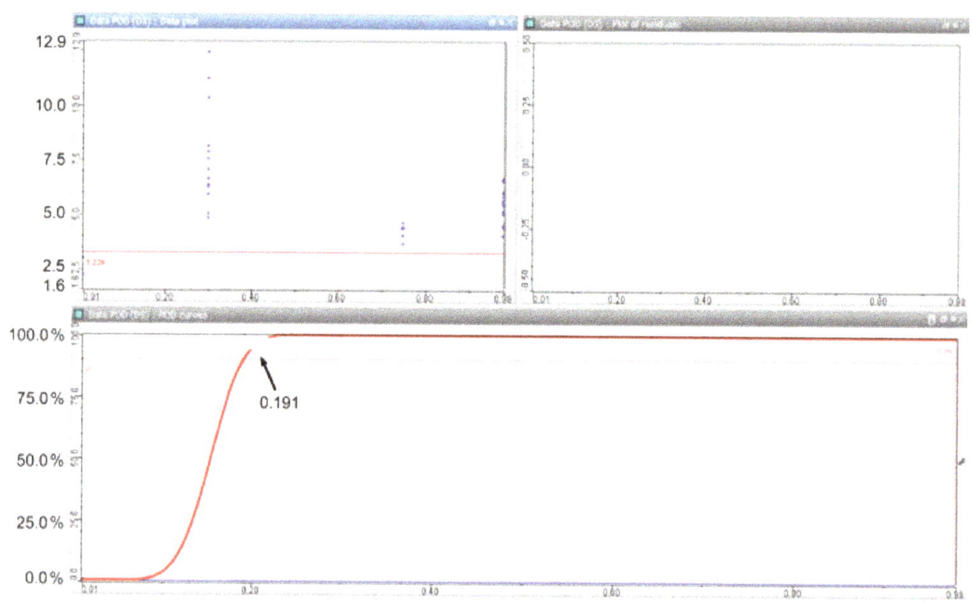

Fig. 5.60 POD curve for the NLUS datasets from the MO contamination scenario, applying a hit/miss model approach

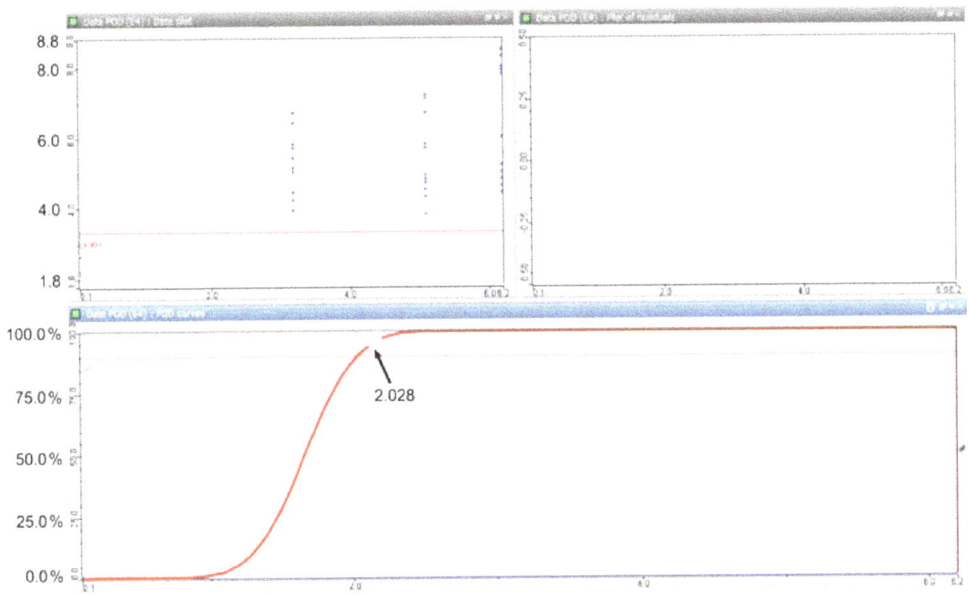

Fig. 5.61 POD curve for the NLUS datasets from the RA contamination scenario, applying a hit/miss model approach

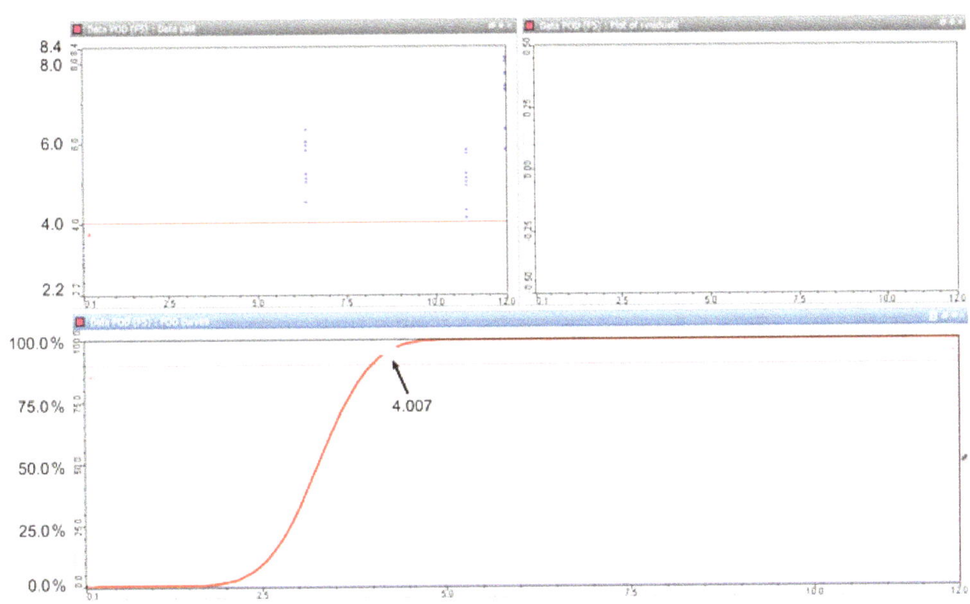

Fig. 5.62 POD curve for applying NLUS as an ENDT technique for the DI contamination scenario, applying a hit/miss model approach

After examining the NLUS datasets for each scenario, this approach was motivated on the one hand by the finding that the correlation between the feature signal amplitude and the applied contamination level was never monotonous or linear. However, the values obtained for the adhesive joints prepared from intentionally contaminated

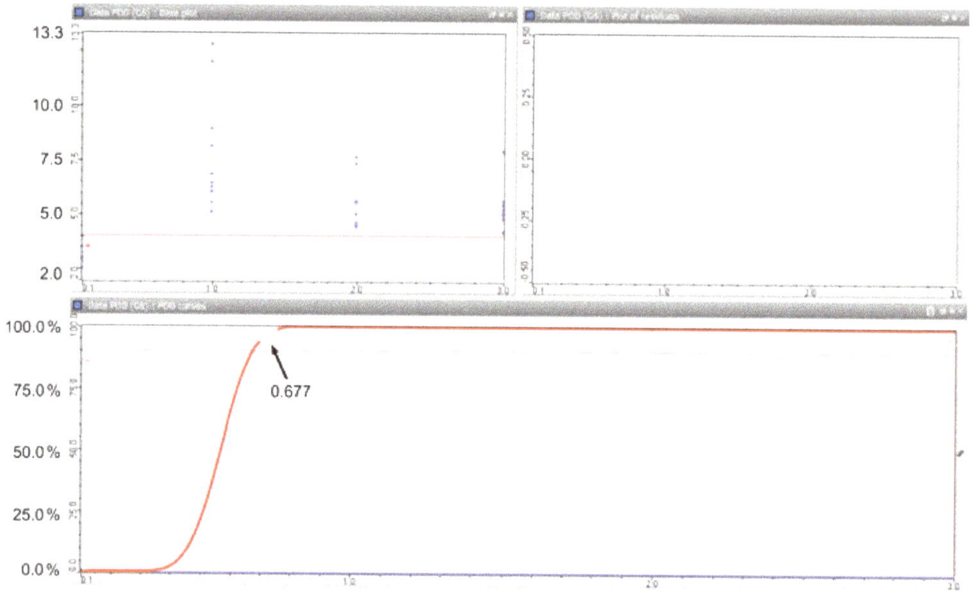

Fig. 5.63 POD curve for applying NLUS as an ENDT technique for the FC contamination scenario, applying a hit/miss model and tentatively excluding an assumed outlying data point

adherends were well separated from those found for the reference specimens, and this finding enables an NLUS expert to make a diagnosis since, based on their experience, such a signal pattern is typical for ultrasound diagnostics. Effectively, when the signal amplitude exceeds a given threshold (called a "gate" in the case of conventional ultrasound), then a defect is indicated. This behavior is thus well suited to a procedure based on the hit/miss model approach.

The results are detailed in the captions of the respective figures and are summarized comprehensively below. We would like to remind the reader here that applying the hit/miss model requires 60 data points per contamination degree. As during our orienting approach we used only ten data points, we would like to highlight that some care shall be taken before drawing in-depth conclusions.

- For the P-FP scenario, the data points for contamination level 1 were in the range of the values obtained for the reference specimen state. For a given contamination level, this decreased the probability of detection. The $a90/95$ is about 0.48 at.% of Na (as measured by XPS), which roughly corresponds to contamination level 2. However, the probability of false detection is high. As shown in Chap. 2 for the P-FP-2 contamination scenario, depositing an amount of Na-containing fluid that corresponds to 0.5 at.% Na (as measured by XPS) to a loss of the G_{IC} and G_{IIC} values by 8% and 61%, respectively, as compared to the values observed for joints prepared with CFRP adherends with surfaces treated according to a qualification process.
- For the MO scenario, better discrimination between each contaminated sample and the reference specimen was obtained, leading to $a90/95 = 19\%$ of relative humidity. As shown in Chap. 2 for the MO-1 scenario, incorporating an amount

of water that corresponds to 30% r.h. is not related to a loss of the G_{IC} value but to a significant loss of the G_{IIC} values by 45%.

- Plausible data were also obtained for the RA scenario. The calculated value was $a90/95 = 2.0$ at.% of Si. As shown in Chap. 2, this surface concentration is smaller than the one obtained and measured by XPS in case of the most exiguous contamination level investigated following the RA-1 contamination scenario. For this scenario, a loss of the G_{IC} and G_{IIC} values by almost 18% and 37%, respectively, was found.
- For the R-DI scenario, the POD calculation also showed good performance. The estimated $a90/95$ was about 4 at.% of K, which was low enough to be below contamination level 1. This uncovers a technologically relevant and interesting potential since, as shown in Chap. 2 following the R-DI-1 contamination scenario, depositing an amount of K-containing fluid that corresponds to 6 at.% K (as measured by XPS) is related to a loss of the G_{IC} and G_{IIC} values by almost 30% and 56%, respectively.

The NLUS data evaluation for the FC scenario required slightly more attention than the other scenarios. Indeed, in the case of the R-FC scenario, only one point from contamination level 3 was below the detection threshold that was placed above the reference data range. Since very few points were available, this point alone had a strong influence on the POD calculation as it had more weight. The $a90/95$ is equal to 2.5, meaning it lies between bond strength levels 2 and 3. If we consider this point an outlier and exclude it from the data range, with plausible justification, we may estimate what the POD would be without it or with more data (assuming additional points above the threshold). The result of such a consideration is given in Fig. 5.63. This leads to $a90/95 = 0.677$, i.e., below level 1, which is much better than when considering this assumed outlier to be significant. From a data-driven perspective, we thus conclude that it appears to be implausible to weigh and use this value unless an appropriate justification is given, e.g., through experimental replication or a physics-based explication. In essence, we recommend providing a sufficiently substantial and consolidated set of measured data points for any technically relevant POD approach.

5.8 The Results of the Full-Scale Demonstration: An Overall Synopsis of the Technology's Performance

As a final summary of this chapter, we propose Tables 5.10 and 5.11 as a synthesis of the ENDT performance that was evaluated during the three-day full-scale demonstration during the ComBoNDT project [1]. We considered both production and repair user cases that were based on realistic and real CFRP parts, respectively. We highlight that our retrospective perception that the implementation of the full-scale demonstration with an aeronautic scope of application, of which we summarize the discrimination performances here, was largely paved by the coupon test phase described in Chaps. 2, 3, and 4 of this book.

Table 5.10 Synthesis table of the performance demonstration for extended non-destructive testing (ENDT) tools and procedures that were advanced and applied during the ComBoNDT [1] project for CFRP specimens with flat (coupon samples) or complex (pilot and full-scale parts) shapes. The materials represent technologically relevant production or repair user cases for which different contamination scenarios had been identified, namely the exposure to moisture (MO), release agent (RA), fingerprint (FP), de-icing fluid (DI), false curing (FC), and thermal impact (TD) scenarios, as well as scenarios realized by simultaneously applying a combination of intentional contaminations. The ENDT techniques comprise Optically Stimulated Electron Emission (OSEE), Laser-induced breakdown spectroscopy (LIBS), two differently implemented approaches based on electronic noses (e-nose), laser vibrometry, Laser-Induced Fluorescence (LIF), an aerosol wetting test (AWT), Fourier transform infrared spectroscopy (FTIR), electromechanical impedance (EMI), magnetostrictive sensors (MGSS), nonlinear ultrasound (NLUS), and laser shock adhesion testing (LASAT). Further composite characterization was performed by ultrasound (US), micro-computed tomography (μCT), and mechanical tests focusing on fracture toughness

ENDT	COUPONS						PILOT & FULL SCALE	
	PRODUCTION			REPAIR			PRODUCTION	REPAIR
	MO	RA	FP	DI	FP / FC	TD	RA+FP	R-DI+FP
OSEE	V	V	V	V	V	X		
LIBS	X	V+	V	V+	V+	X	V	
E-nose	V	V	V	V	V	V	V+	V+
Vibrometry	V	X	X	X	X	X		X
LIF	X	X	X	X	X	$V_{220°C}$		X
AWT	V+	V+	V	V	X	V	V	V+
FTIR	V+	X	X	V+	X	V+	X	N more data needed
Mech. test	V+	V+	V+	V+	V+	V+		
US	X	X	X	X	V+	V+		
μCT	V	V	V	V	V	V		
EMI	V	X	X	X	V	V	N	X
MGSS	X	X	V	X	V	V	X	V^{FP}
NLUS	V	V+	V+	V+	V+	V+	N more data needed	N more data needed
LASAT	V	V	V	V	V	V+	V	V
CAPTION	V+: Detection & discrimination – V: detection – X: no detection/ not conclusive – N: Partial/ doubts remain							

As we present the survey of our findings graphically, we would like to give some hints and details to ease the reading and to enhance the clarity of these summarizing presentations. Basically, we used a color code, marking the ENDT outcomes for the surface quality assessment in blue, the ENDT findings for the bonding quality assessment in green, and the respective laboratory-based analytical reference method findings for the bonded specimens in red. Then, for each cell, we use a code following the subsequently listed captions to estimate the performance of the inspection tools and procedures that we advanced and applied during the ComBoNDT project:

- "V+" shows that both the detection of and discrimination between different contamination levels or different contamination types in the case of multiple contaminations were achieved;
- "V" indicates that the detection of a contamination-based deviation from the reference specimen was managed;
- "X" denotes that a detection was not accomplished or was not conclusive;
- "N" means that only a part of the contamination scenario was captured or encompassed with some doubts remaining (see Table 5.10).

Table 5.11 Synthesis table of the *a*90/95 results obtained from the preliminary POD evaluations performed for the example ENDT techniques, the features assessed by data evaluation, and the contamination scenarios applied to CFRP adherends. The ENDT techniques examined here comprise the aerosol wetting test (AWT), Fourier transform infrared spectroscopy (FTIR), Laser-induced breakdown spectroscopy (LIBS), and nonlinear ultrasound (NLUS)

ENDT technique	Contamination/evaluated feature	Calculated *a*90/95
AWT	P-RA/droplet diameter	3.0 at.% of Si
	P-RA/wettability	2.9 at.% of Si
FTIR	P-MO	Moisture uptake upon exposure to 30.4% r.h.
	R-DI	4.5 at.% of K
	R-TD	25.5 °C of overheating
LIBS	P-RA	3.2 at.% of Si
NLUS	P-FP	0.5 at.% of Na
	P-MO	Moisture uptake upon exposure to 19.1% r.h.
	P-RA	2.0 at.% of Si
	R-DI	4.0 at.% of K
	R-FC	2.542

To provide quantitative indications of the ENDT performance, we evaluated example datasets obtained using different ENDT techniques and for different scenarios. For this purpose, we applied a POD approach that was reduced in complexity as compared to more formal and far more comprehensive technologically relevant procedures. The performances were quantified using the *a*90/95 values. When available, the values are listed in Table 5.11. We highlight here again that this was only a first estimation since we did not perform a strict POD procedure to obtain these figures. We suggest that the next step should be to cross-check these values with the results from mechanically testing the adhesively bonded joints comprising intentionally contaminated adherends in order to correlate the POD with the expected decrease in the design-relevant mechanical material properties.

Furthermore, our simplified and preliminary POD investigations provided useful information. In some cases, a direct output of our POD approach was a refinement of ENDT testing procedures. For example, it was shown that evaluating the wettability feature was more promising than relying on the droplet diameter feature in the case of applying AWT for RA detection within the RA scenario. Similarly, the POD results also quantified the effect of the actually performed stepwise testing procedure on the probability of detection in the case of applying LIBS for this scenario. Indeed, better performances were obtained in this case by increasing the number of spot-wise measurements to an extent that was acceptable both from a performance and from a time-demand point of view.

In other cases, our POD approach showed that the assessed testing procedure was not yet sufficiently adapted as decreasing POD curves were obtained. That was the case, for example, when applying laser vibrometry for assessing specimens prepared following the MO scenario. However, our POD approach offered some clues on how to change the testing procedure to enhance the detection performance.

We conclude that the POD work introduced in this chapter represents the first step, and taking all the promising results into account, it would be interesting to push these investigations further as they offer much promise for the application of ENDT datasets obtained in the frame of a comprehensive real and adapted POD procedure.

References

1. ComBoNDT "Quality assurance concepts for adhesive bonding of aircraft composite structures by advanced NDT" (2015–2018) Project funded from the European Union's Horizon 2020 research and innovation programme under grant agreement No 636494
2. Berens AP (1989) NDE reliability data analysis. In: Nondestructive evaluation and quality control, 9th edn. ASM International, OH
3. PICASSO "imProved reliability inspeCtion pf Aeronautic structure through Simulation Supported POD" (2009–2012) Project funded under FP7-TRANSPORT and grant agreement No 234117
4. Pavlovic M, Zoëga A, Zanotelli C et al (2017) Investigations to introduce the probability of detection method for ultrasonic inspection of hollow axles at Deutsche Bahn. Procedia Struct Integr 4:79–86
5. Chung H-Y, Manuel L, Frank KH (2006) Optimal inspection scheduling of steel bridges using nondestructive testing techniques. J Bridge Eng 11:305–319. https://doi.org/10.1061/(ASCE)1084-0702(2006)11:3(305)

6

Bonded Products Life Cycle Management

Welchy Leite Cavalcanti, Elli Moutsompegka, Konstantinos Tserpes, Paweł H. Malinowski, Wiesław M. Ostachowicz, Romain Ecault, Neele Grundmann, Christian Tornow, Michael Noeske, Peter Schiffels, and Bernd Mayer

Abstract In this chapter, we outline some perspectives on embracing the datasets gathered using Extended Non-destructive Testing (ENDT) during manufacturing or repair process steps within the life cycle of bonded products. Ensuring that the ENDT data and metadata are FAIR, i.e. findable, accessible, interoperable and re-usable, will support the relevant stakeholders in exploiting the contained material-related information far beyond a stop/go decision, while a shorter time-to-information will facilitate a prompter time-to-decision in process and product management. Exploiting the value of ENDT (meta)data will contribute to increased performance by integrating all defined, measured, analyzed and controlled aspects of material transformation across process and company boundaries. This will facilitate the optimization of manufacturing and repair operations, boosting their energy efficiency and productivity. In this regard, some aspects that are currently driving activities in the field of pre-process, in-process and post-process quality assessment will be addressed in the following. Furthermore, some requirements will be contemplated for harmonized and conjoint data transfer ranging from a bonded product's beginning-of-life through its end-of-life, the customization of stand-alone or linked ENDT tools, and the implementation of sensor arrays and networks in joints, devices and structural parts to gather material-related data during a product's middle-of-life application phase, thereby fostering structural health monitoring (SHM).

W. Leite Cavalcanti (✉) · N. Grundmann · C. Tornow · M. Noeske · P. Schiffels · B. Mayer
Fraunhofer Institute for Manufacturing Technology and Advanced Materials IFAM, Wiener Str. 12, 28359 Bremen, Germany
e-mail: welchy.leite.cavalcanti@ifam.fraunhofer.de

E. Moutsompegka · K. Tserpes
Laboratory of Technology & Strength of Materials, Department of Mechanical Engineering & Aeronautics, University of Patras, 26500 Patras, Greece

P. H. Malinowski · W. M. Ostachowicz
Institute of Fluid–Flow Machinery, Polish Academy of Sciences, Fiszera 14, Gdańsk 80231, Poland

R. Ecault
Airbus Operations S.A.S., 316, Route de Bayonne, B.P. D4101, 31060 Cedex 9 Toulouse, France

Keywords Life cycle management · FAIR ENDT data and metadata · Harmonized characterization data (CHADA) documentation · Customisation · Standardization · Sensor network

6.1 Introduction

One of the foremost ongoing trends and initiatives is not only that data are collected and exchanged for documentation purposes along a process or value chain, but also that value can be added if the substantial information contained in these data is extracted by the providers and owners. This typically means evaluating several sets of interconnected metadata, measured characterization data or simulation-based modelling data that constitute knowledge sources.

In the first of the subsequent sections, we accentuate the technical benefits of consistently linking joints with their associated material-related data. We base our selection of potential prospects on our perception of the distinctive prevailing global transport of material goods, combination of logistic processes and transfer of ideas along with the life cycle of material products such as adhesive joints. From this network point of view, the urgent need for composite product data interoperability for life cycle management and the sustainment of aircraft fleets was recently highlighted by McMillan et al. [1]. Next, we highlight how the balancing of customization and efficiency will profit from a certain readiness for standardization and from establishing the consistency of (meta)data formats. This will help overcome the need for data translation at each interface [2]. In this context, we present the outline of a research project that resulted in the successful development and implementation of a customized hand-held aerosol wetting test (AWT) device relying on the core elements presented in the ComBoNDT project [3]. After that, we introduce stepwise approaches for integrating ENDT investigations using optically stimulated electron emission (OSEE) and the space-resolved findings in, e.g., material data management systems while safeguarding data integrity. We show how the concept developed in the ComBoNDT project [3] facilitates the linking of ENDT data with design-relevant features that are significant along the life cycle of adhesive joints. Finally, we complement the latter two surface quality-related approaches through the prospect of material-integrated sensor networks and fibre-based sensors, thereby facilitating the structural health monitoring (SHM) of adhesive joints.

6.2 Data Transfer Along the Product Life Cycle

In Chap. 1, we introduced the image of an enhanced triangular industrial automation pyramid, in which only two of the three lateral faces of the pyramid were presented. The third face, which was hidden in the sketch due to the limited angle of view rather than a limited awareness, represents the environmental impact during the manufacturing phase. Therefore, we highlight the environmental aspects during the complete product life cycle in Fig. 6.1, showing the top view of a tetrahedron and sketching the incorporation of the manufacturing process as a part of a product's beginning-of-life (BoL) period. In this way, the relevance of the manufacturing characteristics for the sustainability of a product is illustrated.

In the previous chapters of this book, we demonstrated that using advanced ENDT tools and procedures facilitates the gathering of high-quality data from production or repair processes based on adhesive bonding technology to obtain high-quality products. From an overall material-related and process-comprehensive perspective, the outcomes of the United Nations Conference on Environment and Development (UNCED) in Rio de Janeiro, 1992, stressed that sustainability is the principal worldwide political goal for the future development of humankind, and thus, shall be respected during any product life cycle. This challenge encompasses environmental, economic and social (or societal) aspects. Related tools supporting researchers and developers in tackling these challenges are the internationally standardized life cycle assessment (LCA) for the environmental part, life cycle costing (LCC) for the economic assessment, which also accounts for the use and end-of-life phases, and social life cycle assessment (SLCA), which may be applied after identifying the system boundaries [4]. Product sustainability may also be captured following the holistic concept of life cycle management (LCM). As pointed out in the framework of the United Nations Environment Programme [5], "understanding, quantifying and communicating the environmental impacts and resource consumption of products" are essential steps on the way towards a "resource-efficient twenty-first century

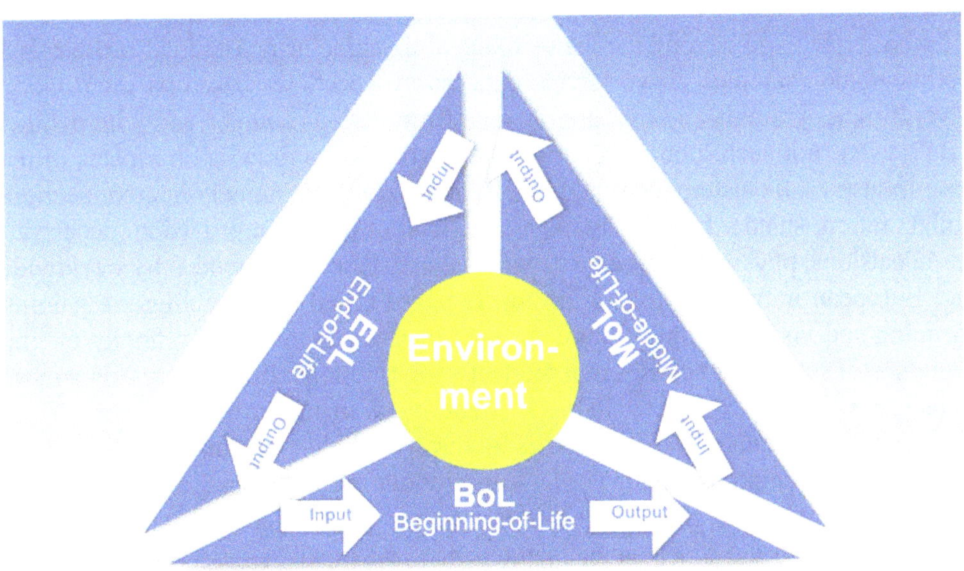

Fig. 6.1 Triangular pyramid highlighting the relevance of all product life cycle phases on a product's environmental impact

green economy". Aggregation of the involved data at the product system level and their management in databases support an intelligent understanding and a trajectory towards sustainable development. Therefore, it is not surprising that LCA and LCC are used in the early phases of aircraft structural design based on CFRP parts [6]. Calado et al. demonstrated in the framework of a combined LCA and LCC approach that the quality control (QC) process step following the demoulding and finishing of a CFRP part may be evaluated based on the QC cycle time and the reject rate of the quality assurance step [6]. This approach throws light on the relevance of reliable and accurate ENDT procedures not only for the quality assessment of CFRP parts before the subsequent process step, but also for the ecological and economic impacts of quality assurance when designing an adhesive bonding process. Considering merely the surface quality of a CFRP adherend, space-resolving ENDT tools facilitating a local cleaning on demand, while also largely avoiding false negatives within the quality assessment, will guide and support process engineers in choosing an appropriate and sustainable pre-treatment in the case of a reject during QC. Beyond that, we infer that the life cycle-based sustainability assessment of adhesively bonded products will largely profit from ENDT procedures and interoperable ENDT datasets, for instance, for optimizing future material-related process steps or due to the availability of joint-specific data that can guide individualized decisions during maintenance, repair and overhaul (MRO) and the end-of-life phase.

Avoiding waste during all the life cycle phases of a product will contribute to greatly enhancing its performance from a twenty-first century perspective. As Otte et al. point out with respect to manufacturing, the "ability to quickly access and share data is key to optimizing and streamlining any industrial production process" [7]. Following the apparently linear time axis in Fig. 1.4 (Chap. 1), which points from input towards output in the material transformation process during adhesive bonding, we would like to go beyond the view of a local manufacturing or repair process and outline some examples of worldwide data-based workflows based on interlinking several such pyramids and exchanging data from various sources and with distinct data quality. For such global scenarios, we would like to draw upon aspects of the nine Industry 4.0 clusters identified by Hermann et al. [8], namely interconnection, collaboration, standards, security, data analytics, information provision, decentralized decisions, physical assistance and virtual assistance. Referenced by worldwide and European initiatives, holistic life cycle-based thinking and acting are gaining meaning and visibility, thus embracing the putative perception of a finally ending sequence of subsequent phases in a product's life through a life cycle within a more holistic conception, as depicted in Fig. 6.1. Therefore, in the more general perception, such circular reasoning is, on the one hand, throwing light on connecting the end-of-life (EoL) and beginning-of-life (BoL) scenarios of products from the material-based technical, social, ecological and economic points of view comprising sustainable development. On the other hand, maintenance and repair, as essential user cases during the middle-of-life (MoL) life cycle phase [7], are gaining attention in light of product life cycle assessment (LCA) procedures. Pragmatically, this means that for any form of EoL recycling, material-related information on the BoL or MoL actions and their effects shall be seminal. Material-related data that meet FAIR data principles and that are straightforwardly gathered during manufacturing or repair processes are pathbreaking in this sense. The interoperability of exchanged data facilitates knowledge management based on necessary ontologies like Basic Formal Ontology (BFO) [7] or the recently developed European Materials Modelling Ontology (EMMO) for applied sciences [9, 10]. EMMO is integrated into virtual marketplaces like VIMMP [11, 12] or MarketPlace [13] and allows end-users to represent their manufacturing process and innovation challenges in a standard ontological form. This is supported by smart targeted guidance through the whole translation process to “deliver a full complete experience to companies”, profiting from an ontology-based open translation environment grounded in EMMO [2].

From the material point of view, we would like to accentuate that in adhesive bonding technology, the quality and durability of joints are crucially determined by the scale-comprehensive effects of molecular interactions and mesoscopic transport processes during the formation of adhesion between adherends and the adhesive system [14]. On the one hand, we would like to express our expectation that ontologies will be the foundation for combining physics-based and data-driven approaches in order to assess more deeply the formation of adhesion and its time-dependent development during the BoL and MoL phases of bonded joints and the respective products, thus contributing to more sustainable material development. On the other hand, we would also like to exemplify that evaluating the ENDT datasets of pre-bond adherend surfaces and adhesively bonded joints continues to advance and is increasingly assisting in cleaning and joining processes in a growing number of user cases.

From the process point of view, we estimate that the abundance of adherend materials and their geometrical shapes, surface pre-treatment process steps, adhesive systems (multi-layered and multi-component), application tools and implementations of the bonded joints involving distinct environments in production, maintenance and repair processes will drive the need for technological adaptation and the demand for information exchange. We expect further customization of ENDT tools and an increasingly individual definition and realization of their interfaces with the surrounding network of catenated integrants of quality assurance systems, e.g. with further NDT tools or by having control personnel organized within the automation pyramid. From our point of view, a systematic common language based on shared top-level or domain-specific ontologies will greatly support the directed and unambiguous communication and exchange of findable, accessible, interoperable and reusable data and information.

6.3 Customization and Further Advancement of ENDT Tools and Procedures

Based on the available knowledge, e.g. that achieved within the ENCOMB and ComBoNDT research projects, and driven by market reviews and customer feedback for detailing the technical and economic user case aspects, interested stakeholders may instigate further developments along with the initiated ideas, thereby contributing to innovative ENDT procedures in the third decade of the twenty-first century. Here, we would like to show through examples that the customization of ENDT tools and procedures is, on the one hand, based on the elaborated, documented and available outcomes of the last decade, at the end of which we performed the developments described in the following. On the other hand, such customization further facilitates tailored and improved customer interaction and increased customer satisfaction.

In the previous chapters, we presented how customizing the operands was achieved in the ComBoNDT research project by configuring a set of CFRP adherends with different surface states. In addition, both the ENDT tools and the respective procedures were technologically advanced. Beyond that, their customization while following standardized approaches is a frequent challenge for studies focusing on operator-related aspects. We would like to highlight here that within the consortium of the ComBoNDT joint research project, this line of action was followed not only by the contributing research institutions, but especially by the small and medium-sized enterprise (SME) partners involved. Significantly, a fast and targeted response to the increasing market demand for NDT technologies is being provided by innovative SMEs [15].

In the following, we highlight the findings of the recently completed HANOB joint research project, involving partners from institutional and SME research and development (R&D), which developed a portable ENDT device used to perform a handheld wetting test [16, 17]. Hereby, a portable measuring system to monitor the wettability of surfaces was designed and developed, and a prototype setup was constructed. The basic components comprise an aerosol nozzle for generating a well-defined spray pattern of pure water, LED-based illumination, and a compact camera for capturing the approx. 10 cm wide droplet pattern resulting in the substrate being inspected, as sketched in Fig. 6.2.

With this easy-to-use setup, it is possible to detect, analyze and objectively evaluate changes in the wettability of components. For technologically relevant user cases, it was also demonstrated that the effectiveness of the performed surface pre-treatment procedures can be mapped and the presence of contaminants can be detected. The developed compact prototype enabled a reliable surface quality inspection to be performed on intricately accessible surface regions and locally pre-treated substrate surfaces. The manipulation and operation of the device, as well as user prompting through a newly developed intuitive graphical interface operated via a touchscreen, were designed to facilitate easy operation after brief operator training.

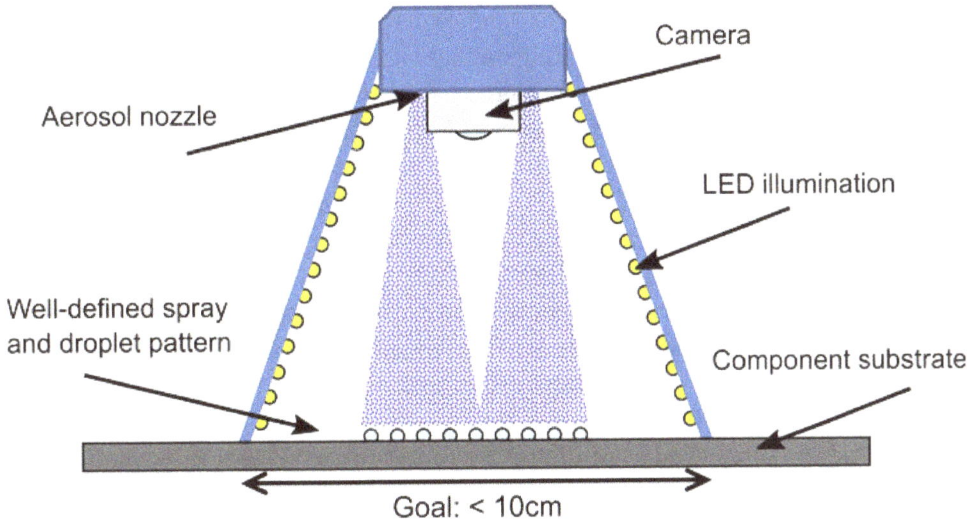

Fig. 6.2 Sketch of a hand-held customized aerosol wetting test device

Once it has been further developed to achieve the readiness level of an industrial device, this portable setup will provide an advantageous cost/performance ratio and will be deployable not only in production, e.g. for the inspection of target areas, but also in the field during repair processes in applications far exceeding aeronautical user cases. Nonetheless, the example application we highlight in Fig. 6.3 shows the HANOB demonstration setup during the inspection of a wing flap. Applied in such demanding environments, this portable device will bridge visual inspection with the naked eye and automated quality assessment.

As both the portable prototype and the industrial robot-aided AWT devices for automated large area inspection yield complex yet similar datasets, we anticipate that standardizing the data acquisition and analysis and further developing data post-processing options for AWT findings will amplify the application portfolio of this

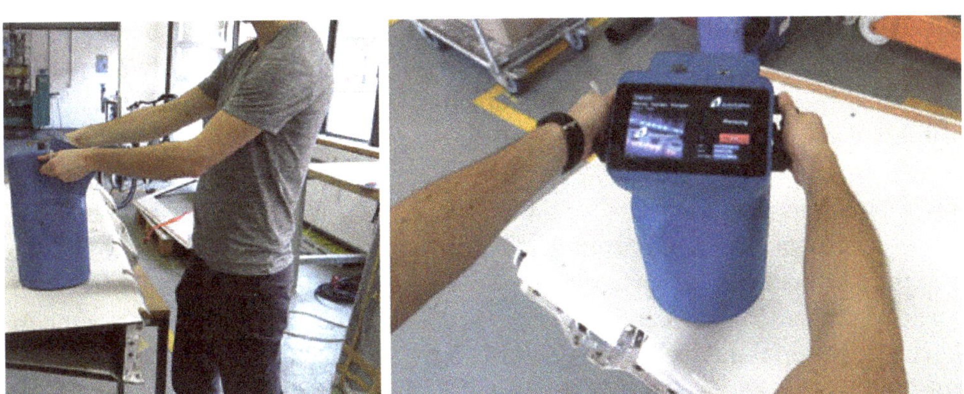

Fig. 6.3 HANOB demonstration setup operated for the inspection of a wing flap from Airbus Germany

ENDT method. Based on the insights and experience gained in the ComBoNDT project reported in Chap. 5, we anticipate that robust physics-based or data-driven post-processing procedures will contribute to increasing the process reliability in coating, painting and bonding applications. For example, our probability of detection (POD) investigations of the AWT datasets obtained for CFRP surfaces showed that in the release agent contamination scenario, the performance of the testing procedure depended on the selection of the characteristic image feature to be evaluated. We revealed a better detection performance when using the feature of wettability instead of using the feature of droplet diameter. As a next step, from the data post-processing point of view, we suggest performing a POD calculation by coupling the two approaches to determine whether the performance can be advanced even further. From the test setup point of view, the area density of the applied aerosol droplets might be further adapted to be optimum for this scenario. Thereby, an automatized optimization of the monitoring process, e.g. by tuning both the aerosol deposition and the feature-dependent detection performance, could facilitate reproducibly adjusted and customized inspection settings. We would like to highlight that at the end of the day, the multi-dimensional optimization of several parameters would be desirable, which would require extended modelling resources. Thereby, the advantageous interplay between quality assurance and digitalization is showcased.

6.4 Harmonized Presentation of ENDT Data and Metadata

Currently, the European Commission is orchestrating several ongoing European initiatives to establish an open innovation environment that, from our point of view, will also have a substantial impact on product quality control. We base these prospects on, for instance, the discernible progress in establishing a standard data structure for materials characterization (i.e. CHAracterization DAta (CHADA) [18]) and materials modelling (i.e. MOdellingDAta (MODA) [19]) with a range of potentially very different approaches and methods. We highlight here examples of those activities under the umbrella of the European Commission, which are embodied by the interacting [20] European Materials Modelling Council (EMMC) [21, 22] and the European Materials Characterisation Council [23], both of which have gained high visibility due to their agile web presence.

As an example, in Fig. 6.4, we suggest an approach for a CHADA documentation implying metadata aspects for the inspection of a CFRP part using the ENDT technique known as optically stimulated electron emission (OSEE). We would like to highlight that such an approach exceeds that described in Chap. 1 due to its more comprehensive and standardization-oriented setting. Moreover, embracing the physics of material interactions, as described by Romanos et al. [18] for impacting probes (here: ultraviolet light) and detected probes (here: photoelectrons), may be perceived as an expression of a physics-based consideration of the material properties and the characterization process itself. We may consider a physics-based approach

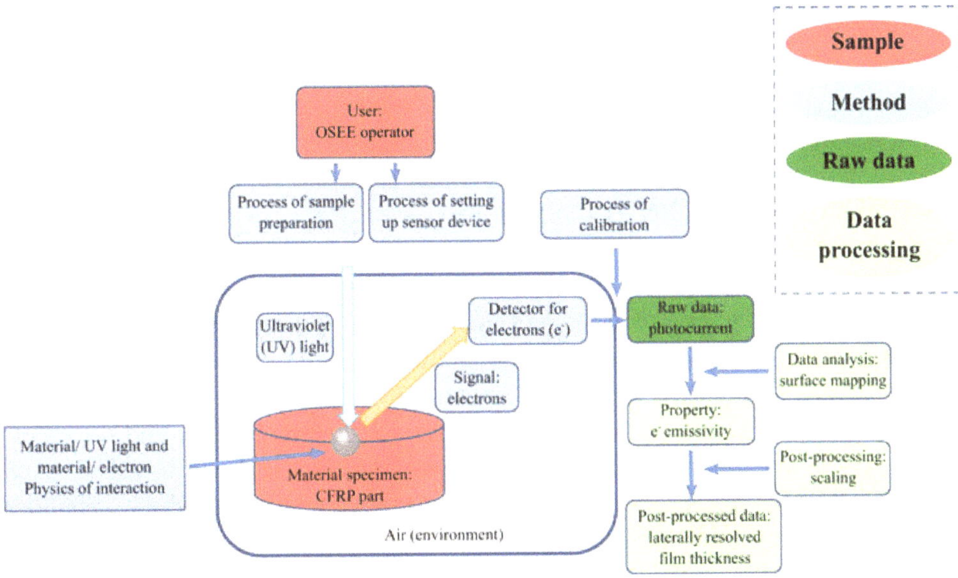

Fig. 6.4 Example of a CHADA (CHAracterization DAta) documentation implying metadata aspects, probe/surface interactions, and data processing steps for the inspection of a CFRP part using the optically stimulated electron emission (OSEE) ENDT technique; composed following the approach described by Romanos et al. [18]

complementary to rather data-driven regression methods that aim to find mathematical structures for the analysis of experimental information [24]. Combining both approaches systematically will be greatly facilitated by the interoperability of models and data.

We start by assessing further details of the recently presented systematic CHADA documentation shown in Fig. 6.4 for the required user responsibilities, comprising several aspects common to multiple-stage analytical processes. Both material-oriented, i.e. operand-related, and instrumental, i.e. operator-related, competencies are highlighted. In addition to setting up a specimen to be analyzed and a sensor before a surface inspection, the CHADA approach stresses the importance of calibration in accessing the characteristic material-related properties from the measured properties of analytical probes. The concept detailed in Chap. 1 and implemented in the ensuing chapters is based on preparatively adjusting finely graduated and technologically relevant surface states to purposefully vary operand features. Starting from such a set of customized operands, we effectively established both the particular sensor response and the response of a standardized technologically approved test for assessing design-relevant (mechanical) features. In this regard, we would like to refer to the discussion in Chap. 3 on the OSEE findings for CFRP coupon specimens. We inferred from our investigations that the OSEE response is affected by both deposited contaminants and the lateral inhomogeneities of the CFRP surface following the manual grinding process. Furthermore, we discussed that de-coupling these two aspects might broaden the application scope of this ENDT technique. We accentuate here that such a de-coupling might be achieved through sophisticated data analysis or the investigation of materials with a well-known composition and structure. In a certain manner, this reasoning places a focus on monitoring and describing the progress and effects of the manual grinding process used for CFRP surface pre-treatment. One lesson learned during the ComBoNDT research project was that the CFRP surface curvature imposes not only challenges related to the accessibility of the CFRP surface for the ENDT tool, but also issues related to a higher lateral inhomogeneity of the abrasive surface pre-treatment. In a physics-based approach, visualizing the latter might be achieved by locally accessing the distribution of exposed carbon fibres and capping by the matrix material as a position-dependent feature of the CFRP surface. Following the more pragmatic concept described in this book, such local variations intrinsically rather contributed to variations in a signal background or to a signal width, and we considered it essential to characterize and document these implications for each user case and contamination scenario based on carefully manufactured reference specimens. This is one of the reasons why we included their consideration within the ten heuristic quality assessment principles outlined in Chap. 1.

From the material application perspective, on the one hand, the ENDT datasets obtained for adherend surfaces in a harmonized way can (from a technical point of view) be communicated and immediately assessed along the process chain in those production and repair user cases that profit from adhesive bonding. On the other hand, the consequences of such relevant material-related information on the behaviour of an adhesive joint are not intuitively evident. Therefore, it is essential to link the inspection data to the material features that are relevant to the design, and eventually the durability, of the device or structural part incorporating the inspected adherend. As outlined in Chap. 1, this link is provided by applying the concept developed in the ComBoNDT project. For example, enhancing the procedure described by Tornow et al. [25], Moutsompegka et al. demonstrated this conceptual approach by linking OSEE-based ENDT data for one scenario of one user case [26] with design-relevant fracture toughness findings. In Fig. 6.5, which is based on test results obtained in technologically relevant user cases during the ComBoNDT, we exemplarily demonstrate that findings from the fracture toughness tests qualitatively and even quantitatively go along with the scenario-specific ENDT inspection datasets obtained from surface monitoring of CFRP adherends with OSEE before the bonding process.

6.5 Sensor Systems, Arrays and Networks for Assessing MoL Data

Following a life cycle approach, a product's beginning-of-life (BoL) period comprises the manufacturing phase, which is succeeded by the middle-of-life (MoL) period. It is attractive to apply non-destructive evaluation (NDE) and ENDT techniques developed for adhesive bonding processes not only during manufacture, but

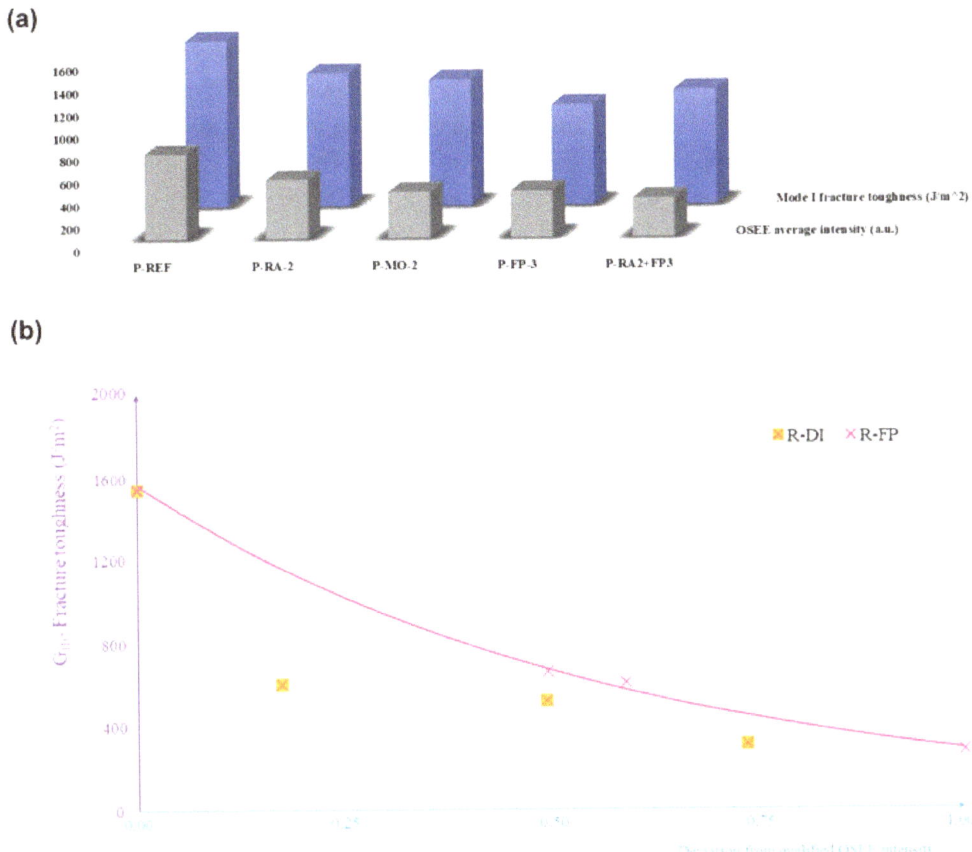

Fig. 6.5 Exemplary representation of outcomes from the ComBoNDT research project showing links between the fracture toughness (G_{IC} or G_{IIC}) measured for CFRP adhesive joints and normalized OSEE intensities measured for the surface of one of the adherends introduced in the bonding process. Distinct scenarios representing a deviation from a qualified surface pre-treatment process were considered for a production (P) and a repair (R) user case. **a** Reduced G_{IC} values qualitatively go along with reduced OSEE intensities; **b** characteristic curves quantitatively depict scenario-specific connections between the design-relevant mechanical joint feature G_{IIC} and the ENDT signal measured using OSEE

also during repair, which is a conceivable user case during a product's MoL. Procedures for monitoring the surface quality of adherends are tailored to open adhesive joints, i.e. the process steps before the application of the adhesive. Beyond that, the approaches sketched in Fig. 1.4 (in Chap. 1) with respect to manufacturing processes encompass bonding quality assessment, which is immediately relevant for all life cycle phases that are governed by a closed adhesive joint, for instance, in the case of composite parts. Accordingly, cyber-physical systems (CPS) relevant for ENDT during the BoL period are also technically relevant for the MoL of a product, facilitating the networking between the machines and equipment as part of the physical world and the cyberspace or internet as part of the digitalized virtual world. As outlined in the previous chapters, sensing systems based on sensors facilitate the observation and documentation of processes or the characteristics of a material object that are not visible to the human eye and cannot be recorded with other human senses [27]. As highlighted by Grosse and discussed in the previous chapters, a sensor acts as a filter that can significantly influence the way humans look at processes and material properties, e.g. by introducing more or fewer distortions and alterations to the data, and thus, the transmission of information. These considerations are featured in Fig. 1.4 (Chap. 1) by highlighting how relevant the quality of cyber-physical interaction is for the quality (assessment) of the output provided by a bonding process.

Moreover, the application of ENDT systems is conceptually closely related to the user cases expected during the MoL period of adhesive joints, such as those comprising the service aspects of MRO. In this context, conventional NDT procedures, such as ultrasonic or X-ray-based inspection or thermography, are rather mature technologies, yet they are intermittently applied. Their reliability has been extensively demonstrated, for instance, in testing aerospace components to detect defects, cracks, corrosion or other degradation features as part of quality assurance and in-service inspection [28]. Following this statement, and also in accordance with the previous chapters, recent directions comprise automated, fast and cost-effective in-service and in-situ monitoring, e.g. of aircraft. However, such conventional NDT procedures rely on access to structural parts with external probes or equipment, thus they are not suitable for condition-based maintenance concepts, and they require human intervention [29]. Beyond that and over an extended period of time, permanently installed SHM systems facilitate the application of non-destructive evaluation and testing methods to survey the state and condition of load-bearing structural components, primarily motivated by further enhancing safety [30, 31]. Following SAE Standard ARP6461, SHM is defined as "the process of acquiring and analyzing data from on-board sensors to evaluate the health of a structure" [29]. To achieve this, SHM systems include three key elements [29] and three steps [32]. Following Güemes et al., these elements comprise first, a network of sensors that are permanently attached to the structure, which is essential for performing automated inspections; second, on-board data handling and computing facilities for processing data in real-time; and third, algorithms that compare recently acquired data with stored data on the pristine structure. In this way, after correcting for environmental factors, a damage index may be calculated and information about damage existence, localization and type may be provided. Furthermore, damage severity may be quantified, and the residual life may be predicted [32]. Such prognosis is closely linked to diagnosis, or the evaluation of structural health, which is the main objective of SHM; prognosis and diagnosis conceptually differ in the tools they use: prognosis is mainly focused on statistical analysis, while diagnosis is more related to sensors, signal processing and algorithms for damage identification [29]. Güemes et al. record that some authors consider prognosis to be the fifth level of SHM, with levels one to four being related to damage detection, localization, classification and quantification of damage, respectively. Accordingly, Haldar [33] attributes the latter two development levels, with respect to the technical sophistication of SHM systems, to the assessment of damage and the prediction of development. Augustin et al. underline that the development and integration of a suitable sensor network is a major challenge facing scientists and engineers of SHM systems [34]. They list eligible methods for SHM systems, including fibre optic sensors, piezoceramic transducers for Lamb wave excitation and collection, vibration-based methods, acoustic emission techniques and electrical conductivity methods. In the following, we highlight recent achievements in diagnosis and provide insights for guided wave propagation, electromechanical impedance or fibre Bragg grating sensors.

An aspect that is inherent to the development and advancement of sensor systems and networks for NDT is considering the comprehensive applicability of such procedures in the distinct processes and environments of various technological sectors. For instance, approaches based on guided wave propagation, as well as electromechanical impedance, in composite materials and material joints have been widely investigated for the purpose of damage assessment. Here, we consider damage as not only an extensive loss of material functionality, but also an implied degradation of relevant material features as compared to the starting state. Widening this scope even further towards assessing this starting state, as well as the material surface, we proposed, tested and established both of these techniques for the non-destructive evaluation of adherends and adhesively bonded CFRP parts following the production and repair processes investigated within the frame of the ENCOMB and ComBoNDT joint research projects. Chapter 3 presented the ENDT results of the guided wave-based studies for surface characterization, and Chap. 4 described the electromechanical impedance results for adhesive bond evaluation. Taking this into consideration, the approaches of damage detection and the assessment of the adhesive bond state may be combined using the same sensor systems and networks. However, in this context, we must still answer the question of how to deploy the sensors. As one answer, designing a correct—and admittedly complex—network is the first step to success. Numerous sensor network solutions are being developed in the academic environment, and over the last decade, some of the results have been further developed and commercialized by companies. One of the better-known examples is the Stanford Multiactuator–Receiver Transduction (SMART) layer concept, which was conceived at Stanford University and later commercialized by Acellent Technologies, Inc. [35]. The SMART layer concept allows for the deployment of a sensor network on a surface, as well as embedding it in a layered composite as one of the plies in the layerwise composite structure. Quite recently, a new approach to sensor networks was proposed: the stretchable sensor network [36]. The leading idea behind this is to have a network that can be deployed across a large area yet remains compact prior to its deployment. Although such ideas do not provide an answer to the problem of how the sensors should be deployed on the structure, the placement of the sensors can be tackled by various approaches. Some of the popular strategies are based on evolutionary algorithms, such as genetic algorithms [37], whereby elliptic-based damage localisation algorithms may be used as a basis for optimisation. Here, the locations of actuating and recording transducers correspond to the ellipse foci, and the circumference of the ellipse informs about the source of wave reflection. If there is no damage or other discontinuity, then no reflection occurs. The type of transducer network determines the positioning of these ellipses. A different approach to the concept of a sensor network involves placing two transducers at each network node—an actuating and a receiving transducer [38]. Such transducer placement causes the resulting ellipses to be almost circular. Distributed systems are used not only for damage location, but also in damage identification for crack shape reconstruction [39]. A very informative review of the problem of optimal sensor placement was presented by Ostachowicz et al. [40]. Meanwhile, the sensor placement oriented towards hot spot monitoring is based on a tomography approach in which a localized area surrounded by sensors is monitored [41]. This was successfully applied for the debonding of stiffening elements from the skin of a composite wing [42], and here we can find a good link between damage detection and the assessment of adhesive bonds. In contrast to distributed networks and tomography approaches, there have also been attempts to employ a concentrated network of transducers. Researchers have analyzed various sensor placements in order to identify the optimum solutions. One of the most popular solutions has a circular shape, and there are examples with 13 [43] or 16 [44] sensors forming the circle. A special case of concentrated arrays is represented by the phased arrays that use a special method of signal processing to simulate wave interference. This effect allows for amplifying waves reflecting from an area of damage [30]. The simple phased array is linear in shape and has been successfully used for locating damage in aluminium specimens both with and without surface curvature [45], as well as in composites [46]. In summary, based on the achievements presented within the scope of this book and in published studies on sensor networks [35–40], we suggest that the same sensor networks may be used for both damage and adhesive bond assessment. Only the signal processing approaches are different for both applications and need to be tailored to the specific task. Such customisation will thus require knowledge of the respective technological environment, as well as standards for communication within the greater supervisory control and data acquisition (SCADA) network. SCADA has become the standard and is currently in operation on wind farms to gather measurements of wind speed, bearing and oil temperatures, voltage, power produced and other variables [47]. Therefore, the integration of new damage and adhesive bonding monitoring techniques should be integrated into the SCADA scheme.

In SHM applications ranging from aircraft to civil infrastructure, fibre optic sensors (FOS) are potentially very well suited to measuring variables such as temperature, electrical current, strain and pressure, even when working in electromagnetic fields, at high temperature or humidity, or in an aggressive chemical environment; also, they have demonstrated surprisingly high reliability [48]. During their life cycle, many adhesively bonded structures experience long-term dynamic loads in the MoL phase, which might lead to the preliminary performance reduction of bonded joints exposed to relatively low load levels [34]. Recently, we reported on the intrusive effect of adhesive bond strength and adhesive layer thickness and quality when embedding optical glass fibres carrying fibre Bragg grating sensors directly in the 0.1 mm thin adhesive bond formed by the structural and repair film adhesives applied in CFRP joints [49]. We showed that influences on the fatigue limit of the tested film adhesive joints were negligible whenever optical glass fibres with total diameters smaller than

100 μm were embedded. Having information on structural events or states available at arbitrary times may reduce the occurrence of unscheduled events, facilitate preventive maintenance [50], and pave the way towards condition-based structural maintenance [51].

Summing up, the research on and development of reliable SHM systems is of great importance. Conventional NDT provides non-destructive and effective tools for detecting and localizing damage in a predefined area and can form part of an SHM system; however, SHM is much more than NDT [52]. For SHM purposes, fibre-based sensors and smart materials may be integrated permanently into the monitored structures and the data may be collected continuously or at least periodically [53]. The collected data are then transferred to a computer system, where they are processed and analyzed for possible defects or changes in the material characteristics, e.g. damage accumulation, in almost real-time [52]. This is crucial not solely for FOS-based sensing because the collected data have little value without proper analysis and the extraction of information about the health state of the structure [33]. In fact, one of the major challenges of SHM systems is related to damage identification and assessment taking place in a mostly unsupervised learning mode [54]. Moreover, the scope of an SHM system enfolds different monitoring and learning levels. On the lowest level, the SHM system only monitors a certain structure or the assembly of a system (e.g. the wing of an aircraft), while on an advanced level the whole structural system is involved (e.g. the whole aircraft). On an even higher level, the data from all individual structural systems (e.g. all aircraft) are pooled and analyzed for recurring damage patterns or principles that can be applied to the whole fleet [52], as outlined in Fig. 6.6.

The successful integration of an SHM system leads to prolonged maintenance intervals, a reduction in maintenance costs, a resource-conserving exchange of wear parts, and a further increase in safety [34].

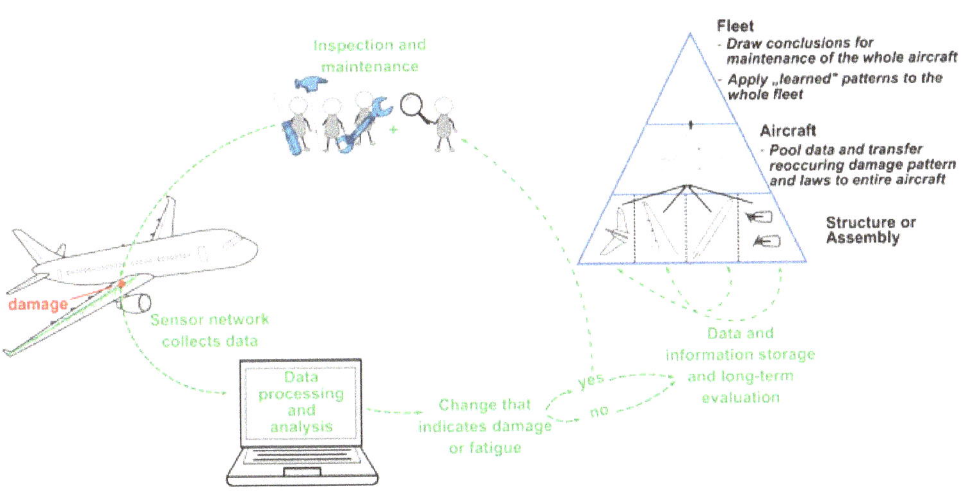

Fig. 6.6 Different monitoring and learning levels in the scope of an SHM system facilitated by data integration

6.6 Synopsis

Encouraged by the foreseeable availability of high-quality data from the advanced ENDT procedures presented in the previous chapters of this book and inspired by the numerous impulses from fruitfully cooperating in a joint research project, we have indicated in this last chapter that we might already project some facets of possible answers onto central future questions, such as:

- What are the drivers in worldwide digitalization and data exchange along the life cycle of a (material) product?
  For instance, sustainable and holistic life cycle product management will benefit from high-quality and relevant information provided to managers because there is some truth in the saying "You can't manage what you can't measure" [55].
- How can we contribute ENDT datasets from quality assessment as an input to worldwide workflows among the involved stakeholders?
  On the one hand, after a bonding process during BoL (e.g. manufacturing), material-related ENDT metadata and data may be available during MoL preventive maintenance and thereby allow for an evaluation of changes in the material state. On the other hand, in the case of MoL repair, a tailored device facilitating local ENDT may guide decisions by remote experts for cases of required cleaning on demand.
- How can quality assurance of local material-related transformation processes be promoted by receiving from external sources an input of material-related data gathered during earlier or subsequent life cycle phases?
  For example, in a bonding process during BoL or MoL (e.g. repair), the quality assessment may be guided by data obtained during the manufacturing of the adherends or during the application of the bonded joint (e.g. using SHM).

All of these questions may be posed for various types of material operands that are changed (e.g., joined) by distinct operators during a broad range of, e.g. adhesion-based, processes within the life cycle of products in different applications. This means that there is still a wealth of aspects and phenomena to be explored. Notwithstanding this, we are convinced that in facing these challenges, the advancement of ENDT tools and procedures achieved in the ComBoNDT research project, as well as FAIR data structures, lay a good technological and knowledge basis that is open for application!

References

1. McMillan AJ, Swindells N, Archer E et al (2017) A review of composite product data interoperability and product life-cycle management challenges in the composites industry. Adv Manuf: Polym Compos Sci 3:130–147. https://doi.org/10.1080/20550340.2017.1389047
2. OntoTRANS "Ontology driven Open Translation Environment" (2020–2024) Project funded from the European Union's Horizon 2020 research and innovation programme under grant agreement No 862136. https://ontotrans.eu/

3. ComBoNDT "Quality assurance concepts for adhesive bonding of aircraft composite structures by advanced NDT" (2015–2018) Project funded from the European Union's Horizon 2020 research and innovation programme under grant agreement No 636494. https://cordis.europa.eu/project/id/636494
4. Klöpffer W, Renner I (2008) Life-cycle based sustainability assessment of products. In: Schaltegger S, Bennett M, Burritt RL et al (eds) Environmental management accounting for cleaner production, vol 24, 1st edn. Springer, Netherlands, pp 91–102. https://doi.org/10.1007/978-1-4020-8913-8_5
5. Sonnemann G, Vigon B (eds) (2011) Global guidance principles for life cycle assessment databases—a basis for greener processes and products. United Nations Environment Programme
6. Calado EA, Leite M, Silva A (2019) Integrating life cycle assessment (LCA) and life cycle costing (LCC) in the early phases of aircraft structural design: an elevator case study. Int J Life Cycle Assess 24:2091–2110. https://doi.org/10.1007/s11367-019-01632-8
7. Otte JN, Kiritsi D, Ali MM et al (2019) An ontological approach to representing the product life cycle. AO 14:179–197. https://doi.org/10.3233/AO-190210
8. Hermann M, Pentek T, Otto B (2016) Design principles for industrie 4.0 scenarios. In: 2016 49th Hawaii international conference on system sciences (HICSS). IEEE, pp 3928–3937. https://doi.org/10.1109/HICSS.2016.488
9. Ghedini E, Friis J, Hashibon A et al (2020) EMMO an ontology for applied sciences. http://emmc.info/emmo-info/
10. Ghedini E, Friis J, Hashibon A et al (2020) European materials & modelling ontology (EMMO). http://github.com/emmo-repo/EMMO/
11. VIMMP "Virtual Materials Market Place" (2018–2021) Project funded from the European Union's Horizon 2020 research and innovation programme under grant agreement No 760907. https://www.vimmp.eu/
12. Horsch MT, Chiacchiera S, Seaton MA et al (2020) Ontologies for the virtual materials marketplace. Künstl Intell 34:423–428. https://doi.org/10.1007/s13218-020-00648-9
13. MarketPlace "Materials Modelling Marketplace for Increased Industrial Innovation" (2018–2022) Project funded from the European Union's Horizon 2020 research and innovation programme under grant agreement No 760173. https://www.the-marketplace-project.eu/
14. Cavalcanti WL, Noeske P-LM (2014) Investigating dynamic interactions by multi-scale modelling: from theory to applications. In: Springborg M, Joswig J-O (eds) Chemical modelling, vol 11. Royal Society of Chemistry, Cambridge, pp 175–200. https://doi.org/10.1039/9781782620112-00175
15. Rottenkohlber M (2010) Industrielle 3D Sensorik – Wo geht die Reise hin? INSPECT 7:12–14
16. Tornow C, Brune K (2019) Vorhabenbezeichnung: HANOB - "Entwicklung Referenzsystem zur handgeführten Oberflächen-Benetzungsprüfung": Abschlussbericht gem. NKBF 98: Berichtszeitraum: 01.05.2016 bis 31.10.2018. Fraunhofer-Institut für Fertigungstechnik und Angewandte Materialforschung (IFAM)
17. Cherrier C, Hanning T (2019) Vorhabenbezeichnung: HANOB - "Entwicklung Referenzsystem zur handgeführten Oberflächen-Benetzungsprüfung" : Abschlussbericht gem. NKBF 98 : Laufzeit des Vorhabens: 01.05.2016 bis 30.04.2018 verlängert zum 30.10.2018. Automation W+R GmbH
18. Romanos N, Kalogerini M, Koumoulos EP et al (2019) Innovative data management in advanced characterization: implications for materials design. Mater Today Commun 20:100541. https://doi.org/10.1016/j.mtcomm.2019.100541
19. de Baas AF (2017) What makes a material function? Let me compute the ways. Note: This document is commonly referred to as the Review of Materials Modelling (RoMM). EU Publications Office, Luxembourg. https://doi.org/10.2777/21919
20. Simperler A, Goldbeck G, Sebastiani M (2019) MODA and CHADA: terminology and standardized documentation for materials modelling and characterization; EuroNanoForum 2019 (Nanotechnology and Advanced Materials Progress Under Horizon2020 and Beyond), Bucharest, Romania, 12–14 June

21. EMMC ASBL, The European Materials Modelling Council (2019). https://emmc.eu/
22. EMMC-CSA "European Materials Modelling Council" (2016–2019) Project funded from the European Union's Horizon 2020 research and innovation programme under grant agreement No 723867. https://emmc.info/
23. EMCC, European Materials Characterisation Council (2020). http://www.characterisation.eu/
24. Gomes FA, de O Assis A, R. da C Reis M et al (2017) Proposal of heuristic regression method applied in descriptive data analysis: case studies. TEEE 2:51. https://doi.org/10.22149/teee.v2i2.105
25. Tornow C, Schlag M, Lima LCM et al (2015) Quality assurance concepts for adhesive bonding of composite aircraft structures—characterisation of adherent surfaces by extended NDT. J Adhes Sci Technol 29:2281–2294. https://doi.org/10.1080/01694243.2015.1055062
26. Moutsompegka E, Tserpes K, Noeske M et al (2019) Experimental investigation of the effect of pre-bond contamination with fingerprints and ageing on the fracture toughness of composite bonded joints. Appl Compos Mater 26:1001–1019. https://doi.org/10.1007/s10443-019-09763-9
27. Grosse CU (2013) Evolution of NDT methods for structures and materials: some successes and failures. In: Büyüköztürk O, Taşdemir MA, Güneş O et al (eds) Non-destructive testing of materials and structures. Springer, Netherlands, Dordrecht, pp 3–18. https://doi.org/10.1007/978-94-007-0723-8_1
28. Purna Chandra Rao B (2017) Non-destructive testing and damage detection. In: Prasad NE, Wanhill RJH (eds) Aerospace materials and material technologies. Springer Singapore, Singapore, pp 209–228. https://doi.org/10.1007/978-981-10-2143-5_11
29. Güemes A, Fernandez-Lopez A, Pozo AR et al (2020) Structural health monitoring for advanced composite structures: a review. J Compos Sci 4:13. https://doi.org/10.3390/jcs4010013
30. Giurgiutiu V (2014) Structural health monitoring with piezoelectric wafer active sensors, 2nd edn. AP Academic Press/Elsevier, Amsterdam. https://doi.org/10.1016/C2013-0-00155-7
31. Lehmhus D, Busse M (2018) Structural health monitoring (SHM). In: Bosse S, Lehmhus D, Lang W et al (eds) Material-integrated intelligent systems—technology and applications. Wiley-VCH Verlag GmbH & Co, KGaA, Weinheim, Germany, pp 529–570. https://doi.org/10.1002/9783527679249.ch22
32. Askaripour K, Zak A (2019) A survey of scrutinizing delaminated composites via various categories of sensing apparatus. J Compos Sci 3:95. https://doi.org/10.3390/jcs3040095
33. Haldar A (ed) (2013) Health assessment of engineered structures: bridges, buildings and other infrastructures. World Scientific, New Jersey, London, Singapore, Beijing, Shanghai, Hong Kong, Taipei, Chennai. https://doi.org/10.1142/8628
34. Augustin T, Karsten J, Kötter B et al (2018) Health monitoring of scarfed CFRP joints under cyclic loading via electrical resistance measurements using carbon nanotube modified adhesive films. Compos A Appl Sci Manuf 105:150–155. https://doi.org/10.1016/j.compositesa.2017.11.015
35. Qing XP, Beard SJ, Kumar A et al (2007) Built-in sensor network for structural health monitoring of composite structure. J Intell Mater Syst Struct 18:39–49. https://doi.org/10.1177/1045389X06064353
36. Guo Z, Kim K, Salowitz N et al (2018) Functionalization of stretchable networks with sensors and switches for composite materials. Struct Health Monit 17:598–623. https://doi.org/10.1177/1475921717709632
37. Soman R, Malinowski P (2019) A real-valued genetic algorithm for optimization of sensor placement for guided wave-based structural health monitoring. J Sens 2019:1–10. https://doi.org/10.1155/2019/9614630
38. Qiang W, Shenfang Y (2009) Baseline-free imaging method based on new PZT sensor arrangements. J Intell Mater Syst Struct 20:1663–1673. https://doi.org/10.1177/1045389X09105232
39. Quek ST, Tua PS, Jin J (2007) Comparison of plain piezoceramics and inter-digital transducer for crack detection in plates. J Intell Mater Syst Struct 18:949–961. https://doi.org/10.1177/1045389X06071435

40. Ostachowicz W, Soman R, Malinowski P (2019) Optimization of sensor placement for structural health monitoring: a review. Struct Health Monit 18:963–988. https://doi.org/10.1177/1475921719825601
41. Wang D, Ye L, Lu Y et al (2010) A damage diagnostic imaging algorithm based on the quantitative comparison of Lamb wave signals. Smart Mater Struct 19:65008. https://doi.org/10.1088/0964-1726/19/6/065008
42. Fasel TR, Todd MD (2010) An adhesive bond state classification method for a composite skin-to-spar joint using chaotic insonification. J Sound Vib 329:3218–3232. https://doi.org/10.1016/j.jsv.2010.02.009
43. Kudela P, Ostachowicz W, Żak A (2008) Damage detection in composite plates with embedded PZT transducers. Mech Syst Signal Process 22:1327–1335. https://doi.org/10.1016/j.ymssp.2007.07.008
44. Stepinski T, Engholm M (2009) Piezoelectric circular array for structural health monitoring using plate waves. In: Proceedings of the 7th international workshop on structural health monitoring, pp 1050–1056
45. Yu L, Santoni-Bottai G, Xu B et al (2008) Piezoelectric wafer active sensors for in situ ultrasonic–guided wave SHM. Fatigue Fract Eng Mater Struct 31:611–628. https://doi.org/doi:10.1111/j.1460-2695.2008.01256.x
46. Sundararaman A, Adams DE, Rigas EJ (2005) Biologically inspired structural diagnostics through beamforming with phased transducers arrays. Int J Eng Sci 43:756–778. https://doi.org/10.1016/j.ijengsci.2004.12.008
47. Antoniadou I, Dervilis N, Papatheou E et al (2015) Aspects of structural health and condition monitoring of offshore wind turbines. Philos Trans R Soc A: Math Phys Eng Sci 373. https://doi.org/10.1098/rsta.2014.0075
48. Meier U, Brönnimann R, Anderegg P et al (2013) 20 years of experience with structural health monitoring of objects with CFRP components. In: Büyüköztürk O, Taşdemir MA, Güneş O et al (eds) Non-destructive testing of materials and structures. Springer, Netherlands, Dordrecht, pp 959–976. https://doi.org/10.1007/978-94-007-0723-8_136
49. Grundmann N, Brüning H, Tserpes K et al (2020) Influence of embedding fiber optical sensors in CFRP film adhesive joints on bond strength. Sensors (Basel) 20. https://doi.org/10.3390/s20061665
50. Speckmann H (2008) Structural health monitoring (SHM). In: International maintenance review board policy board (IMRBPB). EASA
51. Lewis A (2008) Structural health management/monitoring. In: International maintenance review board policy board (IMRBPB). EASA
52. Balageas D, Fritzen C-P, Güemes A (2006) Structural health monitoring. Wiley Online Library. https://doi.org/10.1002/9780470612071
53. Trilaksono A, Watanabe N, Hoshi H et al (2013) Continuous damage monitoring of a thin composite structural with mismatched stiffener in a combined joint using fiber Bragg grating under tension and three-point loading. OJCM 03:63–87. https://doi.org/10.4236/ojcm.2013.33008
54. Farrar CR, Worden K (2010) An introduction to structural health monitoring. In: Deraemaeker A (ed) New trends in vibration based structural health monitoring. Springer, Wien. https://doi.org/10.1098/rsta.2006.1928
55. Baumast A (2017) Geschichte und aktuelle Herausforderungen des betrieblichen Nachhaltigkeitsmanagements. In: Pape J, Baumast A, Weihofen S et al (eds) Betriebliche Nachhaltigkeitsleistung messen und steuern: Grundlagen und Praxisbeispiele. Verlag Eugen Ulmer, Stuttgart, pp 18–30

Permissions

All chapters in this book were first published in ABACSNDTQAC, by Springer; hereby published with permission under the Creative Commons Attribution License or equivalent. Every chapter published in this book has been scrutinized by our experts. Their significance has been extensively debated. The topics covered herein carry significant information for a comprehensive understanding. They may even be implemented as practical applications or may be referred to as a beginning point for further studies.

The contributors of this book come from diverse backgrounds, making this book a truly international effort. We would like to thank all the contributing authors for lending their expertise to make the book truly unique. They have played a crucial role in the development of this book. Without their invaluable contributions this book wouldn't have been possible. They have made vital efforts to compile up to date information on the varied aspects of this subject to make this book a valuable addition to the collection of many professionals and students.

This book was conceptualized with the vision of imparting up-to-date and integrated information in this field. To ensure the same, a matchless editorial board was set up. Every individual on the board went through rigorous rounds of assessment to prove their worth. After which they invested a large part of their time researching and compiling the most relevant data for our readers.

The editorial board has been involved in producing this book since its inception. They have spent rigorous hours researching and exploring the diverse topics which have resulted in the successful publishing of this book. They have passed on their knowledge of decades through this book. To expedite this challenging task, the publisher supported the team at every step. A small team of assistant editors was also appointed to further simplify the editing procedure and attain best results for the readers.

Apart from the editorial board, the designing team has also invested a significant amount of their time in understanding the subject and creating the most relevant covers. They scrutinized every image to scout for the most suitable representation of the subject and create an appropriate cover for the book.

The publishing team has been an ardent support to the editorial, designing and production team. Their endless efforts to recruit the best for this project, has resulted in the accomplishment of this book. They are a veteran in the field of academics and their pool of knowledge is as vast as their experience in printing. Their expertise and guidance has proved useful at every step. Their uncompromising quality standards have made this book an exceptional effort. Their encouragement from time to time has been an inspiration for everyone.

The publisher and the editorial board hope that this book will prove to be a valuable piece of knowledge for students, practitioners and scholars across the globe.

Index

www.ingramcontent.com/pod-product-compliance
Lightning Source LLC
LaVergne TN
LVHW080643210926
848485LV00003B/313
* 9 7 8 1 6 4 7 2 8 4 7 9 4 *